Sustainable Development Goals Series

The **Sustainable Development Goals Series** is Springer Nature's inaugural cross-imprint book series that addresses and supports the United Nations' seventeen Sustainable Development Goals. The series fosters comprehensive research focused on these global targets and endeavours to address some of society's greatest grand challenges. The SDGs are inherently multidisciplinary, and they bring people working across different fields together and working towards a common goal. In this spirit, the Sustainable Development Goals series is the first at Springer Nature to publish books under both the Springer and Palgrave Macmillan imprints, bringing the strengths of our imprints together.

The Sustainable Development Goals Series is organized into eighteen subseries: one subseries based around each of the seventeen respective Sustainable Development Goals, and an eighteenth subseries, "Connecting the Goals", which serves as a home for volumes addressing multiple goals or studying the SDGs as a whole. Each subseries is guided by an expert Subseries Advisor with years or decades of experience studying and addressing core components of their respective Goal.

The SDG Series has a remit as broad as the SDGs themselves, and contributions are welcome from scientists, academics, policymakers, and researchers working in fields related to any of the seventeen goals. If you are interested in contributing a monograph or curated volume to the series, please contact the Publishers: Zachary Romano [Springer; zachary.romano@springer.com] and Rachael Ballard [Palgrave Macmillan; rachael.ballard@palgrave.com].

Priscilla Nyadoi
Fulgensia Kamugisha Mbabazi
Fredrick Ssali
Editors

Sustainable Development Goals in Africa

The Planet, People, and Perspectives

Springer

Editors
Priscilla Nyadoi
Uganda Martyrs University
Kampala, Uganda

Fulgensia Kamugisha Mbabazi
Busitema University
Tororo, Uganda

Fredrick Ssali
Uganda Martyrs University
Kampala, Uganda

ISSN 2523-3084 ISSN 2523-3092 (electronic)
Sustainable Development Goals Series
ISBN 978-3-032-20741-8 ISBN 978-3-032-20742-5 (eBook)
https://doi.org/10.1007/978-3-032-20742-5

This Springer imprint is published by the registered company Springer Nature Switzerland AG
The registered company address is: Gewerbestrasse 11, 6330 Cham, Switzerland

Foreword

Uganda Martyrs University (UMU) opened in 1993 and has over the years been at the forefront of providing quality higher education, training, and research that responds to national, regional, and global needs. The university is committed to nurturing staff and students capable of *conducting futuristic research* that can drive innovation, stimulate economic growth, and create a wide range of jobs as envisioned by the United Nations 2030 Agenda for Sustainable Development. The world is at a stage where science is driving rapid changes very fast and, thus, researchers and theorists need to address problems at hand and as well anticipate future developments and provide solutions for future problems. It is important that we not only work to provide evidence that supports or refutes existing theories but also endeavor to translate the findings into practical applications that can be applied for societal transformation. The need for strict observance of ethical research practices by all stakeholders and at all stages of the process cannot be overemphasized.

Universities in Africa and other regions in the Global South have been challenged to embrace transformational change and to be relevant to the twenty-first entury demands by, for instance, exploring different approaches to curriculum design and delivery. This will entail, among others, ensuring that facilitators and learners are open to learning and the distance between professors and other faculty is reduced. Uganda Martyrs University has partnered with local, national, regional, and global players in ensuring active participation on the SDGs agenda. For example, the university has, through its World Bank-funded African Centers of Excellence (ACE II) project and the African Center of Agro-ecology and Livelihood Systems (ACALISE), developed programs and trained graduates from eastern, southern, and western Africa in agro-ecological practices. The university has also supported student and community initiatives in collaboration with the Ministry of Agriculture and the National Environment Management Authority (NEMA) to raise awareness and plant trees across the equator in Uganda.

As per SDG 17, there is also a need to strengthen partnerships and linkages in research to enable the attainment of goals that are otherwise challenging to attain by working individually or as one institution. Therefore, senior researchers and graduate students at universities must find ways to work in teams. If one is good at conceptualization, another may be good at formal analysis and investigation, while another may be better at funding acquisition. Working together, the team would produce much better research output. Such meaningful partnerships ensure that we grow together, remembering the

African Ubuntu philosophy that emphasizes that one does not grow in isolation.

Therefore, I encourage African researchers to work together and generate new knowledge in Africa for African and global development needs.

Uganda Marytrs University
Kampala, Uganda
October 11th 2025

Patrick Edrin Kyamanywa

Preface

Africa has a vast diversity of genomes of humans and other animals. The continent has the youngest human population globally and is facing a significant disease burden that requires ingenious solutions. The role of science and technology (S&T) in providing these solutions cannot be overstated. In Uganda, the Uganda National Council for Science and Technology (UNCST) is responsible for developing strategies for integrating S&T into the national development planning process. The agency advises government of Uganda on policy matters necessary for advancing S&T and oversees and coordinates research and development activities in Uganda. The UNCST receives and reviews research protocols, and with support from the Research Secretariat in the Office of the President, clears all research intended to be carried out in Uganda. Researchers are required to submit the necessary research application forms in print or use the UNCST online research registration platform.

National governments and other stakeholders should invest more in science, research, technology, and innovation so as to achieve faster growth and higher levels of economic and social prosperity. Our researchers need diverse resources including personnel, time, and equipment to carefully plan and execute their studies. We should support them. I urge universities and other tertiary institutions to implement strategies that can effectively bridge the "innovation valley of death" and transform research outputs into tangible economic and social benefits. These include: investment in modern facilities; encouraging a culture of entrepreneurship; aligning research priorities with identified market needs; and establishing partnerships with private sector companies. As UNCST, we are glad to partner with universities and research institutions in promoting and coordinating research, science, and technology to support national development and improve the quality of life.

Kampala, Uganda
October 11th 2025

Martin Patrick Ongol

Introduction

The Sustainable Development Goals (SDGs) were adopted by the United Nations in 2015 as the 2030 Agenda for Sustainable Development. This global initiative, which replaced the Millennium Development Goals (MDGs), outlined 17 integrated goals to address pressing worldwide challenges including poverty, inequality, climate change, and environmental degradation by balancing social, economic, and environmental needs. Senior researchers and students from various disciplines can advance inclusive development within the framework of the SDGs by generating knowledge, informing policy, and driving innovation. This can be achieved by conducting research to understand societal challenges, designing innovative solutions for a more just and equitable world, and advocating for transformative changes that benefit all people. With the aim of providing a platform for fostering interdisciplinary approaches, disseminating research findings to various stakeholders, and building capacity for sustainable development in Africa, the Uganda Martyrs University in November 2024 convened its 15th Annual Research Conference. Its theme was "Beyond the Sustainable Goals in Africa: The People, Planet, Markets and Education." Conference submissions were subjected to peer-review, and the revised manuscripts are presented here as chapters of this book.

Chapter 1 in Part I advances the state of knowledge on the role of higher educational and research institutions. It gives an overview of how these institutions are contributing toward achievement of the SDGs in Africa through teaching, research, and community engagement. In Chaps. 2, 3, 4, and 5 in Part II, authors tackle issues concerning the environment, climate change and biodiversity, food security, and governance in natural resource management. Africa is experiencing significant biodiversity loss, climate change, and land and environment degradation. These impact its ecosystems, economic stability, and water resources. The authors discuss food security, climate finance, environmental governance, conservation of wildlife resources, among others. They propose a host of interventions to drive sustainable development in Africa. In Chaps. 6, 7, 8, 9, 10, and 11 in Part III, topics on people, places, accountability, violence, and intergenerational justice are discussed. The chapters feature migrant workers, human trafficking, female genital mutilation, women empowerment, peace and justice promotion, among others. Chapters 12, 13, 14, and 15 in Part IV highlight public health issues in Uganda and other African countries. These include HIV AIDS, preventive care, maternal health, and child health and nutrition, among others. Chapters 16,

17, 18, 19, and 20 in Part V focus on poverty and markets in rural Africa, youth unemployment, performance of agricultural cooperatives, and livelihood dynamics of urban refugees. Chapters 21, 22, 23, 24, and 25 in Part VI focus on education and knowledge generation, highlighting interventions to enhance students learning experiences and the completion of study programs.

The editors and contributing authors would like to thank management for supporting the preparation and publication of this book. We thank all the reviewers and session chairs, including Professor Patrick Kyamanywa, Assoc. Prof. Jimmy Spire Ssentongo, Assoc. Prof. David Ngendo Tshimba, Assoc. Prof. Mark Olweny, Assoc. Prof. Laura Ariko Otaala, Rev. Dr. Marie Gorretti Nakitende, Dr. Robinah Nakabo, Dr. Kizito Omona, Dr. Joseph Ssekandi, Dr. Modest Kayi O'dama, Dr. Denis Musinguzi, and Simon Esibo Omaada. We are grateful to the entire staff of the Directorate of Graduate Studies Research and Enterprise for dedicating time and effort to the 15th UMU Annual Research Conference. We thank Aaron Schiller and the team at Springer for their kind help during the preparation of this book.

Priscilla Nyadoi
Fulgensia Kamugisha Mbabazi
Fredrick Ssali

Contents

Contributors

Sarah Specioza Abeja Department of Governance, Peace and International Studies, Uganda Martyrs University, Nkozi, Uganda

Christopher Mukidi Acaali Uganda Martyrs University, Kampala, Uganda

Mary Goretti Acila Makerere University, Kampala, Uganda

Faculty of Agriculture, Uganda Martyrs University, Kampala, Uganda

Sarah Akello Makerere University, Kampala, Uganda

Augustin Nkundimana Alfred Université Catholique la Sapientia de Goma, Goma, Democratic Republic of the Congo

Susan Arayo Department of Governance, Peace and International Studies, Uganda Martyrs University, Nkozi, Uganda

Florence Munyonyo Asiimwe Uganda Martyrs University, Kampala, Uganda

Samuel Ssanyu Balamaga Faculty of Health Sciences, Uganda Martyrs University, Kampala, Uganda

Jacinta Bwegyeme Uganda Martyrs University, Kampala, Uganda

Paulinus Byabasaija Uganda Martyrs University, Kampala, Uganda

Aloysius Byaruhanga Uganda Martyrs University, Kampala, Uganda

Morris Egesa Uganda Martyrs University, Kampala, Uganda

Richard Ekodeu Agency for Cooperation in Research and Development (ACORD) Uganda, Kampala, Uganda

Simon Omaada Esibo Uganda Martyrs University, Kampala, Uganda

Nathan Hagobi Hasahya Faculty of Business Administration and Management, Uganda Martyrs University (UMU), Kampala, Uganda

Seidu Abubakari Jeremiah Jaksally Development Organisation, Bole, Ghana

Seniwoliba Attia Joseph Jaksally Development Organisation, Bole, Ghana

Bendicto Kabiito Uganda Martyrs University, Kampala, Uganda

James Kagaari Kyambogo University, Kampala, Uganda

Dushime Kashori Université Catholique la Sapientia de Goma, Goma, Democratic Republic of the Congo

Andrew Kigongo Faculty of Health Sciences, Uganda Martyrs University, Kampala, Uganda

Mark Kiiza Uganda Martyrs University, Kampala, Uganda

Hellen Mghoi Kishuke Mshilla Uganda Martyrs University, Kampala, Uganda

Daniel Kizza Makerere University, Kampala, Uganda

Christopher Kugonza Uganda Martyrs University, Kampala, Uganda

Donald Rugira Kugonza Makerere University, Kampala, Uganda

Patrick Kyamanywa Uganda Martyrs University, Kampala, Uganda

John Bosco Lokolimoi Uganda Martyrs University, Kampala, Uganda

Samuel Mafabi Makerere University, Kampala, Uganda

Yusuf Mayega School of Arts and Social Sciences, Uganda Martyrs University, Kampala, Uganda

Denis Rwabiita Mugizi Makerere University, Kampala, Uganda

John Chrysostom Kigozi Munene Makerere University, Kampala, Uganda

Veronica Munuve Uganda Martyrs University, Kampala, Uganda

Denis Musinguzi Uganda Martyrs University, Kampala, Uganda

Peter Mwesigwa Uganda Martyrs University, Kampala, Uganda

Robinah Nakabo Uganda Martyrs University, Kampala, Uganda

Juliet Musoke Nakabuye Faculty of Law and Criminology, Ghent University, Ghent, Belgium

Uganda Martyrs University, Kampala, Uganda

Marie G. Nakitende Faculty of Business Administration and Management, Uganda Martyrs University (UMU), Kampala, Uganda

Charles Patrick Namisi Uganda Martyrs University, Kampala, Uganda

Dorothy Kalule Nampanzira Makerere University, Kampala, Uganda

Josephine Namugumya Uganda Martyrs University, Kampala, Uganda

Makerere University, Kampala, Uganda

Josephine Namuli Uganda Martyrs University, Kampala, Uganda

Maria G. Nassuna Musoke Makerere University, Kampala, Uganda

Helen Nambalirwa Nkabala Uganda Martyrs University, Kampala, Uganda

Gideon Nkurunziza Department of Business and Management Sciences, Makerere University Business School (MUBS), Kampala, Uganda

Priscilla Nyadoi Uganda Martyrs University, Kampala, Uganda

Uganda Wildlife Society, Kampala, Uganda

Modest Kayi O'dama Uganda Martyrs University, Kampala, Uganda

Lwanga Okello Uganda Martyrs University, Kampala, Uganda

Samuel Okello Makerere University, Kampala, Uganda

Kizito Omona Faculty of Health Sciences, Uganda Martyrs University, Kampala, Uganda

Fred Felix Ongom Uganda Martyrs University, Kampala, Uganda

Laura Ariko Otaala Uganda Martyrs University, Kampala, Uganda

Pascal Ndyanabo Rutazihana Dschang School of Economics and Management, University of Dschang, Dschang, Cameroon

Association des chercheurs en Economie de Développement Appliquée (ACEDA), Yaoundé, Cameroon

Université Catholique la Sapientia de Goma, Goma, Democratic Republic of the Congo

Peter Celestine Safari Uganda Martyrs University, Kampala, Uganda

Fredrick Ssali Uganda Martyrs University, Kampala, Uganda

Edward K. Ssegawa Uganda Martyrs University, Kampala, Uganda

Joseph Ssekandi Uganda Martyrs University, Kampala, Uganda

Jimmy Spire Ssentongo Uganda Martyrs University, Kampala, Uganda

Gilbert Tayebwa Uganda Wildlife Society, Kampala, Uganda

The International Crane Foundation, Baraboo, WI, USA

Alex Alison Tongu Uganda Martyrs University, Kampala, Uganda

Simeon Wanyama Faculty of Business Administration and Management, Uganda Martyrs University (UMU), Kampala, Uganda

Part I

Sustainable Development Goals in Perspective

1 Promoting SDGs Through Community Engagement: The Case of Uganda Martyrs University

Christopher Mukidi Acaali

Abstract

Community engagement at Uganda Martyrs University (UMU), one of the three traditional pillars, alongside teaching, learning and research, serves as a vehicle for providing mutually beneficial services to the UMU community, either as a single group or in collaboration. This paper presents an account of some community engagement interventions at Uganda Martyrs University and the lessons drawn from these over the last three decades. With examples, the paper shows how the University, through community empowering interventions, has promoted and demonstrated activities contributing toward social and economic development of the communities and the realization of Sustainable Development Goals and targets.

Keywords

Community engagement · Farmer cooperatives · Education · Malnutrition · Equator Valley Farm · Teaching and learning · Sub-Saharan Africa

C. M. Acaali (✉)
Uganda Martyrs University, Kampala, Uganda
e-mail: cmukidi@umu.ac.ug

Introduction

Like any other university, Uganda Martyrs University (UMU) has three traditional functions that regularly guide its operations and strategic direction. They are: teaching, learning, research, and community engagement. Of late, innovation and entrepreneurship have been added to the three functions. Aware that the university's vision and mission are shaped, permeated, and informed by these functions, the chapter explores how these, in conjunction with the community engagement policy, have guided UMU in implementing several community interventions. Thus, the chapter focuses on the engagement policy, among others, by distinguishing Internal Community Engagement (ICE) from External Community Engagement (ECE). Furthermore, the chapter examines the community engagement activities UMU has implemented in its neighbourhood communities. The chapter argues that such interventions which, for example, targeted the reduction of poverty, hunger, and malnutrition, contributed to the implementation of the SDGs. But above all, they contributed to the social and economic transformation of the communities. This research was done by consulting online resources, several UMU and partner reports, and baseline surveys that underscored UMU's engagement activities.

P. Nyadoi et al. (eds.), *Sustainable Development Goals in Africa*, Sustainable Development Goals Series, https://doi.org/10.1007/978-3-032-20742-5_1

Uganda Martyrs University Engagement Policy

Community engagement activities at Uganda Martyrs University (UMU) are traceable to its establishment in 1993. Several community engagement activities were implemented in the Nkozi sub-county, for example, the provision of water to Nkozi Hospital and households of Nkozi Trading Centre at a subsidized cost. The university also implemented other engagements to support educational activities in the Nkozi Trading Centre. Uganda Martyrs University Nursery School, later renamed St. Patrick Primary School, started in 1997 (originally as a Nursery School) to provide nursery education services for the children of UMU's staff and the people of Nkozi. The Nursery School relocated to its present site in 2003, and in 2014, it was handed over to the Banakalooli Brothers of Kiteredde for proper management.

While the provision of water to the community members was coordinated by the Estates Department, the nursery school was coordinated by Management. With the establishment of the Centre for Extra-Mural Studies (CEMS) in 1996, UMU formally started its journey of community engagement activities. Initially, the centre was not established to run community engagement activities but rather to coordinate Distance Learning Programs and act as an incubator of new programs and short courses. However, during the implementation of these programs, it was discovered that another unit was necessary to coordinate community activities. Thus, the university established this unit in 1999 and later converted it into a Directorate for Community Outreach Programs Coordination. However, in 2015, the directorate was transformed into the Department of Education for Sustainable Development and Community Engagement (ESD & CE, 2022). In 2022, the department relocated from the Office of the Deputy Vice Chancellor to the Directorate of Graduate Studies, Research, and Enterprise. Although the Office continued to coordinate community engagement activities, the task of mainstreaming education for sustainable development in the university and coordinating the Regional Centre of Expertise in Greater Masaka was added. Meanwhile, as early as 2016, the ESD&CE embarked on a community engagement policy for the university. Through a Senate committee headed by Dr. Salome Najjuka, formerly a lecturer in the School of Arts and Social Sciences (SASS), a community engagement policy was written and presented to the Senate in 2017, duly approved, and subsequently promulgated for the university's use (ESD & CE, 2022). The Community Engagement Policy (2017) introduces its framework through five foundational elements: the policy background, its objectives, categories and forms of engagement, and the composition of university-community membership. Within its background section, the policy situates community engagement as a core function of the university, alongside teaching and learning, and research. This triad is presented as a mutually reinforcing set of mandates through which the university systematically addresses societal needs. The document also traces the institutional evolution of the present Department of Education for Sustainable Development and Community Engagement, charting its transformation from the Extra-Mural Office to the Directorate of Outreach, to its current departmental status.

One of the central emphases of the Community Engagement Policy (2017) is its overarching objective, which aligns with and operationalizes the university's vision, mission, and strategic goals. Notably, the policy asserts that community engagement serves as a vehicle through which these institutional aims are realized. Specifically, it seeks to "build life and improve the standards of the university environment, plus the quality of life of the university community members" (CEP, 2017). This objective encompasses three key dimensions through which the university contributes to community vitality, notably via educational initiatives that enhance livelihoods, such as agricultural programs offering skills training to farmers and providing student internship opportunities. The notion of "building life" is conceptualized as reciprocal: while the university imparts knowledge, skills, and resources to the community, the community, in turn, enriches the university through experiential learning opportunities for students and staff. Accordingly, this symbiotic exchange fosters a transformative improvement in the quality of life for both the institution and its surroundings.

A related objective extends this impact beyond the immediate environment, envisioning contributions to the development of neighbourhoods, the nation, and the global community. As Mukidi (2024) observes, this is achieved when students from diverse faculties identify community problems, investigate them, and share evidence-based recommendations, thereby generating collaborative solutions. Such problem identification and solution provision catalyse both social and economic transformation. Guided by its religious ethos, the policy also prioritizes outreach to vulnerable, poor, and marginalized groups, emphasizing inclusivity so that "no one is left behind" in the development process. Additionally, it affirms the principle of equity, ensuring all stakeholders are treated with fairness and dignity. Ultimately, these objectives seek to position the university not as an isolated "ivory tower," but as "an academic community characterized by the non-negotiable values of service, solidarity, justice, and respect for others" (CEP, 2017).

In outlining the categories of community engagement, the Community Engagement Policy (2017) begins by clarifying its use of the two terms: community and engagement. The term community refers primarily to the stakeholders and geographical areas surrounding the Main Campus at Nkozi. The policy implies that this definition was not extended to other provincial campuses at the time of drafting, as these had not yet been established. However, this rationale does not adequately explain the omission of Lubaga, Masaka, and Nsambya campuses, which were already operational by the time the policy was promulgated. The term engagement is defined as "an active and participatory process involving university members and the community which can influence and shape policy, services, and the sharing of resources using different methods and techniques and for varied objectives" (CEP, 2017). The policy further specifies that the community may include any three or more individuals associated with Uganda Martyrs University (UMU). Notably, this encompasses university staff, students, alumni, administrators (including the University Council and its committees), associations and societies within the institution, professional and funding bodies affiliated with the university, and non-governmental organizations connected to UMU. It also extends to partner educational institutions, local neighbourhoods, trading centre, parishes, hospitals, prisons, police stations, schools, local government units, and village settlements, as well as the wider network of associates, partners, and collaborators at national and international levels. The above notwithstanding, the policy conceptualizes community engagement as a "vehicle" for delivering mutually beneficial services to members of the UMU community, either individually or collaboratively. In this metaphorical sense, just as a vehicle transports people, goods, and services, community engagement facilitates the advancement of the social, cultural, spiritual, and economic well-being of the university, its stakeholders, and the beneficiaries of specific engagement initiatives (CEP, 2017).

Within the policy, several terms are used interchangeably with community engagement, namely outreach, service learning, community service, and volunteerism. Outreach is described as "an extension of services, programs, and other forms of engagement to the community outside the university environs" (CEP, 2017), emphasizing reciprocity and equality in benefits for both the university and the external community. Similarly, service learning refers to the integration of service activities with academic coursework, with the results contributing toward a student's graduation requirements. In the policy, this approach acknowledges that learning is a multi-faceted process in which both the university and the community contribute knowledge and skills through structured projects and engagement beyond the classroom. As in the other two terms, community service encompasses all activities in which staff and students dedicate their time, expertise, and effort for the benefit of the community, with the community likewise contributing knowledge to academia, thereby enhancing the university's credibility in addressing community needs. Lastly, but not least, volunteerism is characterized by the provision of services with the primary intent to benefit the community, without expectation of personal gain by the volunteer. Granted that incidental benefits may accrue to the volunteer, they are secondary to the altruistic, extracurricular, and non-credit-bearing nature of the engagement.

Having established the conceptual scope of community engagement and the interchangeable terms associated with it, the Community Engagement Policy (2017) further distinguishes between two principal categories: Internal Community Engagement (ICE) and External Community Engagement (ECE). On the one hand, the ICE is defined as active and purposeful participation in one or more activities within the university environment that contribute to the welfare and development of the university community, carried out in a spirit of self-giving and service. It encompasses the work of staff in organizing institutional events such as graduations, workshops, exhibitions, open days, fundraising campaigns, and environmental conservation initiatives, as well as providing mentorship to staff and students, and offering guidance to clubs and associations. The policy notes that the range of activities may vary according to the creativity and initiative of individual staff members. On the other hand, the ECE, also referred to in certain contexts as outreach or placement, is defined as active and purposeful participation in activities beyond the physical boundaries of the university, aimed at enhancing the well-being and development of the institution, its surrounding neighbourhoods, and other stakeholders. These engagements seek to address, mitigate, or alleviate various social, economic, and environmental challenges faced by the communities in which the university operates. Importantly, the policy emphasizes that such engagements are not one-sided; they also enrich pedagogy and research by integrating classroom learning with practical applications in real-world contexts. One key form of ECE is student placement or internship, whereby students undertake projects aligned with community needs. These projects may include initiatives in sanitation, health, food security, water access, housing, leadership training, educational support, and advocacy for peace and justice. Such experiences enable students to apply theoretical knowledge in the field, develop practical problem-solving skills, and cultivate a community-oriented mindset. The policy argues that graduates who undergo this form of experiential learning possess a competitive advantage in the job market, as they are more attuned to the social realities and developmental needs of the communities they serve.

The objectives underpinning both ICE and ECE are closely aligned with the university's vision and mission, as well as its triad traditional functions: teaching and learning, research, and community engagement. The policy commits the university to improving the standards of the campus environment, enhancing the quality of life for its members, and contributing to the development of surrounding neighbourhoods, the nation, and the global community. Special emphasis is placed on outreach to vulnerable, marginalized, and disadvantaged groups, thereby fostering inclusivity and equity. By promoting values such as service, solidarity, justice, and respect for others, the university positions itself as a socially responsive institution rather than an isolated academic enclave. With the policy framework established, it becomes pertinent to assess how Uganda Martyrs University has operationalized these provisions, mostly retrospectively, in implementing interventions within its neighbouring communities.

Community Engagement Interventions

Since its inception in 1993, Uganda Martyrs University has made several significant interventions in the communities surrounding Nkozi Sub-County, Mpigi District. Several of these interventions were made with the support of partners like the University of Notre Dame, USA, UPFORD, and AP-TRIAS, a Belgian funding organization. The purpose of this section is to explore the nature of these interventions and how far they went in changing the social and economic lives of the people. Between 1999 and 2002, Uganda Martyrs University, with the support of UPFORD, implemented a nutrition project for the malnourished children of Buwama, Kituntu, and Nkozi sub-counties. Specifically, the project provided nutritious soya food that contained high-quality protein, which supplied children with essential amino acids for building and repairing tissues. In addition, soya provided the much-needed sources of vitamins B and E. Through this project, therefore, UMU provided the basics for children's growth and development. Meanwhile, farmers were encouraged to grow soya on their farms as a guarantee of the project's sustainability. Specifically, UMU provided extension workers who provided expertise to farmers growing soya. It was out of these initiatives that the famous Equator Valley Farm of the university, now extinct, was established (ESD & CE, 2022).

While the nutrition project was coming to an end, UMU's other project, funded by AP-TRIAS, a Belgian funding organization, started in Nindye, Katonga, Masaka, between 2003 and 2008 to increase farmers' household incomes. Farmers were trained in business management, entrepreneurship, and marketing skills. Additionally, they were taught skills of crop handling and how to avoid post-harvest losses, as well as how to collaborate within their groups and communities. With the acquired skills, maize productivity increased for consumption and the market. Collective marketing centres and market linkages were established where their produce was sold to Uganda Grain Traders and the World Food Program. With these efforts, the farmers' household incomes increased due to increased production and quality of produce. But above all, the major contributing factor to this success was the ability to control post-harvest losses (ESD & CE, 2022). Inadvertently, the project improved the interpersonal relationships among farmers, as they had to work together to aggregate their produce and establish common markets. Consequently, collaborations among farmer groups and communities increased exponentially.

As the AP-TRIAS concluded, another project supported by the SIEBEN Foundation, Belgium, and the University of Notre Dame, USA, began between 2009 and 2010. The initiative to promote commercial agriculture through cooperatives and improve nutrition in households in Nkozi Sub-County, Mpigi District, was successfully achieved. This project supplemented the efforts of an earlier project implemented between 1999 and 2002, during which households were sensitized about nutrition. Thus, by the end of the project in 2010, three objectives of the project were accomplished, namely, food security in the area had increased, households could afford and access nutritional foods, and farmers' outputs had increased such that they had to aggregate their produce for sale through cooperatives (Mukidi, 2024). Accordingly, UMU's intervention improved people's nutrition, created food security in the area, and increased people's incomes.

To date, UMU's visible success with community engagement activities has been in Nindye, a village in the Nkozi Sub-County, Mpigi District, where several interventions have been made. The most prominent of these interventions was the one spearheaded by the University Partnership for Research and Development (UPFORD) and the University of Notre Dame, between 2009 and 2015, in which UMU intervened in the community by helping farmers improve their agricultural production through the adoption of modern farming methods. Furthermore, UMU intervened by enabling farmers to access better health services through the acquisition of a mobile motorcycle (boda-boda) that transported pregnant women to the nearby clinic (ESD & CE, 2022). In addition, the project sensitized the farmer communities to observe hygiene and ensure the availability of water and sanitation in the community. Inevitably, improved agricultural production and health automatically led to improved incomes for the community. With the support of UPFORD, UMU initiated the formation of microsavings groups branded "Savings and Internal Lending Communities" (SILC) (Ssekijjo, 2018). As Mukidi (2024) points out, "The primary purpose of these groups was to encourage a saving culture for the people but also create possible financial opportunities for their businesses." Consequently, the spirit of saving culture was adopted by the groups, and household finances at the community and individual levels tremendously improved the farmers' incomes. More interestingly, it became more possible for the children of these farmers to attend good schools beyond their village. Due to these interventions, the Nindye community became a social and educational laboratory where staff and students alike from the Faculties of Agriculture, Business Administration, and Management, and the School of Arts and Social Studies benefited from the empirical research and internships by observing the life and activities of the people (Mukidi, 2024).

To complement the efforts of UPFORD, UMU, in collaboration with VEROZON, implemented a health project in Nindye between 2012 and 2013, aimed at improving the community's health. Working with Nindye Health Centre, the project focused on providing health information to the population. To access health services conveniently, the project gifted some households with mobile phones to contact health workers for treatment (ESD & CE, 2022). In addition to agriculture and business projects, UMU's health sector interventions improved people's lives in Nindye. Another project, with the support of UPFORD, named the Social Enterprise Project (SEP), was implemented by UMU between 2015 and 2019. This project was premised on the understanding that it was difficult for farmers in Nindye to access credit lines from commercial banks, largely due to their prohibitive conditions, one of which included security of titled land. To overcome this obstacle, STEP, a continuation of the SILC project, devised a financing loan facility that farmers could easily access to start local business enterprises (Guzman, 2018). The significance of this intervention was that the SILC project extended longer loan repayment schedules than commercial banks. Moreover, the loan facility could be obtained at any time, in contrast to the earlier practice where it was tied to the agricultural season. In addition, STEP provided on-site business/entrepreneurial skills training for the participating members (ESD & CE, 2022). According to Emily Pool and Danice Guzman (2020), the impact of these initiatives was reflected in farmers' expanded enterprises, increased output, large capital purchases, and employment of more workers.

Between 2014 and 2019, the School of Arts and Social Sciences implemented the Karamoja Cluster Project. By doing so, UMU extended its outreach to communities beyond Mpigi District. This project focused on the peace-to-peace approach among the Karamojong and the Turkana people. The project was premised on the prevailing violence among the pastoral communities of the Karamojong and Turkana peoples due to cattle rustling. The channels used for promoting peace were sports and community dialogue. By the time the project ended, several successes had been accomplished. First and foremost was the restoration of peace within the communities and their neighbours. In addition to these efforts, the school participated in the Equator Peace Academy program, where some of its staff regularly engaged in scholarly discourses, with themes of cultural identity and religious tolerance, all aimed at promoting and building peace. Such scholarly discourses were held in Kapchorwa, Karamoja, Acholi, and Rwanda.

One of the upbeat interventions of UMU in Nkozi Sub-County was between 2020 and 2021. The project, supported by UPFORD and implemented by UMU, was the Community Lending and Outside Capital (CLOC) research project. This CLOC project arose from research findings regarding the challenges encountered by the SILCs' and SEPs' projects. Key among the challenges was that funds were insufficient to support the farmers. More crucially, since farmers were paid money from their farms on a seasonal basis, they all tended to borrow funds at about the same time, thus straining the resource envelope. Due to these findings, CLOC was established to address these challenges by providing funds to SILC groups at the beginning of the agricultural season, allowing farmers to borrow at any given time. By the end of the project, CLOC established that farmers could easily access loans due to the project's intervention. Using the loans, farmers invested in poultry, businesses, and agriculture (ESD & CE, 2022).

Additionally, the university has also planned for more interventions through more faculties. This section explores some of the planned interventions, although several of them are yet to be implemented. The Faculty of Law, for example, has had several planning meetings with Kayabwe Town Council. Through these meetings, the Council has expressed interest in the faculty writing ordinances on matters that concern the people, such as land and domestic violence. The council has also expressed interest in the faculty providing training and legal literacy sessions on several issues regarding the people. Staff and students from the Faculty of Law are already working on some of these projects. In addition to legal matters, the Council has also been teaming with the Faculty of the Built Environment on the prospects of developing a master plan for the town council. The faculty intends to deploy its students to interact with town Council leaders to capture their ideas on their concept of the master plan.

Since 2022, the Centre for African Studies has been implementing a Heritage Education Program in eight schools in Nkozi, Buwama, and Kamengo, all in Mpigi District. The program's intention is to train young people to appreciate their cultural heritage and traditional activities. In the schools covered, cultural heritage clubs have been formed, and several competitions have been organized in debates, speeches, music, dance, drama, and essay writing on culture. One of the objectives of this program is to encourage young people to embrace their cultural heritage instead of copying Western cultures and values that are mostly a contradiction of their African values.

Currently, the ICT Department of UMU is implementing an ICT research project in Nkozi and surrounding areas, funded by RENU. Working with the Faculty of Agriculture, the project aims to establish soil moisture and disseminate that knowledge to the farmers using ICT technology. Thus, the project gives farmers the skills to know when and what to plant during different seasons. When fully implemented, this engagement with farmers is projected to increase their yields as farmers will circumvent the poor soils and unfavourable seasons. Besides, farmers will plant only seeds they know will perform better in certain soils. This engagement is likely to increase production and consequent incomes for the farmers. The other initiative the Faculty of Agriculture is implementing is the establishment of seed banks to help farmers improve their farms, especially in organic farming. In this, indigenous varieties of seeds have been multiplied and supplied to farmers in the villages of Nkozi and its neighbourhood. Staff and students from the faculty regularly visit and give farmers advice on organic farming practices. This initiative is expected to make Nkozi and surrounding areas more food secure and increase opportunities for raising their incomes. Along with this project, the Faculty of Agriculture requires students of the Bachelor of Agriculture Year 3 Semester Two to be attached to farmers. This practice is intended to achieve two related objectives. The first one is for students to identify problems that farmers face in their farming practices. Students will relate what they have learned in theory classes to what farmers experience. This intercourse between theory and practice will increase the learner's knowledge. The second objective is for the learner to provide a solution to the problems farmers have experienced. The learner will research more about the challenges and, in the process, acquire more knowledge about them. By providing a solution to farmers, the student will project UMU in a more favourable perception that it teaches courses relevant to society.

Another recent UMU's intervention, with the support of the RUFFORD Foundation, UK, is the promotion of aquaculture in Mpigi District. The project was premised on the assumption that farmers deplete lakes of fish to support their livelihoods. Farmers' fishing methods have tended to collect young and old fish, thus threatening the conservation of some species. Consequently, UMU, through the Faculty of Agriculture, has encouraged farmers to rear fish in cages instead of catching them from the lakes. Farmers have been sensitized about the practice and taught to rear fish in cages. When fully embraced, the intervention will conserve the species of fish but also increase farmers' incomes from direct sales from the cages.

To sum up, UMU's community interventions during the past years have had several benefits for the university and the community. One of the key benefits is UMU's positive contribution to the physical, social, and economic development of the communities where external community engagements have been implemented. Many of UMU's engagements have resulted in food security, increased incomes, and improved health benefits. Due to improved living conditions, some members of communities have had the luxury of sending their children to good schools. On this note, UMU's contribution to communities must be further studied against the parameters of the Sustainable Development Goals.

UMU Community Engagement Interventions: Contribution to Sustainable Development Goals

The relationship between Uganda Martyrs University and the Sustainable Development Goals can be scrutinized through the university's implementation of community engagement. Although some of UMU's community engagement programs were implemented long before the UN SDGs agenda was promulgated, the spirit of their implementation was in tandem with it, as they targeted uplifting households and communities largely in the Nkozi Sub-County area from the social and economic obstacles to development.

With the UN's and member states' adoption and promulgation of the 17 SDGs in 2015 (United Nations, 2024a), community engagements in universities earned themselves a backbone and a platform upon which their activities would be aligned. For the record, the SDGs covered several key topics under the social, economic, environmental, and technological developments (Chankseliani & McCowan, 2021). In the same breath, community engagements at UMU touched similar areas of focus, albeit on a community level (ESD & CE, 2022). It should be noted that some of the community engagement activities cut across several of the SDGs. Therefore, this section's objective is to highlight how UMU's community engagement activities have contributed to the implementation of the SDGs.

Goal number 1 of the 17 SDGs was premised on the prevailing poverty in the world, where 670 million people still live in abject poverty. These poor people, mostly in Sub-Saharan Africa and Southern Asia, live on less than 2.15 dollars per day (United Nations, 2024b). UMU's main interventions from 1999 to date have targeted increasing household incomes in Nindye, Buwama, Nkozi, Kituntu, and Nkozi sub-counties in Mpigi District. This was accomplished through training farmers in business and management skills and entrepreneurship, embracing modern farming methods such as crop handling, post-harvesting, acquiring processing and marketing skills, forming cooperatives, and working collaboratively with other farmers (Bakiika, 2001). Programs initiated by UMU in collaboration with UPFORD trained farmers to save in groups and acquire loans from the SILC and CLOC groups to start enterprises (Guzman, 2018). The promotion of aquaculture, the enterprise of cage fishing, was another initiative where farmers improved their incomes. All these initiatives contributed to fighting poverty in the areas where UMU's community engagements improved household incomes.

Although the Equator Valley Farm started in 1994 and targeted generating income for the university and serving as a demonstration unit for agricultural students, it employed individuals from the surrounding communities. Those employed were paid salaries to support their households. Besides, the Farm provided advisory services and farm inputs, resulting in increased production. Furthermore, coordinating several training workshops in crops such as maize, coffee, and bananas improved farmers' production and income. Other livestock training that focused on the use of feeds, dairy management, poultry management, record keeping, and piggery management all resulted in increased production and income. On top of this, UMU spearheaded the coordination of markets for farmers' produce, resulting in farmers' increased incomes (Bakiika,

2001). Accordingly, UMU's initiatives reduced poverty in households through community engagement programs, thus contributing to the implementation of SDG 1.

Goal 2 of the SDGs, which targets ending hunger, attaining food security in households, and improving and promoting sustainable agriculture (Ramona & Monica 2023), was based on the prevailing hunger problem in the world. Sadly, a large percentage of the world's population that is hungry lives in developing countries, with many malnourished, and most of them in Sub-Saharan Africa. Uganda is not an exception to these alarming statistics. Owing to the prevailing situation as described above, the UN identified hunger in SDG 2 as one of its 17 goals. UMU's several community interventions since 1999 have contributed to the efforts to reduce hunger within the communities surrounding it. Such interventions included the provision of soya to the communities living in the sub-counties of Buwama, Kituntu, and Nkozi. Also, UMU promoted the growing of soya by the farmers by providing extension workers. Realizing that farmers were struggling to get reasonable production from their maize and bananas, UMU trained farmers in modern farming methods such as post-harvest management, crop handling, processing, and marketing techniques to enable them to become food secure. The university additionally trained farmers and mobilized them to form cooperatives where they jointly sold their produce to organizations such as the World Food Program. In all these efforts, UMU contributed to reducing hunger and malnutrition within the communities and hence contributed towards the implementation of SDG 2.

Goal 3 of the SDGs aims to guarantee healthy lives and promote the well-being of people of all ages. Goal 3 was informed by the dire world health situation, where child mortality, maternal mortality, and HIV/AIDS are still high, especially in Sub-Saharan Africa (Batliwala, 2019). UMU's contribution to this SDG was to sensitize communities to observe basic hygiene and give households health information. Later, UMU supplied mobile phones to households, through which health workers would be contacted to bring health services to the sick. Pregnant mothers were provided with a motorcycle (boda-boda) service to transport them to the clinics. Through these initiatives, UMU implemented SDG 3.

Goal 4 of the SDGs focuses on inclusive and equitable quality education that promotes lifelong learning opportunities for all. The key words to underscore here are inclusivity (where no one is excluded), equity (fair and impartial, meaning equal and fair opportunities are accessible by all without favour), quality education (accessed by all and not being a preserve by some), and lifelong learning (where learning is accessible by all, irrespective of one's age). This SDG was informed by the UN's statistics that although pupils' enrolment had increased, millions of children remained out of school, particularly in Sub-Saharan Africa (United Nations, 2024c). In many developing countries, equality between boys and girls accessing primary education was still a dream. UMU's community engagement interventions on SDG 1 (on poverty), SDG 2 (on hunger), and SDG 3 (on guaranteeing the health and well-being of all ages) resulted in improved incomes for households to pay school fees for their children. By so doing, UMU implemented SDG 4.

Goal 5 aims at gender equality and the empowerment of all women and girls. This is against the inequality that exists in society regarding opportunities and services extended to males and females. For example, globally, women in employment earn less than men, and they also spend three times as much of their time on domestic and care labour as men, which is unpaid (Department of Global Communications, 2023). In Sub-Saharan Africa and Asia, for example, girls face more barriers than boys in enrolling in primary and secondary schools. In UMU's environs, the same inequalities that often arise from cultural biases persist. Though there were no specific interventions on gender, several initiatives that UMU implemented benefited the female gender. The several savings groups of SILC, CLOC, and SEP consisted mostly of the female gender and, consequently, the loans accrued from such groups benefited the female gender more than the male one. The majority of the members in these groups were women. UMU, by supporting several initiatives within the communities, such as the formation of savings groups, implemented SDG 5.

The target of Goal 6 is to provide safe, sufficient, and affordable clean water to the world's population. Safe and affordable clean water would, in effect, improve people's sanitation, hygiene, and health (Batliwala, 2022). UMU's intervention of sensitizing the community, especially in boiling and drinking clean water, contributed to the community's good health. Besides, drilling a borehole in Kikoota and serving residents in Kayunga, UMU ensured the provision of clean water in the area. Moreover, before the National Water reached Nkozi Sub-county, UMU provided water to the community. In all these initiatives, UMU implemented Goal 6 of the SDGs.

Goal 8 focuses on sustained, inclusive, sustainable economic growth, full and productive employment, and decent work for everybody. Thus, Goal 8 is addressing two interlinked themes of economic growth and decent work. Behind these themes is the idea of spurring growth by creating jobs and reducing poverty among the population (United Nations, 2019). Indirectly, this goal has been achieved by the implementation of several interventions. UMU's interventions by training farmers, forming cooperatives and SACCO groups, establishing common markets, growing soya, and improving maize and banana farming all created decent and sustainable jobs for farmers. With the acquired skills, farmers employed themselves on their farms; others created enterprises to sell their produce. Due to these efforts, farmers have diversified their income by venturing into other enterprises in addition to farming. For example, some households ventured into poultry farming and selling several goods in shops (Guzman, 2018). However, the most visible employment for some youth was the Equator Valley Farm by the university, where they worked as herdsmen and labourers in the university gardens. To date, the university employs about 60 casual workers at the Farm. By these initiatives, UMU implemented Goal 8.

Goal 11 targets cities and human settlements to make them more inclusive, safe, resilient, and sustainable, among other reasons, by reducing slums, improving urban governance, enhancing dwellers' participation, addressing climate change issues, and improving urban management by controlling pollution and waste management through rules and regulations. This goal presumes sectoral interlinkages and collaborations (Küfeoğlu, 2022). UMU implemented this goal through other interventions as mentioned above, affirming the role of interlinkages and cooperation with other sectors. Through its several programs that address poverty, hunger, health, water and sanitation, education for all, and provision of employment, UMU has contributed to building safe, resilient, and sustainable communities. Such interventions have created safe environments where youth can employ themselves and reduce temptations to engage in anti-social behaviours against society. But in addition to these initiatives, the Faculty of Law is working with the community to sensitize them on legal matters regarding land and domestic violence, and coming up with ordinances that can help the Town Council's governance. Similarly, the Faculty of the Built Environment (FoBE) is drawing up a Master Plan for the Town Council. Like any other city or town council, the planning for such requires the participation of stakeholders so that together with experts, they can create an ecosystem where everybody's views are considered for building a safe and sustainable community (Kosalraman, 2024). Through these engagements, UMU has contributed to the implementation of Goal 11.

Goal 13 targets mitigating the impacts of climate change by encouraging the promulgation of policies that would curb behavioural hazards to the climate at the international and national levels (Owusu-Sekyere et al., 2024). Additionally, countries would cooperate in mobilizing their populations against behaviours that would fuel climate change. With climate change already on our doorsteps, Goal 13 calls for resilience and the development of adaptive capacity to cope. In its reliance and adaptive measures, UMU initiated the planting of trees at every celebrated event within its environment. More importantly, UMU started a program of planting trees following the path of the Equator throughout the country. This initiative is expected to contribute to the country's forest cover and mitigate climate hazards. Through these initiatives, therefore, UMU has implemented Goal 13.

Goal 14 targets life below water with special consideration of the protection of oceans and marine resources within them. Notably, oceans are a significant component of the planet, as they occupy 70% of the Earth's surface and, consequently, play a crucial role in regulating the planet's climate (Huck, 2022). Goal 14, therefore, aims to conserve the oceans, seas, lakes, and rivers. Although Uganda is a landlocked country, its lakes and rivers empty into the oceans, implying that what happens on land will automatically affect the oceans. In Uganda's case, oceans can be affected by its pollution of its lakes and rivers mostly through dumping of plastic materials (plastic bottles, fishing nets), chemical pollution through use of pesticides, oil spills from cars as they are washed in swamps that empty into rivers and lakes and industrial chemicals which end up in rivers and lakes. Some pollutions originate from the use of agricultural chemicals and inputs that contain excess nitrogen and phosphorus, which reduce oxygen in water. As UMU, given its proximity to Lake Victoria, the dangers of such pollution reaching the lake are not far-fetched. Thus, to mitigate such hazards, UMU enforced the collection of plastic materials and polythene bags on all its campuses. Additionally, UMU has opted for organic farming to distance itself from chemicals that would otherwise affect soil and eventually filter into rivers and lakes. Additionally, UMU is encouraging farmers to engage in cage farming in lakes and rivers. Although UMU's objective is to provide a convenient source of income and also allow the local fish species to grow and multiply, the benefits of the enterprise have far-reaching consequences on the oceans' ecosystems. Fish that would otherwise be obtained from the oceans to supply several markets would be obtained from local rivers and lakes through cage farming. In effect, fishing pressure on the oceans would be lessened with the consequent protection of ocean habitats such as coral reefs and seagrass beds. This is because there would be less fishing gear in the oceans to disrupt their ecosystems. UMU, through the Faculty of Agriculture, has been encouraging farmers to engage in cage farming, indirectly, promoting the sustainability of the oceans by farmers' production of seafood that would be obtained from the oceans. In other words, by shifting to cage farming, oceans would be spared the effects of fishing gear, including but not limited to the use of plastic nets. By so doing, UMU is implementing Goal 14 of the SDGs.

Goal 15, which focuses on life on land, aims to preserve and promote the responsible and sustainable use of the Earth's ecosystems by encouraging, among other things, reforestation and the mitigation of desertification (Sayer et al., 2020). Given the widespread global practices of land degradation and biodiversity loss, this goal aims to halt and reverse the trend. UMU has implemented this goal through its implementation of reforestation programs. In all its campuses, UMU initiated a practice where every key event is celebrated with the planting of several trees. The Main Campus, for example, located at the Equator, has increasingly become a green campus with several species of birds hibernating there. Additionally, in collaboration with the Wildlife Authority, since 2024, UMU has initiated a tree-planting project along the Equator path, from the Kenya border to the Democratic Republic of the Congo border. When the project is completed, the livelihoods of the people living along the path of the Equator will be improved. But more importantly, the people will also be sensitized about the importance of protecting and promoting the planting of trees within their localities. Furthermore, the University, at the Main Campus, has protected and preserved the swamp that connects with River Katonga. Through these several initiatives, UMU has implemented Goal 15.

Goal 16, which emphasizes building strong communities through peace and justice (DCAF, 2021), was implemented by UMU through its several community engagement activities. Sensitizing communities to domestic violence and legal matters, such as land, which is often a source of conflict in communities, UMU is at the forefront of contributing to the building of peaceful and inclusive societies. UMU spearheaded peace-to-peace dialogues in Karamoja, Kapchorwa, and Turkana. In these regions, communities were reminded of the significance of peace and its interlinkage with development. In addition to these efforts, through the Equator Peace Academy, UMU implemented Goal 16 by engaging communities on themes of cultural identity and religious tolerance, all for building and promoting sustainable peace.

Goal 17 focuses on partnerships and collaborations among stakeholders, particularly in the implementation of all the SDGs. From governments, civil society, non-governmental organizations, and private partnerships, to communities and individuals, the implementation of all SDGs requires the cooperation and collaboration of all partners (Larissa et al., 2021). Therefore, aware of the significance of these partnerships, the goal advocates for their strengthening and revitalization for sustainable development. In implementing this goal, UMU identified global and local partners with whom several SDGs are implemented. As indicated above, several interventions and projects on poverty, hunger, and health were initiated and implemented through collaborations with international and local partners. Some of these partners included the University of Notre Dame, USA; UPFORD, USA; and AP-TRIAS, Belgium. Locally, UMU's partners have been and continue to be the communities where these projects are implemented. Thus, by working with these partners, UMU is implementing Goal 17.

To sum up, UMU's community engagements have enabled it to implement the UN's SDGs. But more importantly, they have made UMU more relevant and focused on the needs of communities. Through them, the university and the community have benefited from the interventions. Staff and students have related what is learned in lectures to what happens in communities. Teaching, learning, and research are ipso facto related to what occurs in the community. At the same time, students benefit from the exposure and opportunity to engage in internships. The initiatives enable UMU to implement the UN's SDGs and make UMU more vibrant and relevant to the community. The significance of UMU's community engagements in contributing to the implementation of SDGs cannot be overstated. The lessons UMU has learned and which can also be adopted by other institutions are several, as the next section highlights.

The Lessons

One of the three core functions of Uganda Martyrs University is community engagement; the other are teaching, learning, and research. Lately, universities worldwide have also adopted innovation and entrepreneurship as strategies for making their graduates more employable and thus more relevant to the needs of society. That aside, the community engagement function, as clearly defined by the university's mission, should be placed at the same level as the other two functions. Thus, community engagement is as important as teaching, learning, and research. The three in this triangular or *trinitarian* relationship mutually need and support each other. When one function is not adequately present in the life of the university, the centre cannot hold.

Although community engagement is an important function in the life of the university, as evidenced by its recognition in university policies, such as the appraisal of academic staff, the attention it receives in the deliberate planning and execution in several universities is inadequate. Worse still, it is given a back seat against the other competing needs of the university that relate to teaching, learning, and research. For some universities, including UMU, community engagement activities are funded mostly through projects. Without project budgets, community engagement activities would hardly be financed. The few community engagement activities financed by universities are implemented as an afterthought. Thus, a deliberate plan and strategy backed by an intentional budget to execute community engagement activities should be the strategy to undertake if UMU and other Higher Education Institutions (HEIs) are to strive for relevance in society.

Community engagement should reclaim its rightful position among the three traditional functions of a university; that is, its role should be perceived as equal and supportive of the roles of the other two functions, namely, teaching, learning, and research. For example, as Erjavec (2021) points out, community engagement activities can play a significant role in curriculum development, as the faculty would be in a position to establish the actual needs of society through feasibility studies. Besides curriculum development, community engagement activities can play a role in reviewing curricula. Notably, community engagement activities in HEIs can inform the mode of lecturers' teaching in lecture halls, as it would be practical and learning carried out onsite. This is important on three accounts: One, community engagement would mean moving away from a theoretical to a practical-oriented approach that is meaningful to learners, lecturers, and the community. Secondly, but more importantly, learning backed by community engagement activities would move from a lecturer-centred to a student-centred approach (Freire, 1993). This type of learning focuses on acquiring skills and expertise, rather than accumulating theories that are irrelevant to learners and the community. Thirdly, the community's participation in the university's engagement activities would contribute to its social and economic development.

Community engagement can support research (Nina et al., 2023). Students would identify problems in communities and embark on them to provide solutions. At the same time, the community would serve as a field laboratory against which students would test their hypotheses as they embark on research. Students' and staff's disseminated findings would contribute to the community's social and economic development. While the university would benefit from these research engagements, the community would also benefit from the university's activities. In other words, staff and students' research would be informed by society's needs, and the findings and recommendations would spur growth and transformational change in the community.

For years, universities have often been perceived for their knowledge production and development, with less emphasis on their impact on society. In Africa and Uganda, in particular, due to this perspective, the only university we had before the entry of other public and private universities was perceived as an ivory tower. This implied it could not be easily reached, nor would its stakeholders freely reach out to the communities. Increasingly, however, with several universities and with a new understanding that universities should address society's needs and consequently contribute to its social and economic development, this mentality of the ivory tower has slowly but surely given way to one of universities contributing to society's social and economic development. The channel through which this can be implemented is community engagement. Accordingly, if this is to occur, HEIs should intentionally plan and budget for community engagement, as it is this that will contribute to the community's social and economic growth (Mukidi, 2024).

Although community engagement in some universities is side-lined and emphasis is placed on other functions, it is the pathway through which universities can be relevant to society. Moreover, it is through community engagement activities that HEIs can play a role in the implementation of SDGs. By implementing SDGs, HEIs will contribute to society's economic development. Besides, they will overcome the generally negative perception of society that universities are ivory towers without meaningful and tangible contributions to society's development (Mukidi, 2024). Moreover, as Ekene and Oluoch-Suleh (2015) affirm, if nothing is done, HEIs will appear irrelevant and inconsequential to society.

As stated above, almost all HEIs espouse community engagement as a key function in their missions. They even require it for the academic staff's appraisal in their renewal. In most cases for most HEIs, the emphasis remains at the policy level, without allowing it to permeate all HEIs' activities and getting all the support. Thus, community engagement should be prioritized in university plans. Secondly, HEIs must establish a conscious and intentional relationship between community engagement and the SDGs. This will require deliberate planning of community engagement activities at the top management and faculty board levels, with a supportive budget. In contrast, the current practice of community engagement is implemented by default. Finally, this study contends that community engagement should be aligned with HEIs' strategic plans. In the same breath, and more importantly, HEIs' strategic plans should be aligned with the national development plans if African countries are to develop.

Conclusion

This chapter focuses on UMU's community engagement as the key vehicle for implementing the UN's SDGs. Beginning with the community engagement policy, approved and promulgated by the Senate in 2017, the chapter explores several topics within the policy: objectives, university membership, categories, and types of community engagement, distinguishing between Internal Community Engagement (ICE) and External Community Engagement (ECE). Additionally, the chapter explores how community engagement activities initially began at UMU. Furthermore, the chapter highlights some of the community engagement activities UMU has implemented in its neighbourhood communities. The chapter demonstrates that by deploying a participatory approach that included students, staff, external partners, and community members, UMU contributed to the reduction of poverty, hunger, and improvement of people's health and well-being. The chapter also argues that UMU's interventions had far-reaching impacts, for example, the promotion of gender equality, especially by empowering women and girls, and the creation of a conducive environment where human settlements are inclusive, safe, resilient, and sustainable. The chapter also contends that Uganda Martyrs University, through its community engagement, has created an environment for partnership building where several households worked together in cooperatives. In addition, communities partnered with the university and partners from the funding agencies. Through these collaborations, all stakeholders benefited: UMU through the improvement of its teaching and learning, and staff and students had communities for their engagements. But above all, community livelihoods were improved, and external partners' dreams of transforming communities were fulfilled. To sum up, UMU's community engagements contributed to the social and economic development of the people and also abetted the implementation of several SDGs, albeit some retrospectively.

References

Bakiika C.S. (Cord Consult LTD (2001). Equator Valley Farm: Baseline Survey in Six Sub-Counties, Mpigi District.

Batliwala, Y. (2019). *SDG 3: Good health and well-being. A Legal Guide*. Retrieved December 13, 2024, from https://www.a4id.org/wp-content/uploads/2022/02/A4ID-Legal-Guide_Chapter-6.pdf

Batliwala, Y. MBE (2022). SDG 6: Clean Water and Sanitation. A Legal Guide. Accessed from https://www.a4id.org/wp-content/uploads/2022/02/A4ID-SDG-Legal-Guide_Chapter-6.pdf on 10-12-2024.

Chankseliani, M., & McCowan, T. (2021). Higher education and the sustainable development goals in higher education. *Higher Education., 81*, 1–8. https://doi.org/10.1007/s10734-020-00652-w

DCAF. (2021). *Sustainable Development Goal 16: The importance of good security sector governance for the achievement of the 2030 Agenda*. Retrieved December 2, 2024, from https://www.dcaf.ch/sites/default/files/publications/documents/DCAF_BG_21_SustainableDevelopement_0.pd

Department of Global Communications. (2023). *What is Goal 5—Gender Equality?* Retrieved December 8, 2024, from https://www.un.org/sustainabledevelopment/wp-content/uploads/2023/09/Goal-5_Fast-Facts.pdf

Education for Sustainable Development and Community Engagement-ESD & CE. (2022). A brief to the Vice Chancellor. Uganda Martyrs University.

Ekene, O. G., & Oluoch-Suleh, E. (2015). Role of institutions of higher learning in enhancing sustainable development in Kenya. *Journal of Education and Practice, 6*, 91. Retrieved December 7, 2023, from https://files.eric.ed.gov/fulltext/EJ1079963.pdf

Erjavec, J., (2021). Stakeholders in curriculum development: Case of Supply Chain and Logistics programme. In: 7th International Conference on Higher Education Advances (HEAd'21). Valencia, Spain, 2021. Valencia: Universitat Politècnica de València, pp. 397–403. https://doi.org/10.4995/HEAd21.2021.12875.

Freire, P. (1993). *The pedagogy of the oppressed*. The Continuum International Publishing Group.

Guzman, B. D. (2018). *Rural Uganda: Nurturing Social Enterprises in the Ford Program in human development and solidarity*. University of Notre Dame, Keough School of Global Affairs.

Huck, W. (2022). Goal 14: Conserve and sustainably use the oceans, seas and marine resources for sustainable development. In *Sustainable development goals: Article-by-article commentary* (1st ed., p. 743). Nomos, C.H. Beck, Hart Publishing. ISBN: 9783848760770. Retrieved August 14, 2025, from https://ssrn.com/abstract=4035316

Kosalraman, A. (2024). *Community engagement in sustainable development*. Faculty of Architecture, MEASI Academy of Architecture. Retrieved December 6, 2024, from https://www.ijser.in/conf/NDCSAPT/NDCSAPT11.pdf

Küfeoğlu, S. (2022). SDG-11: Sustainable Cities and Communities. In: Emerging Technologies. Sustainable Development Goals Series. Springer, Cham. https://doi.org/10.1007/978-3-031-07127-0_13 (Accessed: 6 December 2024).

Larissa, O. D., Diane, A. R., Andre, L. F., Rosana, A. V., Homero, F. F., Mikko, K., & Julia, B. R. (2021). Innovation ecosystem framework directed to Sustainable Development Goal #17 partnerships and implementation. *Sustainable Development*. https://doi.org/10.1002/sd.2191. Retrieved August 14, 2025.

Mukidi, C. A. (2024). Sustainability and the triple Mission of the University: Uganda Martyrs University in perspective in Sub-Saharan Africa. In P. N. Abooki (Ed.), *The sustainability of higher education*. Palgrave Macmillan, Springer Nature Switzerland AG.

Nina, P. M., Kiambati, K., & Mbalassa, M. J. (2023). Using black soldier fly bio-waste conversion to produce commercial animal feed rations. In *A grant proposal submitted to bio-innovate Africa*. Retrieved December 6, 2024, from https://bioinnovate-africa.org/call-for-proposals-for-early-stage-bio-incubation-projects/

Owusu-Sekyere, Y. S. N., Obakeng, T. S., & Dora, A. T. (2024). *Sustainable Development Goal 13: Urgent action to combat climate change and its impacts*. Retrieved August 14, 2025, from https://www.elgaronline.com/edcollchap/book/9781839105449/book-part-9781839105449-35.xml

Pool, E. & Guzman, B. D. (2020). Social enterprise project: End-line report to the Ford Family Program in Human Development and Solidarity Studies. [report] Ford Family Program in Human Development and Solidarity Studies, University of Notre Dame / Kellogg Institute for International Studies, Notre Dame, IN.

Ramona, Lile & Monica, Ocnean & Balan, Ioana. (2023). Challenges for Zero Hunger (SDG 2): Links with Other SDGs. https://doi.org/10.3390/books978-3-03897-863-3-2 (Accessed: 5 December 2024).

Sayer, J., Sheil, D., Galloway, G., Riggs, A. R., Mewett, G., MacDicken, K. G., Arts, B., Boedhihartono, A. K., Langston, J., & Edwards, D. P. (2020). Life on land—the central role of forests in sustainable development. In P. Katilla et al. (Eds.), *Sustainable development goals: Their impacts on forests and people*. Retrieved August 14, 2025, from https://bebassampah.id/public/files/uploads/sustainable-development-goals-their-impacts-on-forests-and-people-compressed.pdf#page=518

Ssekijjo, F. (June 2018). SILC-SEP PROJECT REPORT NINDYE – UGANDA

Uganda Martyrs University. (2017). *Community engagement policy of Uganda Martyrs University*. Uganda Martyrs University.

United Nations. (2019). *Sustainable Development Goal 8: Promote sustained, inclusive and sustainable economic growth, full and productive employment and decent work for all*. Retrieved December 5, 2024, from https://sustainabledevelopment.un.org/content/documents/2144EGM_SDG_8_Concept_Note_15_Feb_2019.pdf

United Nations. (2024a). *Department of Economic and Social Affairs, Sustainable Development*. Retrieved December 12, 2024, from https://sdgs.un.org/goals

United Nations. (2024b) *Sustainable development goals: Goal 1—End poverty in all its forms everywhere*. Retrieved December 6, 2024, from https://www.un.org/sustainabledevelopment/poverty/

United Nations. (2024c). *Sustainable development goals: Quality education*. Retrieved December 8, 2024, from https://www.un.or/sustainabledevelopment/wp-content/uploads/2018/09/Goal-4.pdf

Part II

Land-Use, Environment, Climate and Biodiversity in Africa

2 Food Security and Climate Finance in East Africa

Pascal Ndyanabo Rutazihana, Dushime Kashori, and Augustin Nkundimana Alfred

Abstract

This study examines the effect of climate finance on food security in East Africa, a region grappling with increased vulnerability to the effects of climate change. Recurrent droughts and floods in the region have severely affected agriculture and food security. Data obtained over a nine-year period (2012–2021) was analysed using the fixed effects method. The results indicated that food security is influenced by climate finance, other factors include public spending on health, per capita income and agricultural productivity. Implications: there is a need for targeted allocation of climate finance and institutional policy.

Keywords

Mitigation · Adaptation · Prevalence of undernourishment · Sustainable Development Goal 2

Introduction

The most pressing environmental issue facing the world today is climate change, and its impact on food systems and food security in Africa is among the most feared. Climate change poses many challenges, particularly for African nations, which have contributed the least to environmental degradation but suffer the most from its negative effects. These challenges include changing rainfall patterns, rising temperatures, rising sea levels and deteriorating food security and water supplies. The fact that the majority of African countries lack the institutional, financial and technical means to reduce emissions and increase resilience to climate change exacerbates these difficulties (Doku et al., 2021). Whether funds allocated to East African Community (EAC) countries for climate change adaptation and mitigation improve food security is an important policy concern that this study addresses. The motivation for conducting this study is a combination of policy concerns and knowledge gaps in literature.

According to Nakai (2018), the Sub-Saharan Africa region has the highest prevalence of undernourishment, affecting nearly 21% of the

P. N. Rutazihana (✉)
Dschang School of Economics and Management, University of Dschang, Dschang, Cameroon

Association des chercheurs en Economie de Développement Appliquée (ACEDA), Yaoundé, Cameroon

Université Catholique la Sapientia de Goma, Goma, Democratic Republic of the Congo

D. Kashori · A. N. Alfred
Université Catholique la Sapientia de Goma, Goma, Democratic Republic of the Congo

P. Nyadoi et al. (eds.), *Sustainable Development Goals in Africa*, Sustainable Development Goals Series, https://doi.org/10.1007/978-3-032-20742-5_2

population, and the worst levels of hunger in the world. In contrast, other developing regions such as Asia and South America have prevalence rates of 11.4% and 5%, respectively. The situation is particularly alarming in the East African Community (EAC), where food insecurity remains a persistent crisis. In 2022, around 55 million people were in a situation of acute hunger, according to the FAO, mainly due to repeated droughts, conflicts and the degradation of agricultural land. In Ethiopia, 22.6 million people were affected, while in Kenya, 6.5 million also suffered from severe food insecurity. In addition, 25% of children under the age of five in the region are affected by chronic malnutrition, with alarming rates in Burundi (56%) and Ethiopia (37%). These statistics underline the structural challenges linked to food production in the region, exacerbated by climate change (FAO, 2023).

Agricultural production fell by 7% between 2015 and 2022, mainly due to prolonged droughts. Kenya and Somalia have suffered a 30% drop in cereal production. In addition, food inflation has reached record levels in 2022, with increases of 34% in Rwanda and 25% in Kenya, according to IMF estimates, severely limiting access to food for the poorest households. This crisis is exacerbated by rapid population growth of 2.8% per year, which brings the total population of the EAC to over 300 million people, increasing pressure on natural resources and food systems (IMF, 2023). In addition to climate impacts, armed conflict has exacerbated food insecurity in countries such as South Sudan, where 70% of the population, or nearly eight million people, were facing severe hunger in 2023. A study published in Global Food Security points out that the drought between 2019 and 2023 was one of the most severe in recent decades, affecting more than 12.5 million people in the Horn of Africa. This has led to major losses of livestock, reduced harvests and massive population displacements. The reduction in arable land per capita is exacerbating this situation: the average area per capita has fallen from 0.18 hectares in 1990 to 0.12 hectares in 2022, according to a study by IFPRI (2023). In addition to these few facts, some recent works argue that the current amounts of climate finance paid by donor countries to Africa may not be sufficient to meet the climate change challenges facing the continent. For example, Ngwenya and Simatele (2020) note that while developed countries have pledged to mobilise a total of US$100 billion by 2020, this falls far short of the World Bank's estimate of US$200 billion per year per country needed by African nations. More recently, Timperley (2021) point out that developed countries have already failed to meet their financing target of US$100 billion by 2020, and these authors put forward proposals for commitments of at least US$750 billion a year. For African countries, these developments are of interest because the climate finance received from donor countries is the main source of funding used to circumvent climate-related environmental changes that result in heavy rainfall, droughts, floods and high temperatures. Recently, Stuch et al. (2021) estimated that over the next few decades, climate-related disasters will reduce staple food crop yields by 80–85% in Central and West Africa, 29% in Southern Africa and 32% in East Africa. These figures are worrying because agricultural production is responsible for more than 70% of the continent's food supply (Appiah et al., 2019).

Lipper et al. (2014) shows how climate finance can be used to achieve food security through climate-smart agricultural practices. Junghans and Köhler (2016) highlight the need for a mediator in the administration of climate finance for smallholder farmers in order to more effectively promote food security through climate finance support. Castro-Nunez (2018) demonstrates the compatibility of programmes aimed at forest conservation with efforts related to peace-building and sustainable food systems in Colombia. Rahaman and Rahman (2020) documented how various climate funds influence food security in a single country, Bangladesh. More recently, Sarker and Fagun (2021) examine the role of climate finance governance in hazard-prone vulnerable areas in Bangladesh and find that while the climate finance mechanism is good enough, its implementation is unsatisfactory. The main shortcoming of the previous literature is that it only provides conceptual channels through which climate finance influences food security, but it does not empirically quantify the effect of climate finance on food security or test whether the effect is significant. Previous studies do not also investigate the separate effect of climate change adaptation and mitigation financing. This study thus examined whether climate finance provided to recipient countries in East Africa has contributed to food security in the region. The following hypothesis was used—H1: Climate finance for adaptation and mitigation has a positive impact on food security.

This study makes four significant contributions to the academic discourse on climate finance and food security. First, it is a pioneering study which, to the author's knowledge, marks one of the first comprehensive attempts to empirically examine the effect of climate finance on food security, particularly in East Africa. This makes the study a crucial reference point for further research in this area. Second, the results of the study not only add something new to the empirical literature, but also contribute to the ongoing discussion on how best to improve food security in Africa through the use of climate finance and other related mechanisms. The study contributes to the theoretical and policy discourse on the adequacy of existing research on climate finance in Africa to achieve 'climate justice' in poor regions. Third, the focus on the East African Community is particularly important because of the unique characteristics of the region, including its heightened vulnerability to the impacts of climate change, heavy reliance on rain-fed agriculture, coupled with rapid population growth, putting food systems under pressure.

This study therefore aimed to analyse how climate financing, when directed towards mitigation or adaptation, influences food security in the East African Community. The hypothesis was that adaptation funding leads to a more tangible and immediate improvement in food security. Moreover, very little is known about the effects of climate finance categories on improving food security and food systems in Africa, and even more so in East Africa.

Methodology

To assess the effect of climate finance on food security in East Africa, we used three main data sources. Data on climate finance were obtained from the Organisation for Economic Co-operation and Development (OECD), while data on food security indicators were collected from FAOSTAT. Control variables were obtained from the World Development Indicators (WDI) database. We collected data for seven East African countries including Burundi, Democratic Republic of Congo, Rwanda, Tanzania, Uganda, South Sudan and Ethiopia. The study was conducted between 2012 and 2021. This period was chosen based on the availability of reliable data from reputable sources, to ensure accuracy and reliability of the results. East Africa was chosen as the study region because of its unique environmental, social and economic contexts. The region is particularly vulnerable to climatic hazards, such as droughts and floods, which severely disrupt agriculture, the main source of livelihood for the population. The importance of climate financing in this area, through international funds such as the Green Climate Fund, provides an opportunity to assess the impact of these investments on the resilience of food systems. In addition, the countries of the region have implemented some active climate policies, making East Africa a relevant field of study for examining the concrete effects of adaptation and mitigation in a context characterised by climatic and socio-economic diversity.

Dependent Variable

To analyse the impact of mitigation and adaptation climate finance on food security, it is necessary to clearly define the dependent variable used in our study. Food security is a multidimensional concept that encompasses four essential pillars: availability, access, use and stability of food. According to the classic FAO definition, food security is achieved when 'all people, at all times, have physical, social and economic access to sufficient, safe and nutritious food to meet their dietary needs and preferences for an active and healthy life' (Smith & Haddad, 2015). This implies that food must not only be available in sufficient quantities, but that individuals must also have the economic and social means to access it, while ensuring that this access is sustainable and constant in the face of external shocks. For the purposes of this study, food security will be measured by the prevalence of undernourishment (all other things being equal, if the prevalence of undernourishment decreases, food security increases). The prevalence of undernourishment is an indicator that measures the percentage of the population whose calorie intake is insufficient to meet the minimum energy requirements for a healthy and active life (Ahmad, 2009). This indicator focuses on the quantity of food consumed and assesses whether populations have regular access to adequate food. It is widely used to assess food insecurity, as it directly captures the effects of food crises, climate shocks and economic constraints that limit access to sufficient food. This indicator is relevant to our study for several reasons.

Firstly, the prevalence of undernourishment provides a direct measure of food insecurity, by focusing on populations whose diets are insufficient to meet basic energy needs. This makes it possible to understand the impact of climatic variations, economic crises or disruptions to agricultural production on access to food (Fan & Brzeska, 2012). Secondly, this indicator highlights inequalities in access to food within the population. The most vulnerable households, often located in rural or poor areas, are generally the most affected by undernourishment, which makes it possible to target the necessary interventions. Finally, the prevalence of undernourishment is a commonly measured indicator, which makes it easily comparable across time and countries, facilitating longitudinal and international analyses. In previous studies, the prevalence of undernourishment has been widely used to assess the impact of climate shocks on food security. Smith et al. used this indicator to analyse the effect of climate change on food security in low-income countries. They showed that rising temperatures and reduced rainfall directly increase the proportion of the population suffering from undernourishment, as these phenomena reduce agricultural productivity. Similarly, Fan and Brzeska (2012) studied how agricultural productivity and climate adaptation policies influence the prevalence of undernourishment, concluding that investments in resilient agricultural practices reduce this indicator by stabilising the food supply. The prevalence of undernourishment is a key indicator for measuring food insecurity in studies on the impact of climate financing. This indicator accurately captures the difficulties of access to food in vulnerable populations, taking into account factors linked to production, food prices and climatic crises.

Independent Variable

To capture our main variable of interest, namely climate finance, we follow recent literature (e.g. Lee et al., 2022; Zhao et al., 2022) and use three categories of climate funds: (1) global climate funds, (2) climate mitigation funds and (3) climate adaptation funds. The data comes from the OECD DAC's climate-related development finance in constant 2020 dollars. The UNFCCC defines climate finance as 'local, national or transnational funding—from public, private and alternative sources—to support mitigation and adaptation actions to address climate change'. The Global Climate Finance Fund (GCFF), generally referred to as 'climate finance', can be divided into two groups: The Adaptation Finance Fund (AF) and the Mitigation Finance Fund (MF). The Adaptation Finance Fund maintains the modifications made to predicted or actual climate change and its effects. Mitigation funding focuses on reducing greenhouse gas emissions (Pickering et al., 2017). And, based on these definitions, summarised in Table 2.1, our climate finance data combines funding from bilateral and multilateral sources.

Table 2.1 Definition of variables

Variable	Definition	Source
FS	Prevalence of undernourishment (% of population)	FAOSTAT
GCF	Log of climate-related development finance—commitment (thousands of US dollars)	OCDE-CAD
MCF	Log of mitigation climate-related development finance—commitment (thousands of US dollars)	OCDE-CAD
ACF	Log of adaptation climate-related development finance—commitment (currently in thousands of US dollars)	OCDE-CAD
FDI	Log of climate-related development finance—commitment (thousands of US dollars)	WDI Database
Health_exp	Domestic general government health expenditure (% of GDP)	WDI Database
Income	Log of GDP per capita (constant 2015 USD)	WDI Database
Rur_Pop	Rural population (% of total population)	WDI Database
Agri_Prod	Agriculture, forestry, and fishing, value added (annual % growth)	WDI Database

Control Variable

Following the determinants of food security, we chose a vector of control variables for the robustness of the study (McMichael & Schneider, 2011; Smith & Haddad, 2015; Yao et al., 2023; Dasgupta & Robinson, 2021). The first control variable is total foreign direct investment (FDI) in sub-Saharan Africa, which comes from the World Development Indicators (WDI). We note that while some previous studies have used agricultural FDI as a more appropriate determinant of food security (Santangelo, 2018; Yao et al., 2023), we have opted to use total FDI due to insufficient time series data on agricultural FDI for African countries recorded by FAOSTAT. In addition, previous studies by Mihalache-O'Keef and Li (2011), Slimane et al. (2016) and Adom et al. (2019), generally find that total FDI is positively related to food security, which further motivates the use of this variable as a control variable in this study. The second control variable is public expenditure on health measured by the share of current health expenditure financed by domestic public sources for health. National public sources include national revenues in the form of internal transfers and subsidies, transfers, subsidies to beneficiaries of voluntary health insurance, non-profit institutions serving households (NPISHs) or business financing schemes, as well as compulsory contributions to prepaid and social health insurance. They do not include external resources spent by governments on health. In their study, Smith and Haddad (2015) showed that increased public health spending has a direct effect on reducing malnutrition, particularly among children, by improving access to quality medical care and allowing better use of available food. Alderman et al. (2006) have also shown that countries that invest more in healthcare are more resilient in the face of food crises, as populations are better equipped to deal with the illnesses that accompany periods of food shortage. The third control variable is per capita income, taken as gross domestic product divided by the mid-year population. GDP is the sum of gross value added by all resident producers in the economy, plus taxes on products and minus subsidies not included in the value of products. It is calculated without deducting the depreciation of manufactured assets or the depletion and degradation of natural resources. Data are expressed in constant 2015 US dollars. The work of Ahmad (2009) has shown that per capita income is a key factor in explaining access to food and adaptation to food crises. Households with higher incomes are less affected by food and climate shocks, as they can buy products on world markets and meet their needs even when local production declines. In addition, Dasgupta et al. have pointed out that low incomes are often correlated with increased food insecurity, due to the inability of poor households to absorb increases in food prices. The fourth control variable is the rural population, which refers to people living in rural areas as defined by the national statistical offices. It is calculated as the difference between the total population and the urban population. The rural population plays a key role in food production, particularly in developing countries where a large proportion of agriculture is carried out by smallholders. The resilience of this population to climate shocks has a direct influence on national food security (McMichael & Schneider, 2011). The final control variable is agricultural productivity, measured by the annual growth rate of value added in agriculture, forestry and fisheries in constant local currency. Agricultural productivity is closely linked to food security, in particular to the availability of food, one of the four pillars of food security (availability, access, utilisation, stability). High agricultural productivity means that more food is produced from the same resources, making it possible to meet the food needs of a growing population. In addition, agricultural productivity indirectly influences food prices, a crucial factor in access to food, particularly for low-income households. When productivity falls due to climatic hazards or a lack of investment, food prices rise, which can exacerbate food insecurity for vulnerable populations (Cline, 2007).

Econometric Model and Estimation Strategy

This research aims to provide an analysis of the impact of climate finance on food security. Therefore, previous empirical and theoretical research is used as the basis for the specification of the empirical functional benchmark model. As shown in Eq. (2.1) (with X representing the set of control variables), we investigate the following relationship:

$$\text{Food Security}_{\text{it}} = f\left(\text{Climate Finance}_{\text{it}},\ X_{\text{it}}\right) \quad (2.1)$$

The literature indicates that empirical data obtained by a log-linear specification gives more reliable and efficient results than a simple linear definition (Shahbaz & Lean, 2012). Thus, in order to reduce the variability of the data—which can result in a distribution of the error term that violates the normality assumption—we converted the majority of the variables to natural logarithms. The coefficients are then considered as elasticities as in Eq. (2.2):

$$\text{FS}_{it} = \beta_0 + \beta_1 \ln \text{CF}_{it} + \beta_2 \ln X_{it} + \varepsilon_{it} \quad (2.2)$$

where index i denotes the country and t the year, β_0 is the constant, β_1 and β_2 are the parameters to be estimated, and ε_{it} is a random error term. FS represents the dependent variable (Food Security), ln CF the explanatory variable of interest (representing the different categories of climate finance), X represents the vector of control variables, selected after an extensive review of the existing literature: Foreign Direct Investment (FDI); Public Health Expenditure (Health_Exp); Agricultural Productivity (Agri_Prod), log GDP per capita (lnIncome) and Rural Population (Rural_Pop). Equation (2.3) can be rewritten as follows:

$$\begin{aligned}\ln \text{FS}_{it} &= \beta_0 + \beta_1 \ln \text{CF}_{it} + \beta_2 \text{FDI}_{it} \\ &+ \beta_3 \text{Health\_exp}_{it} + \beta_4 \text{Agri\_Prod}_{it} \\ &+ \beta_5 \ln \text{Income}_{it} + \beta_6 \text{Rural\_Pop}_{it} + \varepsilon_{it}\end{aligned} \quad (2.3)$$

where β_1, β_2, β_3, β_4 and β_5 are the parameters to be estimated.

To ensure robust statistical inference, we use a range of estimation techniques. First, our basic model is estimated using the fixed effects method following recent studies (Adesete et al., 2023). There are a number of methodological and economic reasons for choosing this method, which are directly related to the nature of the variables we are studying and the objective of our research. The fixed-effects model is particularly appropriate when we wish to control for unobserved characteristics that may vary between entities (countries, regions or other units of analysis) but remain constant over time. In our study, it is likely that each country or region has intrinsic characteristics that influence both food security and how they benefit from climate finance. These characteristics may include, for example, the quality of political and economic institutions; agricultural infrastructure; food-related cultural practices; and the general climate of a region. These unobserved factors may be correlated with the explanatory variables in our model. Not controlling for these effects could introduce an omission bias into the estimates. The fixed-effects model neutralises this effect by capturing differences specific to each unit of analysis that are constant over time, thus guaranteeing unbiased estimates (Wooldridge, 2010). The fixed-effects model also focuses on intra-group variability (within countries), by eliminating differences between observed units that are constant over time. In the context of our study, some factors may simultaneously affect climate finance and food security, but these factors may remain constant over time. For example, variables such as political preferences or structural agricultural policies that do not change significantly from 1 year to the next may bias the results if they are not taken into account. The fixed effects model eliminates these constant unobservable factors, focusing only on internal changes within each unit of analysis.

In a study using panel data, problems of heteroscedasticity (non-constant error variance) and autocorrelation (correlation of errors over time) are common. These problems can affect precision of the estimates obtained using fixed effects. To analyse the robustness of our basic results, we employ one of commonly used methods to correct for these problems, namely the Feasible Generalised Least Squares (FGLS) method.

Results and Discussion

The results in Fig. 2.1, suggest a negative relationship between climate finance and food security. This means that, on average, countries with higher levels of mitigation and adaptation climate finance have lower rates of undernourishment/food insecurity.

The results in Tables 2.2 and 2.3 present the fixed effect analysis on the effect of climate finance on food security.

These results (Tables 2.2 and 2.3) show a negative relationship between climate finance indicators and the food security proxy (bearing in mind that the decrease in this proxy reflects an increase in food security in this study, ceteris paribus). These results therefore confirm our hypothesis that the increase in global climate funds, those earmarked for mitigating the effects of climate change and those linked to adaptation,

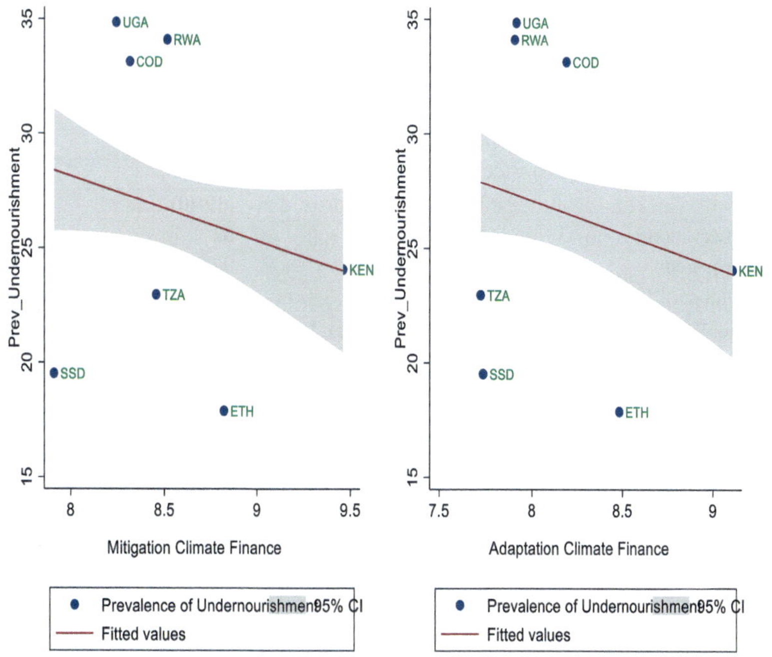

Fig. 2.1 Scatter plot with fit: Mitigation and Adaptation climate finance with the food security (Prev_Undernourishment)

Table 2.2 Effect of climate finance on food security (fixed effect)

	Dependent variable: Food Security (Prev_Undernourishment)		
	(1)	(2)	(3)
LnMCF	−0.105**		
	(0.215)		
LnACF		−0.731**	
		(0.320)	
LnGCF			−0.269**
Constant	25.06***	30.62***	26.52***
	(1.690)	(2.562)	(2.468)
Observations	65	62	65
R-squared	0.04	0.088	0.015
Number of ID	7	7	7

Note: Standard errors in parentheses *** $p < 0.01$, ** $p < 0.05$, * $p < 0.1$

Table 2.3 Baseline results with control variables

	Dependent variable: Food Security (Prev_Undernourishment)		
	(1)	(2)	(3)
LnMCF	−0.123***		
	(0.232)		
lnACF		−0.573***	
		(0.317)	
lnGCF			−0.226**
			(0.293)
FDI	−0.318***	−0.180**	−0.274*
	(0.264)	(0.272)	(0.262)
Health_Exp	−0.475*	−1.125***	−0.558*
	(1.354)	(1.455)	(1.334)
lnIncome	−6.09***	−5.24**	−6.18***
	(3.834)	(3.996)	(3.802)
Rural_Pop	0.261*	0.131*	0.186*
	(0.403)	(0.448)	(0.372)
Agri_Prod	−0.698***	−0.888***	−0.991***
	(0.179)	(0.204)	(0.181)
Constant	82.68**	71.81*	82.08**
	(37.33)	(39.40)	(37.12)
Observations	65	62	65
R-squared	0.604	0.628	0.713
Number of ID	7	7	7

Note: Standard errors in parentheses *** $p < 0.01$, ** $p < 0.05$, * $p < 0.1$

contribute to increasing the food security of the populations of the East African countries (listed in the appendix), but in different proportions. It is interesting to note that our empirical results are comparable to existing studies and, in some cases, provide additional knowledge to the literature. According to the results in Table 2.2, all the coefficients are statistically significant at the 1% level in the first two columns and at the 5% level in the last column. All other things being equal, undernourishment is negatively and significantly affected by the various components of the climate funds, indicating a positive relationship with food security.

The results of the first model (1) stipulate econometrically that a 1% increase in funds earmarked for mitigating the effects of climate change by 10% translates into a 1.05 unit decrease in the prevalence of undernourishment in East African countries over the period studied. Similarly, a 10% increase in funds earmarked for adaptation to climate change results in a 7.31 units reduction in the prevalence of undernourishment in the populations of the region studied (2). Finally, the 10% increase in overall climate funds is linked to a 2.69 unit reduction in the undernourishment rate according to the results of the third specification (3). From an economic point of view, these results corroborate those found by Phiri and Doku (2024), Lipper et al. (2014) and Vermeulen et al. (2012) showing that climate finance, whether for mitigation or adaptation, is more useful for guaranteeing the availability, access, use and stability of food (food security).

Different arguments can be put forward to explain these results. Firstly, in East Africa, funding for climate change mitigation is essential to limit the long-term risks of climate disruption while enhancing food security. Although such funding focuses on reducing greenhouse gas emissions, it also has beneficial effects on food production and the resilience of farming systems (Bryan et al., 2013). Funding mitigation projects, such as reforestation and sustainable land management, helps to restore degraded ecosystems and stabilise hydrological cycles. In Ethiopia, for example, reforestation programmes supported by international funds have reduced the risk of drought and soil erosion, critical factors for food security (Campbell et al., 2021). This makes it possible to maintain more stable levels of agricultural production, thereby reducing the prevalence of malnutrition. Also in East Africa, mitigation funding has supported the introduction of climate-smart agricultural practices, such as land management and soil carbon conservation. Initiatives such as agroforestry in Kenya and Uganda have increased agricultural productivity while reducing greenhouse gas emissions (Cohn et al., 2017). These practices enable farmers to produce more efficiently in a less predictable environment, thereby enhancing food security. Mitigation financing therefore helps to limit the negative effects of climate change while increasing the capacity of East African agricultural systems to withstand climate shocks, thereby improving regional food security. Secondly, with regard to the effect of climate change mitigation funds in East Africa, the latter can help to reduce food insecurity and, by analogy, stabilise and/or increase food security through the introduction of more resilient agricultural practices, such as the use of drought-tolerant crop varieties and improved irrigation systems. In countries such as Kenya, Tanzania and Ethiopia, these initiatives have led to an increase in agricultural yields despite adverse climatic conditions (Thornton et al., 2018). These improvements help to secure food supplies in regions often affected by drought. Moreover, a large proportion of rural communities depend on rain-fed agriculture, which is particularly sensitive to climatic hazards. Funding for adaptation could explain why the diversification of livelihoods is a topical issue, thanks in particular to training projects and support for the development of non-agricultural activities such as livestock farming, crafts and small businesses. These efforts are of great importance in reducing the food vulnerability of households, by increasing their economic resilience in the face of poor harvests (Harvey et al., 2014; Adger et al., 2009). Adaptation funding has also enabled the construction of water infrastructure such as dams, reservoirs and irrigation systems. These initiatives have helped countries such as Uganda and Ethiopia to reduce dependence on rain-fed agriculture and stabilise agricultural production, even during periods of water stress. Such initiatives have led to a significant improvement in local food security (De Pinto et al., 2020). Financing adaptation in East Africa therefore contributes to the fight for food security by making farming systems more resilient to climate hazards and diversifying the livelihoods of rural communities.

Table 2.2 presents the results with the introduction of control variables for robustness to the omission of relevant variables in the study. The signs and significances of the explanatory variables of interest remain unchanged. As for the control variables, the results reveal mixed effects. While foreign direct investment, public spending on health, per capita income and agricultural productivity have a negative effect on the prevalence of undernourishment (and therefore a positive effect on food security), the rural population has a positive effect on the prevalence of undernourishment in East African countries. Specifically, the result that FDI is negatively related to the prevalence of undernourishment (positively related to food security) can be compared to previous findings by Mihalache-O'Keef and Li (2011) and Slimane et al. (2016) who also found a positive effect of FDI on food intake. This finding corroborates those of previous works that FDI through the strengthening of agricultural infrastructure (Dube & Vargas, 2013); the transfer of agricultural technologies (Msuya & Maganza, 2018) and the creation of jobs and improvement of incomes; can contribute to enhancing food security in the region. Agricultural productivity remains a major determinant of food security in the region under study, where the majority of the population depends directly on agriculture for their livelihoods. An increase in agricultural productivity makes it possible to meet growing food demand while reducing rural poverty. Higher agricultural productivity increases food supply, thereby reducing shortages and stabilizing food prices (Dercon & Gollin, 2014). Improved agricultural productivity also leads to higher incomes for farmers, reducing poverty in rural areas and improving access to food. In Kenya, programmes to boost the productivity of small farms have shown positive results in terms of improving food security and reducing poverty (Bezu et al., 2014). Public spending on health plays a key role in improving food security, as it directly influences people's ability to access adequate food and maintain good health. In East Africa, where poverty levels and health challenges are high, public spending on health is crucial to ensuring food security for the most vulnerable populations. Investments in public health services, particularly nutrition programmes, have a direct impact on reducing undernourishment, especially among children, and uptake of food supplementation programmes (WHO, 2021).

The results also show that per capita income is a determining factor in food security, as it directly influences the ability of households to buy food. In East Africa, higher incomes contribute to improved access to food and poverty reduction, both of which are essential for food security. An increase in income enables households to buy not only more food, but also more nutritious food, thereby improving the quality of their diet. Studies have shown that in countries such as Kenya and Uganda, higher incomes have led to improved diets, including greater consumption of fruit, vegetables and animal protein (Hirvonen et al., 2016). The link between income and poverty reduction is also well established, and improved economic conditions enable households to lift themselves out of poverty, thereby reducing their risk of food insecurity (Babatunde & Qaim, 2010). In East Africa, the high proportion of the population living in rural areas contributes to food insecurity, mainly due to increased dependence on subsistence farming and inadequate agricultural infrastructure. The majority of people living in rural areas depend on small farms, which are often vulnerable to climatic hazards and characterised by low-productivity farming practices. This situation limits households' access to regular and adequate food. In addition, rapid population growth in rural areas is putting pressure on available land, reducing farm sizes and hence food production (Bai et al., 2020).

Robustness Check Using Alternative Estimation Method (FGLS)

In a study using panel data, problems of heteroscedasticity (non-constant error variance) and autocorrelation (correlation of errors over time) are common. These problems can affect the precision of estimates obtained using fixed effects. One of the methods commonly used to correct these problems is the feasible generalised least squares method (Kassaye et al., 2021). The feasible generalised least squares (FGLS) method, as Baltagi points out, makes it possible to correct the problems of heteroscedasticity and autocorrelation by adjusting the covariance matrix of the residuals. Unlike fixed effects, which assume

Table 2.4 Effect of climate finance on food security (FGLS)

	Dependent variable: food security (Prev_Undernourishment)		
	(1)	(2)	(3)
lnMCF	−1.322**		
	(0.663)		
lnACF		−1.082***	
		(0.926)	
lnGCF			−1.607***
			(0.846)
FDI	2.759***	2.410***	2.709***
	(0.673)	(0.730)	(0.671)
Health_Exp	−2.496	−2.560	−2.542*
	(1.674)	(1.657)	(1.676)
lnIncome	12.44***	12.04***	12.31***
	(2.223)	(2.180)	(2.218)
Rural_Pop	3.916***	4.058***	3.412***
	(1.044)	(1.156)	(0.981)
Agric_Prod	−0.692***	−0.836***	−0.704***
	(0.261)	(0.296)	(0.262)
Constant	−41.72**	−36.28**	−36.09**
	(16.70)	(17.03)	(16.88)
Observations	65	62	65
Number of ID	7	7	7

Note: Standard errors in parentheses *** $p < 0.01$, ** $p < 0.05$, * $p < 0.1$

errors are homoscedastic and uncorrelated, FGLS modifies estimates to account for unequal variances (heteroscedasticity) and temporal correlations (autocorrelation) in errors. Secondly, in our study, we analyse panel data covering several periods for each country or region. In such data, the residual errors are often correlated over time, making it necessary to use a method capable of correcting these autocorrelations, as the FGLS does. In addition, FGLS also allows heteroscedasticity problems to be dealt with simultaneously, which is important in a study where the variance of errors can vary considerably between countries due to their institutional and economic diversity (Wooldridge, 2010). To ensure validity and reliability, we adopted a more robust alternative estimation technique and our results as presented in Table 2.4.

Overall, the prevalence of undernourishment remains negatively explained by the different climate finance indicators. More specifically, an increase in climate finance for mitigation and adaptation or overall of 1% translates into a decrease in the prevalence of undernourishment of the order of 1322; 1082 and 1607 units respectively. Our basic results remain valid when another more robust estimation technique is used, and also confirm the positive link by analogy between climate finance and food security.

Conclusion

Africa in general, and the East African region in particular, is one of the regions most vulnerable to the effects of climate change, with recurrent droughts, floods and increased climatic variability seriously affecting agricultural production and food security. Rural populations, heavily dependent on rain-fed agriculture, bear the brunt of these climatic disruptions. Faced with these challenges, climate financing, particularly for adaptation and mitigation, is essential to strengthen the resilience of food systems. This study examines the impact of such financing, as well as the influence of other variables such as foreign investment, public spending on health and agricultural productivity on food security in this region. To this end, this study used the fixed effects method for 7 East African countries over the 2012–2021 period to assess the effect of climate finance on food security. The empirical results revealed that climate finance positively and significantly affects food security at conventional thresholds in the region under review. The results also showed that foreign direct investment, public spending on health, per capita income and agricultural productivity were valuable in stabilising if not sustaining the fight for food security and the achievement of SDG2 in the region. The use of a robust alternative estimation technique also demonstrated the nature of the relationship between climate finance and food security.

Based on the results of the study, some key recommendations have been formulated to combat food insecurity in the region. The results of this study underline the need to continue and intensify climate financing for adaptation and mitigation in the agricultural sectors. It is crucial that this funding be integrated into multi-sectoral policies that take account of the economic, health and social dimensions. In addition, particular attention should be paid to good governance and transparency in the management of climate funds, to ensure that they reach the most vulnerable populations. Additional investments in public health and agricultural productivity must also be encouraged to consolidate progress in food security and sustainable development. One of the limitations of this study is its short study period, limited by data availability, which only covers the period 2012–2021. Future research could consider extending the study period to include more historical data, allowing for a more comprehensive analysis of the long-term effects of climate finance on food security. This extension of the time period could provide a better understanding of the trends and effectiveness of climate finance in ensuring food security in the region over a longer period.

References

Adesete, A. A., Olanubi, O. E., & Dauda, R. O. (2023). Climate change and food security in selected sub-Saharan African countries. *Environment, Development and Sustainability, 25*(12), 14623–14641.

Adger, W. N., Dessai, S., Goulden, M., Hulme, M., Lorenzoni, I., Nelson, D. R., et al. (2009). Are there social limits to adaptation to climate change? *Climatic Change, 93*, 335–354.

Adom, P. K., Opoku, E. E. O., & Yan, I. K. M. (2019). Energy demand–FDI nexus in Africa: Do FDIs induce dichotomous paths? *Energy Economics, 81*, 928–941.

Ahmad, F. (2009). Food security in Pakistan. *Pakistan Journal of Agricultural Sciences, 46*(2), 83–89.

Alderman, H., Hoogeveen, H., & Rossi, M. (2006). Reducing child malnutrition in Tanzania: Combined effects of income growth and program interventions. *Economics and Human Biology, 4*(1), 1–23.

Appiah, D. O., Asante, F., & Nketiah, B. (2019). Perspectives on agricultural land use conversion and food security in rural Ghana. *The Sciences, 1*(1), 14.

Babatunde, R. O., & Qaim, M. (2010). Impact of off-farm income on food security and nutrition in Nigeria. *Food Policy, 35*(4), 303–311.

Bai, Z., Schmidt-Traub, G., Xu, J., Liu, L., Jin, X., & Ma, L. (2020). A food system revolution for China in the post-pandemic world. *Resources, Environment and Sustainability, 2*, 100013.

Bezu, S., Kassie, G. T., Shiferaw, B., & Ricker-Gilbert, J. (2014). Impact of improved maize adoption on welfare of farm households in Malawi: A panel data analysis. *World Development, 59*, 120–131.

Bryan, E., Ringler, C., Okoba, B., Roncoli, C., Silvestri, S., & Herrero, M. (2013). Adapting agriculture to climate change in Kenya: Household strategies and determinants. *Journal of Environmental Management, 114*, 26–35.

Campbell, D., Moulton, A. A., Barker, D., Malcolm, T., Scott, L., Spence, A., et al. (2021). Wild food harvest, food security, and biodiversity conservation in Jamaica: A case study of the Millbank farming region. *Frontiers in Sustainable Food Systems, 5*, 663863.

Castro-Nunez, A. (2018). Responding to climate change in tropical countries emerging from armed conflicts: Harnessing climate finance, peace building, and sustainable food. *Forests, 9*(10), 621.

Cline, W. R. (2007). *Global warming and agriculture: Impact estimates by country*. Peterson Institute.

Cohn, A. S., Newton, P., Gil, J. D., Kuhl, L., Samberg, L., Ricciardi, V., et al. (2017). Smallholder agriculture and climate change. *Annual Review of Environment and Resources, 42*(1), 347–375.

Dasgupta, S., & Robinson, E. J. (2021). Improving food policies for a climate in secure world: Evidence from Ethiopia. *National Institute Economic Review, 258*, 66–82.

De Pinto, A., Cenacchi, N., Kwon, H. Y., Koo, J., & Dunston, S. (2020). Climate smart agriculture and global food-crop production. *PLoS One, 15*(4), e0231764.

Dercon, S., & Gollin, D. (2014). Agriculture in African development: Theories and strategies. *Annual Review of Resource Economics, 6*(1), 471–492.

Doku, I., Ncwadi, R., & Phiri, A. (2021). Determinants of climatefinance: Analysis of recipient characteristics in sub-Sahara Africa. *Cogent Economics & Finance, 9*(1), 1964212.

Dube, O., & Vargas, J. F. (2013). Commodity price shocks and civil conflict: Evidence from Colombia. *Review of Economic Studies, 80*(4), 1384–1421.

Fan, S., & Brzeska, J. (2012). In S. Fan & R. Pandya-Lorch (Eds.), *The nexus between agriculture and nutrition: Do growth patterns and conditional factors matter?* (p. 31).

Food and Agriculture Organisation. (2023). *Africa regional overview of food security and nutrition*. Annual Report. Retrieved from https://openknowledge.fao.org/server/api/core/bitstreams/f1ee0c49-04e7-43df-9b83-6820f4f37ca9/content/state-food-security-and-nutrition-2023/food-security-nutrition-indicators.html

Harvey, C. A., Chacon, M., Donatti, C. I., Garen, E., Hannah, L., Andrade, A., et al. (2014). Climate-smart landscapes: Opportunities and challenges for integrating adaptation and mitigation in tropical agriculture. *Conservation Letters, 7*(2), 77–90.

Hirvonen, K., Taffesse, A. S., & Hassen, I. W. (2016). Seasonality and household diets in Ethiopia. *Public Health Nutrition, 19*(10), 1723–1730.

IFPRI. (2023). *Global food policy report: Rethinking food crisis responses—Considerations for Africa*. Retrieved from https://www.ifpri.org/event/africas-discussion-ifpris-2023-global-food-policy-report-rethinking-food-crisis-responses/

International Monetary Fund. (2023). Sub-saharan Africa. *Light on the Horizon?* Retrieved from https://www.imf.org/en/Publications/REO/SSA/Issues/2023/10/16/regional-economic-outlook-for-sub-saharan-africa-october-2023

Junghans, L., & Köhler, M. (2016). Cropping and cashing: Institutional solutions for synergetic climate finance for mitigation and adaptation in agriculture. *Climate and Development, 8*(3), 207–210.

Kassaye, A. Y., Shao, G., Wang, X., Shifaw, E., & Wu, S. (2021). Impact of climate change on the staple food crops yield in Ethiopia: Implications for food security. *Theoretical and Applied Climatology, 145*(1), 327–343.

Lee, C. C., Li, X., Yu, C. H., & Zhao, J. (2022). The contribution of climate finance toward environmental sustainability: New global evidence. *Energy Economics, 111*, 106072.

Lipper, L., Thornton, P., Campbell, B. M., Baedeker, T., Braimoh, A., Bwalya, M., et al. (2014). Climate-smart agriculture for food security. *Nature Climate Change, 4*(12), 1068–1072.

McMichael, P., & Schneider, M. (2011). Food security politics and the millennium development goals. *Third World Quarterly, 32*(1), 119–139.

Mihalache-O'Keef, A., & Li, Q. (2011). Modernization vs. dependency revisited: Effects of foreign direct investment on food security in less developed countries. *International Studies Quarterly, 55*(1), 71–93.

Msuya, E., & Maganza, D. (2018). The impact of foreign direct investment on agricultural productivity and poverty reduction in Tanzania. *Tanzanian Economic Review*.

Nakai, J. (2018). Food and agriculture organisation of the United Nations and the sustainable development goals. *Sustainable Development, 22*, 1–450.

Ngwenya, N., & Simatele, M. D. (2020). Unbundling of the green bond market in the economic hubs of Africa: Case study of Kenya, Nigeria and South Africa. *Development Southern Africa, 37*(6), 888–903.

Pickering, A. J., Arnold, B. F., Dentz, H. N., Colford, J. M., Jr., & Null, C. (2017). Climate and health co-benefits in low-income countries: A case study of carbon financed water filters in Kenya and a call for independent monitoring. *Environmental Health Perspectives, 125*(3), 278–283.

Rahaman, M. A., & Rahman, M. M. (2020). Climate justice and food security: Experience from climate finance in Bangladesh. In *Environmental policy: An economic perspective* (pp. 249–268).

Santangelo, G. D. (2018). The impact of FDI in land in agriculture in developing countries on host country food security. *Journal of World Business, 53*(1), 75–84.

Sarker, M. M. R., & Fagun, A. N. (2021). COVID-19, food security, food prices and urban-rural interrelationship for sustainable food and nutritional security: A study on Dhaka City. *International Journal of Agricultural Economics, 6*(1), 47–58.

Shahbaz, M., & Lean, H. H. (2012). Does financial development increase energy consumption? The role of industrialization and urbanization in Tunisia. *Energy Policy, 40*, 473–479.

Slimane, M. B., Huchet-Bourdon, M., & Zitouna, H. (2016). The role of sectoral FDI in promoting agricultural production and improving food security. *International Economics, 145*, 50–65.

Smith, L. C., & Haddad, L. (2015). Reducing child undernutrition: Past drivers and priorities for the post-MDG era. *World Development, 68*, 180–204.

Stuch, B., Alcamo, J., & Schaldach, R. (2021). Projected climate change impacts on mean and year-to-year variability of yield of key smallholder crops in sub-Saharan Africa. *Climate and Development, 13*(3), 268–282.

Thornton, P. K., Kristjanson, P., Förch, W., Barahona, C., Cramer, L., & Pradhan, S. (2018). Is agricultural adaptation to global change in lower-income countries on track to meet the future food production challenge? *Global Environmental Change, 52*, 37–48.

Timperley, J. (2021). Green tech transfer is too slow: To help mitigate the effects of climate change on a global scale, there are projects in place to promote the transfer of climate technologies in developing countries. So why aren't they working? *Engineering & Technology, 16*(10), 1–2.

Vermeulen, S. J., Campbell, B. M., & Ingram, J. S. (2012). Climate change and foodsystems. *Annual Review of Environment and Resources, 37*(1), 195–222.

Wooldridge, J. M. (2010). *Econometric analysis of cross section and panel data*. MIT Press.

World Health Organisation. (2021). *The State of Food Security and Nutrition in the World 2021: Transforming food systems for food security, improved nutrition and affordable healthy diets for all.* World Health Organisation.

Yao, P., Fan, H., Wu, Q., Ouyang, J., & Li, K. (2023). Compound impact of COVID-19, economy and climate on the spatial distribution of global agriculture and food security. *Science of the Total Environment, 880*, 163105.

Zhao, J., Zhou, B., & Li, X. (2022). Do good intentions bring bad results? Climate finance and economicrisks. *Finance Research Letters, 48*, 103003.

3 Legal Responses to Land and Environmental Governance in Uganda's Albertine Rift: A Review and Perspectives

Juliet Musoke Nakabuye and Priscilla Nyadoi

Abstract

We interrogated Uganda's legal framework to determine whether it effectively addresses land governance and environmental protection in the oil-rich Albertine Graben. The review benchmarks against the legal international agenda on environmental protection established by the 1992 Earth Summit and its subsequent conventions. We used documentary analyses and focused on determining the state of access to justice regarding land rights and the conservation of biodiversity from literature. Our investigation discovered inconsistencies in enforcement practices and legislative gaps in the domestication of international environmental law.

Keywords

Oil developments · Ecologically sensitive areas · Environmental conventions · International standards · National Environment Management Authority (NEMA)

J. M. Nakabuye (✉)
Faculty of Law and Criminology, Ghent University, Ghent, Belgium

Uganda Martyrs University, Kampala, Uganda
e-mail: jnakabuye@umu.ac.ug

P. Nyadoi
Uganda Martyrs University, Kampala, Uganda

Introduction

Environmental conventions mandate parties to implement set standards. Uganda is a signatory to several international environmental agreements, notably the Convention on Biological Diversity (CBD, 1992), the Convention on International Trade in Endangered Species of Wild Fauna and Flora (CITES, 1973), and the Ramsar Convention on Wetlands (Ramsar Convention, 1971). These treaties provide an important legal framework to conserve biodiversity and regulate the exploitation of natural resources. The CBD obligates parties to develop national biodiversity strategies, promote sustainable use, and adopt the precautionary principle especially when scientific knowledge is incomplete (Biermann & Pattberg, 2012). CITES controls cross-border trade of endangered species to prevent overexploitation, with species listed in appendices based on risk assessments (CITES, 1973). The Ramsar Convention protects wetlands of international importance, many of which are habitats for endemic amphibians and birds in the Albertine Graben (Ramsar Convention, 1971).

Uganda's domestication of these treaties occurs primarily through the National Environment Act (2019) and the Uganda Wildlife

P. Nyadoi et al. (eds.), *Sustainable Development Goals in Africa*, Sustainable Development Goals Series, https://doi.org/10.1007/978-3-032-20742-5_3

Act (2019), which provide the statutory basis for environmental protection and wildlife management (Republic of Uganda, 2019, 2020). The National Environment Management Authority (NEMA) and Uganda Wildlife Authority (UWA) are mandated to enforce these laws and regulations. Despite this, several scholars note institutional weaknesses such as limited technical capacity, insufficient funding, and weak interagency coordination which impede full treaty implementation (Twinomugisha, 2007; Muhwezi, 2016). Recently, the country discovered commercially viable oil in the Albertine region. It is expected to produce about 230,000 barrels of crude oil per day over a 20-year period. The oil fields lie underneath one of Africa's most ecologically sensitive landscapes (Ahebwa et al., 2017; IUCN, 2020; Akisiimire, 2021). The developments by China National Oil Offshore Company (CNOOC) and Total Energies in constructing a 1445 km East African Crude Oil Pipeline (EACOP) pose even more threats (Total Energies EP Uganda, 2019; National Environment Management Authority (NEMA), 2020).

Despite Uganda's express commitments to environmental conventions, the Albertine Graben and its inhabitants continually suffer adverse effects arising from oil developments. Environmental and Social Impact Assessments (ESIAs) are often not exhaustive, in favour of investors and not meeting ecological standards and citizens' expectations (Muhwezi, 2016; Weakly & Ssonko, 2019; WWF Uganda, 2020). On the other hand, land governance is marked by ambiguities in ownership and enforcement mechanisms. Oil development has exacerbated conflicts over land, leading to displacement of communities, especially women and indigenous peoples (Muhwezi, 2016; Obbo et al., 2018). For example, the EACOP has displaced 3648 people from ten districts in the Albertine Graben (Kiiza et al., 2011). It is not known how the oil industry development in the region is impacting land and environmental governance in relation to expected international standards. This review aims to provide from literature current information on this. We examined legal and policy frameworks underpinning Uganda's land and environment governance, with regard to international conventions, basing on Environmental and Social Impact Assessments.

Methods

The research adopted a doctrinal legal methodology, reviewing literature about how Uganda is adhering to international conventions underpinning land and environmental governance. This was complemented by reviewing reports from state and non-state actors in the Tilenga and Kingfisher oil project areas (Fig. 3.1). Information obtained was then summarized into themes presented hereafter.

Findings and Discussion

The Uganda Wildlife Act (2019) provides a comprehensive legal framework for the management of protected areas. The Act mandates that any activity likely to affect wildlife habitats must undergo ESIAs and obtain appropriate permits. The National Environment Act (2019) serves as the overarching environmental management law in Uganda. It empowers NEMA to oversee the sustainable use and conservation of natural

Fig. 3.1 Map showing the location of the Albertine Graben in Western Uganda

resources, including the review and approval of ESIAs for projects that may impact the environment (Republic of Uganda, 2020). This law is critical for regulating the oil and gas sector, which has expanded rapidly since the early 2010s. Despite these legal provisions, a study by WWF Uganda (2020) found inappropriate developments in ecologically sensitive areas, inadequate community participation, and weak enforcement of mitigation measures. The Tilenga and Kingfisher oil developments are situated within or near critical wildlife habitats in the Albertine Graben. Although ESIAs were undertaken as required by the law, existing literature indicates that there is inadequate monitoring of the developments (Weakly & Ssonko, 2019; National Environment Management Authority (NEMA), 2020). The literature provides evidence that inappropriate activities are taking place in protected areas and there are governance issues especially relating to community participation and land ownership (WWF Uganda, 2020). Uganda's Access to Information Act (2016) guarantees public access to environmental data, enabling communities to engage in environmental decision-making (Republic of Uganda, 2016). The National Environment Act (2019) and Uganda Wildlife Act (2019) also incorporate provisions for community involvement in natural resource management (Republic of Uganda, 2019, 2020). The governance issues often result from inadequate legal mandates, institutional capacity challenges, and competing economic priorities (Muhwezi, 2016; Kairaba & Nyandwi, 2019). The finding underscores the need to not only enforce existing laws more robustly but also to reformulate them to explicitly provide for community participation in land and environmental governance.

Conclusion

The paper provides an overview of the state of land and environmental governance in the Albertine Graben emerging from the literature. Our findings highlight some of the gaps in the implementation of legal and policy frameworks in relation to expected international standards. We pinpoint the need for a detailed study to determine factors underpinning the performance of ESIAs and other legal and policy frameworks currently.

References

Ahebwa, W. M., van der Duim, R., & Sandbrook, C. (2017). Oil discovery and its impacts on Uganda's tourism and conservation development: Perceptions of local communities. *Journal of Environmental Planning and Management, 60*(10), 1772–1791.

Akisiimire, H. (2021). *The effects of oil and gas exploration on mammalian population distribution and habitat change in Murchison falls National Park, Uganda.* MSc thesis, Kyambogo University, Kampala.

Biermann, F., & Pattberg, P. (2012). *Global environmental governance reconsidered.* MIT Press.

Convention on Biological Diversity. (1992). *Convention on biological diversity.* United Nations. Retrieved from https://www.cbd.int.

Convention on International Trade in Endangered Species of Wild Fauna and Flora (CITES). (1973). *Convention on International Trade in endangered species of wild fauna and flora.* Retrieved from https://cites.org/

International Union for Conservation of Nature (IUCN). (2020). *Protecting biodiversity in Africa's extractive sector.* IUCN.

Kairaba, A., & Nyandwi, E. (2019). Competing claims: Land tenure challenges in Uganda amidst oil development. *African Land Governance Journal, 15*(2), 45–65.

Kiiza, J., Bategeka, L., & Ssewanyana, S. (2011). Righting resources-curse wrongs in Uganda: The case of oil discovery and the management of popular expectations. *The Journal of Humanities and Social Sciences, 10*(3), 183203.

Muhwezi, W. (2016). *Oil and gas governance in Uganda: A critical analysis of the legal and policy framework.* ACODE Policy Briefing Paper Series, No. 31.

National Environment Management Authority (NEMA). (2020). *Environmental and social impact assessment guidelines for the oil and gas sector.* Government of Uganda.

Obbo, E., Nambooze, C., & Kwagala, A. (2018). Women and land rights in Uganda: Impacts of oil discoveries. *Gender and Development, 26*(3), 403–417.

Ramsar Convention Secretariat. (1971). *Convention on wetlands of international importance especially as waterfowl habitat (Ramsar convention).* Ramsar Convention Bureau. Retrieved from https://www.ramsar.org/

Republic of Uganda. (2016). *Access to information act, 2005.* Uganda Legal Information Institute. Retrieved from https://ulii.org

Republic of Uganda. (2019). *National Environment act, 2019*. Uganda Legal Information Institute. Retrieved from https://ulii.org.

Republic of Uganda. (2020). *Third National Development Plan (NDPIII) 2020/21–2024/25*. National Planning Authority.

Total Energies EP Uganda. (2019). *Tilenga project environmental and social impact assessment report*. Retrieved from https://totalenergies.ug.

Twinomugisha, B. (2007). A content analysis of Uganda's environmental impact assessment guidelines. In A. R. Markus (Ed.), *Environmental law in Uganda*. Fountain Publishers.

Uganda Wildlife Authority. (2019). *Uganda wildlife act, 2019*. Government of Uganda.

Weakly, J., & Ssonko, D. (2019). Environmental impact assessments and oil extraction: Lessons for Uganda from East Africa. *East African Environmental Studies Journal, 11*(3), 233–245.

World Wide Fund for Nature (WWF) Uganda. (2020). *Safeguarding Uganda's natural heritage amidst oil and gas development*. WWF Uganda.

Tackling Conflict and Conservation in the Albertine Rift and Lake Victoria Crescent, Uganda

4

Priscilla Nyadoi, Gilbert Tayebwa, and Fredrick Ssali

Abstract

Human-wildlife conflicts (HWC) significantly impede achievement of Sustainable Development Goals (SDGs) by negatively impacting livelihoods, food security and overall well-being. Reducing this problem has remained a significant challenge in protected area management. Moreover, the impacts and types of conflict are dynamic and poorly characterised. This study identified wildlife species involved in HWC and explored the potential strategies for mitigating conflicts in and around protected areas in the Albertine Rift and Lake Victoria Crescent in Uganda. Data from 369 randomly sampled participants show that no fewer than 17 wild vertebrate species, including the African elephant (*Loxodonta africana*), chimpanzee (*Pan troglodytes*), vervet monkey (*Cercopithecus aethiopicus*) and olive baboon (*Papio anubis*), cause HWC in the area. Crop raiding was the most frequently reported type of HWC. All pastoralists stated that they had experienced raiding by wild animals and four fifths of crop farmers reported crop raiding as a conflict they had experienced. The participants mentioned low farm yields, disease transmission and human death as the most common negative effects of HWC and suggested that the negative effects can best be mitigated by erecting physical barriers to deter the raiders. These findings provide valuable insights for protected area management and for the advancement of sustainable development goals.

Keywords

Crop raiding · Mitigation measures · Protected area management · Human-wildlife coexistence · Sustainable livelihoods

P. Nyadoi (✉)
Uganda Martyrs University, Kampala, Uganda

Uganda Wildlife Society, Kampala, Uganda
e-mail: pnyadoi@umu.ac.ug

G. Tayebwa
Uganda Wildlife Society, Kampala, Uganda

The International Crane Foundation, Baraboo, WI, USA

F. Ssali
Uganda Martyrs University, Kampala, Uganda

Introduction

Human-wildlife conflict (HWC) has been defined as struggles that emerge when the presence or behaviour of wildlife poses an actual or perceived direct and recurring threat to human interests, leading to negative impacts on people and/or wildlife (IUCN SSC HWCTF, 2020). This problem can hinder efforts to achieve Sustainable Development Goals (SDGs) including SDG 1

P. Nyadoi et al. (eds.), *Sustainable Development Goals in Africa*, Sustainable Development Goals Series, https://doi.org/10.1007/978-3-032-20742-5_4

('no poverty'), SDG 2 ('zero hunger'), SDG 8 ('decent work and economic growth') and SDG 15 ('life on land') (United Nations, 2015). A wide variety of animals including elephants, hippopotami and lions, often stray out of protected areas into adjacent cultivated fields or grazing areas, consequently impacting lives, livelihoods and food security (Nyhus, 2016; IUCN, 2022; Galley & Anthony, 2024). Wildlife species often damage people's crops leading to economic losses, fear, stress and poor health among affected communities with consequences that the latter develop disinterest in conservation and engage in retaliatory wildlife killings (Mackenzie et al., 2017; Blackie, 2022). Coexistence can be achieved where there are appropriate tools and management, public policies and societal support (Nyhus, 2016; Travers et al., 2019; Braczkowski et al., 2020; Nkansah-Dwamena, 2023). Direct methods such as fencing, translocation, use of repellents and protected area buffer zones as well as indirect methods like compensation and revenue sharing among others may reduce conflicts (Lindsay et al., 2017; Poulsen et al., 2017; van Houdt & Traill, 2022; Shwartz, 2024).

Crop destruction, livestock predation and human attacks are recurrent conflicts in and around many protected areas in Uganda (Tweheyo et al., 2010; Kabuusu et al., 2018; Kolinski & Milich, 2021; Ashepet et al., 2023). The dynamic nature of the issues involved coupled with changes occasioned by climate change call for scientific monitoring and in-depth analyses especially in fragile ecosystems. In this chapter, we report about wildlife species, forms of HWC and the different management and control measures for promoting human-wildlife coexistence around protected areas in the Albertine Rift and Lake Victoria Crescent in Uganda. Findings accruing from methods presented hereafter provide useful information for the advancement of SDG 1, SDG 2, SDG 8, SDG 15 and SDG 17. The specific objectives were to (1) identify wildlife species involved in conflicts with humans and (2) explore the potential strategies for mitigating HWC in and around protected areas in the Albertine Rift and Lake Victoria Crescent.

Materials and Methods

Study Area

The study was conducted among local residents neighbouring four protected areas in the Albertine Rift (i.e., Queen Elizabeth National Park, Bugoma Forest Reserve, Bugungu Wildlife Reserve and Murchison Falls National Park) and three protected areas in the Lake Victoria Crescent (i.e., Lake Victoria, River Malaba and West Bugwe Forest Reserve) (Fig. 4.1).

Data Collection and Analysis

We established two 10-km transects in each protected area, starting at the edge and extending into community land. The transects were placed at least 2 km apart. Along each transect, households that had an adult member (aged 18 years or older) present at home were systematically selected at 1-km intervals to participate in interviews. We excluded households with only children or nobody at all and searched for the households that met eligibility criteria in the area. Systematic random sampling was adopted to make the data collection process easier and to ensure better representation of study participants.

We conducted in-depth interviews with the selected study participants and key informants between December 2016 and March 2019 using questions set a priori (Appendix 1). Key informants were selected purposively and included natural resource officers, environment officers, wildlife officers and community leaders. To enhance data validity and reliability, the questionnaire was pretested during a pilot study. Informed oral consent was obtained from all the participants before data collection commenced. The data collected were kept anonymous and confidential by using unique identifier codes that were distinct for each participant. Data were summarised and analysed using R (R Core Team, 2023). Descriptive and inferential statistics on species mentioned by participants, nature of conflicts and potential strategies were generated.

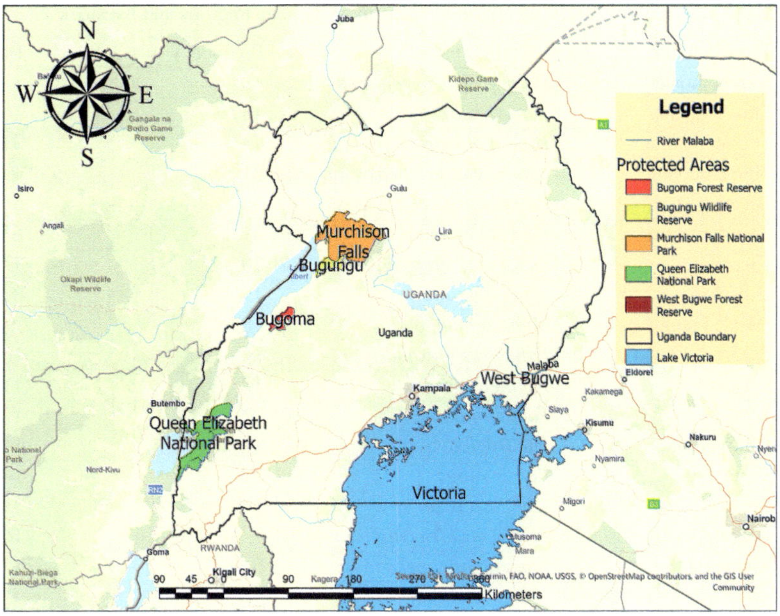

Fig. 4.1 Map of Uganda showing the protected areas where the study was conducted

Results

Demographic and Socio-Economic Characteristics

Of the 369 participants, 75% were men, 26% were aged 18–30 years old, 50% had attained primary level education while 81% owned land (Table 4.1). About 71% were crop farmers, 87.8% were married people and 46.6% had lived in the area for more than 20 years. The vast majority (about 70%) had experienced crop damage, with only a few participants mentioning human and/or livestock attacks (18%) and home invasions (1%) as the other relatively common forms of human-wildlife conflicts occurring up to 10 km from the park boundary. About 76% of the respondents who owned land, and 56% of those who lived on other people's land, reported crop raiding as a conflict they had experienced. All pastoralists stated that they had experienced raiding by wild animals while 79% of crop farmers reported crop raiding as a conflict they had experienced.

Table 4.1 Socio-demographic characteristics of study participants ($n = 369$)

Variable	Category	n (%)
Gender	Male	276 (74.8%)
	Female	93 (25.2%)
Age	18–30 years	95 (25.7%)
	31–40 years	89 (24.1%)
	41–50	81 (22.0%)
	51–60	57 (15.4%)
	61–70	32 (8.7%)
	>70	15 (4.1%)
Level of education	None	60 (16.3%)
	Primary	186 (50.4%)
	Secondary	91 (24.7%)
	Tertiary/ vocational training	24 (6.5%)
	Degree	7 (1.9%)
	Post graduate	1 (0.3%)
Land ownership	Yes	298 (80.8%)
	No	71 (19.2%)

Table 4.2 Wildlife species reported to cause most damage to crops and livestock

English name	Scientific name	Frequency ($n = 369$)	Rank (based on frequency)
Vervet monkey	*Chlorocebus pygerythrus*	187	1
Olive baboon	*Papio anubis*	147	2
Hippopotamus	*Hippopotamus amphibius*	78	3
Nile crocodile	*Crocodylus niloticus*	41	4
Bushpig	*Potamochoerus larvatus*	32	5
Spotted hyena	*Crocuta crocuta*	31	6
Elephant	*Loxodonta africana*	20	7
Buffalo	*Syncerus caffer*	19	8
Banded mongoose	*Mungos mungo*	18	9
Squirrel	*Paraxerus alexandri*	16	10
Leopard	*Panthera pardus*	12	11
Lion	*Panthera leo*	12	12
Chimpanzee	*Pan troglodytes*	11	13
Black rat	*Rattus rattus*	11	14
Serval	*Leptailurus serval*	10	15
Porcupine	*Atherurus africanus*	7	16
Sitatunga	*Tragelaphus spekii*	6	17

Wildlife Species Involved

A total of 17 wild vertebrate species were reported to cause human-wildlife conflicts in the study area (Table 4.2). The vervet monkey (*Cercopithecus aethiopicus*), olive baboon (*Papio anubis*) and hippopotamus (*Hippopotamus amphibius*) were the most frequently mentioned damage-causing species (Table 4.2). Eight of the 17 mentioned species were reported in both regions while the remaining nine including the African elephant (*Loxodonta africana*) ($\chi^2 = 20.0$, $P < 0.001$), lion (*Panthera leo*) ($\chi^2 = 12.0$, $P < 0.001$) and chimpanzee (*Pan troglodytes*) ($\chi^2 = 11.0$, $P < 0.001$) were unique to one of the regions (Appendix 2).

Types of Human-Wildlife Conflicts Experienced

The most prevalent type of human-wildlife conflict reported in the area was physical damage to crops, which accounted for 99.7% of the total incidents. Other forms of HWC, including death of humans and domestic animals (7.9%) and loss of time spent guarding crops (6.5%), were considerably less common (Table 4.3).

Table 4.3 Types of human-wildlife conflicts mentioned by participants

Type of HWC mentioned	Frequency ($n = 369$)	Percentage (%)
Crop damage	368	99.7%
Death of humans and domestic animals	29	7.9%
Loss of time for production during crop guarding	24	6.5%
Disease transmission to livestock and people	5	1.4%

Strategies for Resolving Human-Wildlife Conflicts

The majority of the participants (30%) suggested that erecting physical barriers like fences would deter crop raiding. Other important HWC mitigation measures mentioned include hunting of crop-raiding animals, relocating wildlife species and reporting incidents of crop raiding to authorities (Table 4.4).

Table 4.4 Strategies for mitigating human-wildlife conflicts

Strategy mentioned	Frequency ($n = 369$)	Percentage (%)
Erecting physical barriers like fences	111	30.1%
Hunting crop-raiding animals	63	17.1%
Relocating wildlife	43	11.7%
Reporting to authorities	38	10.3%
Allowing regulated access to park resources	27	7.3%
Scare techniques (scare crows, fires, drumming, throwing stones)	21	5.7%
Guarding farms	9	2.4%

Discussion

Seventeen wildlife species were mentioned by residents neighbouring protected areas in the Albertine Rift and the Lake Victoria Crescent as common species involved in human-wildlife conflict, with monkeys and baboons ranking as the most destructive to crops. While the sitatunga, porcupine and chimpanzee were involved, they were not very common. The most reported forms of HWC are crop raiding and human and livestock attacks. Participants indicated that HWC have affected their livelihoods by leading to low crop yields, deaths of humans and domestic animals, loss of time for production and diseases transmitted to livestock and people. They suggested that local communities should be allowed to access resources from protected areas as a way of mitigating conflicts. When asked who should implement the proposed wildlife conflict reducing strategies, most of the participants suggested the government while a few others suggested non-governmental organisations, local authorities and communities.

Eight vertebrate species were reported to be problem wildlife in both regions while six are unique to the Albertine Rift and three to the Lake Victoria Crescent (Appendix 2). Overall, the forms of HWC experienced across protected areas were similar implying that the proposed mitigation measures may be replicated in several proected areas (PAs). Efforts aimed at combating HWC and contributing to improving the livelihoods of households in frontline villages (for example, revenue sharing, tourism, crop improvement and multiple-use programmes, see Wild & Mutebi, 1996) should be implemented in both the Albertine Rift and the Lake Victoria Crescent. This way, the communities will value and opt for appropriate practices that enable harmonious co-existence with wildlife.

Most of the HWC management approaches and stakeholders proposed in this study call for public-private partnerships and partnering with local communities. The findings reveal the need for Government agencies, in particular the conservation and environmental agencies, to diversify strategies and revise policies guiding management of HWC. Similar HWC and impacts have been reported in the past. For example, wildlife attacks leading to death of people have occurred around Murchison Falls National Park, Queen Elizabeth National Park and other protected areas (Kabuusu et al., 2018; Lacroux et al., 2022; Ashepet et al., 2023). The current study therefore adds knowledge to this widely-studied topic by especially highlighting the need to apply mitigation measures in both the Albertine Rift and Lake Victoria Crescent.

While the effectiveness and sustainability of the proposed strategies may vary across protected areas (due to local variation), it is expected that establishing and/or improving buffer zone management will offer better protection against wildlife damage. A comprehensive study will, however, be required to ascertain and track the effectiveness of each strategy. The current study reveals that the African elephant (*Loxodonta africana*), chimpanzee (*Pan troglodytes*), vervet monkey (*Cercopithecus aethiopicus*) and olive baboon (*Papio anubis*) continue to be involved in HWC despite the many interventions by conservation agencies and civil society (c.f. report by Kolinski & Milich, 2021). The impacts these wild animals cause to communities (such as damaging crops) often lead to economic losses and poor health among the affected people (Mackenzie et al., 2017; Blackie, 2022). It is thus likely that the strategies being used in the management of human-wildlife conflict have not been effective. This again calls for detailed studies to determine the efficacy of existing HWC management approaches and what needs there may be for improvements and/or new approaches.

Studies have shown that where communities participate in park management and benefit sharing from conservation and development projects, the attitudes of people toward wildlife tend to become positive (Travers et al., 2019; Braczkowski et al., 2020; Nkansah-Dwamena, 2023). This, as the findings of this study reveal, may be the case. It may well be that the respondents who called for wildlife translocations and the killing of crop raiding-animals are people who have not been directly involved in benefit sharing and/or in the management of protected areas. Therefore, there is need for government to scale up applicable human-wildlife management strategies currently in use, across regions in Uganda. This should be done, taking into consideration the species involved and the extent and nature of conflicts experienced as these may vary between locations. It will be useful for an in-depth study to be done to help find out who, where and why communities are calling for wildlife killing as a measure to tackle HWC.

Interestingly and of importance to note is that similar conflicts and impacts on livelihoods continue to be reported despite the fact that there are many on-going interventions to address them. For example, Uganda Wildlife Authority (UWA), Fauna & Flora International (FFI), Jane Goodall Institute (JGI), Conservation Through Public Health (CTPH), Uganda Wildlife Society (UWS) and Wildlife Conservation Society (WCS) have had sustained interventions including compensating affected persons and building and supporting communities' capacity to enable them co-exist harmoniously with wildlife. We postulate that strategies in use currently for management of human-wildlife conflicts are not very effective and thus there is need for appraisal not only of these strategies but also the policies and regulatory frameworks underpinning them.

Conclusions

Seventeen wildlife species were reported to cause human-wildlife conflicts in the Albertine Rift and the Lake Victoria Crescent. Of these, the vervet monkey (*Cercopithecus aethiopicus*), olive baboon (*Papio anubis*) and hippopotamus (*Hippopotamus amphibius*) were the most frequently mentioned damage-causing species. Similar HWC were experienced by communities in the two regions and these include crop raiding, attacks on domestic animals and people, and invading homes and farms for rest. The experienced HWC negatively affected livelihoods by leading to low crop yields, death of humans and domestic animals, loss of time during guarding, and diseases transmitted to livestock and people. Erecting physical barriers, hunting crop-raiding animals, relocating wildlife species and reporting incidents of crop raiding to authorities ranked among the most frequently suggested measures for mitigating conflict. The findings of this study provide valuable pointers for improving the coexistence between people and wildlife and for the advancement of sustainable development goals.

Acknowledgements The study was funded by the Royal Dutch Government through the Ecosystem Alliance (a partnership between IUCN Netherlands National Committee, BothEnds and Wetlands International) as part of the programme for empowering communities for natural resource management. We are grateful to Tom Okello, Adonia Bintora, Rose Mutonyi and other UWA staff who participated in the study. We thank the District-Natural Resources, Environment, forests and Community Development Officers of Kasese, Buliisa, Hoima, Tororo, Busia, Bugiri, Namayingo and Mayuge for their kind assistance during fieldwork. We thank the two civil society organisations, AFIEGO and NAPE, for collaborating with us during fieldwork and, thank the local communities who took time to share their experiences. Thanks to Teddy Namirimu, Gilbert Tayebwa, Joel Buyinza, Laster Stoney Ogola, Susan Owino, Noah Jawor and Louis Kyaligonza who conducted the interviews.

Appendix 1: Questionnaire

This questionnaire is intended to identify wildlife species involved in HWC and to explore the potential strategies for mitigating the conflicts in and around protected areas in the Albertine Rift and Lake Victoria Crescent in Uganda. All information obtained is strictly for research purposes and will be treated with confidentiality.

1. Name of interviewer: .... Contact: .... Date: ....

2. Socio-economic characteristics of respondent

(a) Gender: Male ☐ Female ☐

(b) Age: 18-30 ☐ 31-40 ☐ 41-50 ☐ 51- 60 ☐ 61-70 ☐ >70 ☐

(c) Marital status: Single ☐ Married ☐ Divorced ☐ Widowed ☐

(d) Education level: None ☐ Primary ☐ Secondary ☐ Tertiary☐ Degree ☐

(e) Employment/occupation: ....

3. For how many years have you been here? ☐ <5 ☐ 5-10 ☐ 11-15 ☐ 16-20 ☐ >20 yrs

4. Do you own the land on which you have settled? Yes ☐ No ☐

If yes, where is your land located? ☐ Near swamp ☐ Near water body ☐ Near protected area

☐ Other (specify) ....

5. What is the size of your land? ☐ < 1ha ☐ 1-3 ha ☐ 4-7ha ☐ >7ha

6. What type of tenure do you hold for your land?..

7. You have stayed here long enough and you own land in this area, do any wild animals come to your land/home? Yes............ No.........

8. Wild animals come to your land, provide answers to the questions in this Table.

Species involved	Negative activities	Impact on Households	Mitigation measures	Suggested stakeholders

9. Is there any other thing you would like us to know? ....

Thank you

References

Ashepet, M. G., Dahdouh-Guebas, F., Redpath, S., Pooley, S., & Huge, J. (2023). The state and perceptions of human-crocodile interactions around Murchison falls conservation area, Uganda. *Human Dimensions of Wildlife, 29*(2), 194–209. https://doi.org/10.1080/10871209.2023.2212692

Blackie, I. R. (2022). Posttraumatic stress and psychological impacts of human wildlife conflict on victims, their families and caretakers in Botswana. *Human Dimensions of Wildlife, 28*(3), 248–264. https://doi.org/10.1080/10871209.2022.2036394

Braczkowski, A., Fattebert, J., Schenk, R., O'Bryan, C., Biggs, D., & Maron, M. (2020). Evidence for increasing human-wildlife conflict despite a financial compensation scheme on the edge of a Ugandan national park. *Conservation Science and Practice, 2*, 2. https://doi.org/10.1111/csp2.309

Galley, W., & Anthony, P. (2024). Beyond crop-raiding: Unravelling the broader impacts of human-wildlife conflicts on rural communities. *Environmental Management, 74*(3), 590–608.

IUCN. (2022). *Human wildlife conflict*. Retrieved from https://www.iucn.org/resources/issuesbrief/human-wildlife-conflict

IUCN SSC HWCTF. (2020). *What is human-wildlife conflict?* Briefing paper by the IUCN SSC Human-wildlife Conflict Task Force. Retrieved from www.hwctf.org

Kabuusu, R. M., Amuno, J. B., Maseruka, Y., & Macpherson, C. (2018). Incidence and risk factors of wildlife-associated human injuries in queen Elizabeth National Park, Uganda: A retrospective cohort study. *Journal of Zoonotic Diseases and Public Health, 2*(1), 1–5.

Kolinski, L., & Milich, M. K. (2021). Human-wildlife conflict mitigation impacts community perceptions around Kibale National Park, Uganda. *Diversity, 13*(4), 145. https://doi.org/10.3390/d13040145

Lacroux, C., Robira, B., Kane-Maguire, N., Guma, N., & Krief, S. (2022). Between forest and croplands: Nocturnal behavior in wild chimpanzees of Sebitoli, Kibale National Park, Uganda. *PLoS One, 17*(5), e0268132. https://doi.org/10.1371/journal.pone.0268132

Lindsay, K., Chase, M., Landen, K., & Nowak, K. (2017). The shared nature of Africa's elephants. *Biological Conservation, 215*, 260–267. https://doi.org/10.1016/j.biocon.2017.08.021

MacKenzie, C. A., Salerno, J., Hartter, J., Chapman, C. A., Reyna, R., Tumusiime, D. M., & Drake, M. (2017). Changing perceptions of protected area benefits and problems around Kibale National Park, Uganda. *Journal of Environment Management, 200*, 217–228.

Nkansah-Dwamena, E. (2023). Lessons learned from community engagement and participation in fostering coexistence and minimizing human-wildlife conflict in Ghana. *Trees, Forests and People, 14*, 100430. https://doi.org/10.1016/j.tfp.2023.100430

Nyhus, P. J. (2016). Human–wildlife conflict and coexistence. *Annual Review of Environment and Resources, 41*, 143–171.

Poulsen, J. R., Koerner, S. E., Moore, S., Medjibe, V. P., Blake, S., Clark, C. J., Akou, M. E., Fay, M., Meier, A., Okouyi, J., et al. (2017). Poaching empties critical Central African wilderness of forest elephants. *Current Biology, 27*, R134–R135.

R Core Team. (2023). *R: A language and environment for statistical computing*. R Foundation for Statistical Computing. Retreived from https://www.R-project.org/

Shwartz, M. W. (2024). Human-carnivore conflict mitigation and lion population viability in Uganda's queen Elizabeth National Park. *Frontiers of Conservation Science, 5*, 1393399. https://doi.org/10.3389/fcosc.2024.1393399

Travers, H., Archer, L. J., Mwedde, G., Roe, D., Baker, J., Plumptre, A. J., Rwetsiba, A., & Milner-Gulland, E. J. (2019). Understanding complex drivers of wildlife crime to design effective conservation interventions. *Conservation Biology, 33*(6), 1296–1306. https://doi.org/10.1111/cobi.13330

Tweheyo, M., Amanya, B. K., & Turyahabwe, N. (2010). Feeding patterns of sitatunga (*Tragelaphus speki*) in the Rushebeya Kanyabaha wetland, south western Uganda. *African Journal of Ecology, 48*(4), 1045–1052.

United Nations. (2015). *Transforming our world: The 2030 agenda for sustainable development (a/RES/70/1)*. UN General Assembly. Retrieved from https://sdgs.un.org/2030agenda

van Houdt, S., & Traill, L. W. (2022). A synthesis of human conflict with an African megaherbivore; the common hippopotamus. *Frontiers in Conservation Science, 3*, 954722. https://doi.org/10.3389/fcosc.2022.954722

Wild, R. G., & Mutebi, J. (1996). Conservation through community use of plant resources. In *People and plants working paper*. UNECO.

Performance of the Small East African Zebu and Tyrolean Grey Zebu Crosses in Uganda: An Overview

5

Mary Goretti Acila, Samuel Okello, Denis Rwabiita Mugizi, Dorothy Kalule Nampanzira, Maria G. Nassuna Musoke, Daniel Kizza, Sarah Akello, and Donald Rugira Kugonza

Abstract

Livestock production contributes significantly to the economy of Uganda, generating income for farmers and boosting export revenues. The vast majority of livestock breeds, including the Small East African Zebu (SEAZ), are native to Uganda. The Small East African Zebu is highly endangered, despite its desirable quality of being resistant to diseases, parasites and other environmental stresses. In this paper, we give an overview of existing literature about SEAZ and Tyrolean Grey Zebu crosses. Literature was reviewed, and a thematic synthesis was made on these. The findings show that this breed is mainly found in the North, East and Central Uganda. It also emerges that farmers prefer the crossbred Tyrolean Grey Zebu (TGZ), due to their high milk production and fast growth rate. Comprehensive studies on the performance of these animals in different agroecological zones are needed to inform breeding programmes in Uganda.

M. G. Acila (✉)
Makerere University, Kampala, Uganda

Faculty of Agriculture, Uganda Martyrs University, Kampala, Uganda
e-mail: gacila@umu.ac.ug

S. Okello · D. R. Mugizi · D. Kalule Nampanzira
M. G. Nassuna Musoke · D. Kizza · S. Akello ·
D. R. Kugonza
Makerere University, Kampala, Uganda

Keywords

Livestock development · Sustainable Development Goals · Austria · Africa · Livestock

Introduction

The livestock sector is an important component of the Ugandan economy, and contributes up to 3.8% of GDP (Uganda Bureau of Statistics, 2020). Small East African Zebu cattle (SEAZ) are resilient to harsh environmental conditions. However, low productivity has highlighted the need for livestock development. To increase milk production and growth performance, the SEAZ can be crossbred with exotic breeds (Ilatsia et al., 2012). In 2009, the SEAZ was crossed with the Tyrolean Grey from Austria (Acila et al., 2023). This Chapter explores existing knowledge related to cattle breeds, their contribution to development and the morphological characteristics of the SEAZ and the Tyrolean Grey Zebu (TGZ) in Uganda.

P. Nyadoi et al. (eds.), *Sustainable Development Goals in Africa*, Sustainable Development Goals Series, https://doi.org/10.1007/978-3-032-20742-5_5

Livestock production across the world is an important sector, accounting for much of the agricultural development that has made countries prosper (FAO, 2022). In Uganda, the livestock sector accounts for 17% of agricultural GDP and 4.3% of total GDP. The country's total cattle population was 11.4 million in 2008 and 14.5 million in 2024 (UBOS, 2020). According to Waiswa and Akullo (2021), Uganda's cattle industry provides income and employment through the sale of meat, milk and other products, and offers a form of insurance and savings for rural households. The rapidly increasing population in Uganda, especially in the urban areas, has increased the demand for livestock products locally (Matovu & Alcicek, 2023).

Livestock production, however, is not being matched locally. Local demand, gauged by human population and gross domestic product, emerges as a pivotal factor influencing livestock production (Kalungu, 2023). This rapid population growth is correlated with growing consumer requirements for livestock products (FAOSTAT, 2023). A low-income country like Uganda is concerned with livestock productivity to reduce high rates of childhood undernutrition and stunting (Tricarico et al., 2020). Recent FAO publications have illustrated how livestock agriculture practices have contributed to the sustainability of food and agriculture and provided case studies of their contributions to the 17 SDGs (FAOSTAT 2018).

Livestock Contribution to Development

This review is anchored within the broader goals of sustainable livestock development, examining how cattle production can contribute to the Sustainable Development Goals (SDGs). Several of the UN sustainable development goals on zero hunger and environmental health cannot be achieved without livestock and livestock products (Atukunda et al., 2021) as livestock productivity and resilience directly influence household incomes, nutrition and ecological sustainability. The dairy cattle sector is crucial in the food security and agricultural economy of Uganda, providing livelihoods for a significant portion of the rural population (Waiswa & Akullo, 2021).

The importance of livestock in generating income for farmers has been reported in several developing countries (Kugonza et al., 2012). It contributes approximately 90%, and is therefore seen as a major source of livelihood for the people in this area (Tayebwa et al., 2018). Livestock is significant in the economy of Uganda. It constitutes an important source of animal protein for the nutrition of the Ugandan population and contributes 16% to the agricultural Gross Domestic Product (GDP) of Uganda and 4% to the overall GDP (MAAIF, 2020).

Generally, cattle contribute a lot to our lives and well-being through food supply, balancing nutrition and income (Assefa & Hailu, 2018). Livestock is valuable in sustainable agriculture by providing manure and labour for increased productivity and therefore enhancing the well-being and increased income of farmers in SSA (Mugumaarhahama et al., 2020; Acila et al., 2024). Livestock development is an investment including the scaling up of livestock-related enterprises or the development of livestock value chains (FAOSTAT, 2023).

Increased agricultural productivity through research for improved breeds and transition from subsistence towards market-oriented production is foreseen in Uganda's Vision 2040 as the basis for increased rural incomes. In addition to employing Ugandans, the cattle production industry produces economically valuable products such as milk, beef and hides for both local and foreign markets (Waiswa & Akullo, 2021). Besides the provision of food and income, pastoralists keep cattle for traditional purposes and socio-cultural wealth (Akwango et al., 2017; Kabi et al., 2016).

The Tyrolean is a local dual-purpose and high-yielding cattle breed that originates from South Tyrol (Sennereiverband, 2017). Crossbreeding is also regarded as a way of promoting the conservation of local animals. Besides being hardy, Tyrolean Grey cattle are among the few small-bodied temperate breeds that could comfortably be used to cross with the likewise small-bodied SEAZ with no fears of dystocia (difficulty in calving down).

Characteristics of Cattle Breeds in Uganda

Indigenous cattle comprise 77% of the total cattle population, while exotic breeds and crossbreeds make up the remaining 23% (UBOS, 2020). Indigenous breeds are usually preferred because of their dairy and beef characteristics (Kugonza et al., 2012; Sanchez-Molano et al., 2025). The SEAZ and TGZ, two distinct groups of cattle in Uganda, are adapted to hot climates, and they are dual-purpose with high resistance to diseases and environmental stresses, but are reported to be endangered (Mwai et al., 2015). There are four strains of SEAZ cattle, namely the Karamojong, Usuk, Serere and Kyoga, mainly reared under traditional systems (FAO, 2019; Masaba et al., 2024). In 2009, the TGZ from Austria, which is known for producing high-quality milk, was introduced and crossed with the SEAZ in order to have a crossbreed that could produce more milk and have a fast growth rate (Sennereiverband, 2017; Acila et al., 2023). Farmers in Uganda prefer the TGZ because of their high milk production, good mothering ability, hard hooves and fast growth rate.

The authors of this chapter are currently investigating whether the observed morphological characteristics of SEAZ and TGZ vary greatly. The study is being conducted in Aswa Ranch and Lusenke Stock Farm in Uganda (Fig. 5.1). The physical appearance of the SEAZ and TGZ differs with respect to coat colour, hair length and hair type. The TGZ found in Lusenke Stock Farm are mostly grey and light brown while the SEAZ are mainly black, dark brown and brown. In Aswa Ranch, the SEAZ and TGZ have similar characteristics.

Among males, SEAZ cattle are mostly uniform, while TGZ males have a notable presence of shaded patterns, which are absent in SEAZ males. TGZ adults show more shaded and spotted patterns compared to SEAZ. Pied patterns are observed only in SEAZ calves. There is a shift in coat colour patterns across age groups as animals mature. The SEAZ are dominated by straight hair, while TGZ show a higher proportion of curly hair. SEAZ cattle are characterised by short hair, whereas TGZ exhibit a wider variation, including medium and long hair. Curly hair is more frequent among TGZ than SEAZ. TGZ males have medium hair and long hair, whereas SEAZ males have short hair, and none have long hair. In females, the majority have short hair. More TGZ females have medium hair than SEAZ females. TGZ males have longer hair.

Major differences are observed in the dewlap and hump characteristics. SEAZ cattle have small dewlaps and small humps while TGZ animals show variation in structural traits, including medium-sized dewlaps and humps, dropping hump shapes and longer tails, reflecting possibly greater growth potential. The hump position is mostly thoracic in SEAZ and TGZ. The facial profile is uniformly flat across SEAZ and TGZ. There is a compact neck and thoracic region in SEAZ animals compared to their crossbred counterparts.

The back profile shows no major difference, with a majority of animals in both groups having a straight back. The back profile is not different between SEAZ and TGZ. However, TGZ at Aswa Ranch have curved backs compared to SEAZ. At Lusenke Stock Farm, SEAZ have straight backs, while TGZ have curved backs.

Rump shape is generally flat across SEAZ and TGZ, with no major differences. SEAZ have a higher occurrence of flat and roofy rumps, while TGZ have sloping rumps. TGZ cattle at Aswa Ranch have curved backs. A small number of animals at Aswa Ranch exhibited the "rooky rump" type, exclusively among SEAZ cattle.

Both TGZ and SEAZ had horns. The SEAZ mainly had straight horns while the TGZ horns were curved but horn spacing was similar. The horns of TGZ were lateral while SEAZ horns were upward.

The TGZ tails were mostly long while SEAZ had short tails in both the Aswa Ranch and Lusenke Stock Farm. The dewlap size in SEAZ was mostly smaller than in the TGZ, similarly between Lusenke Stock Farm and Aswa Ranch.

Fig. 5.1 SEAZ and TGZ in Aswa Ranch and Lusenke Stock Farm in Uganda. (Photos taken by Mike Okello)

In terms of reproductive features, all males had small testicles, suggesting a uniformity in this trait that could relate to breed management or age category under study. For females, there were no significant differences in udder size, teat length, or navel flap size between SEAZ and TGZ. Testis size among males and traits like udder size, teat length and naval flap size showed no differences between SEAZ and TGZ.

Conclusion

This chapter has provided background information on the SEAZ and TGZ in Uganda, including an account of the morphological characteristics of these animals, particularly those that are relevant for guiding breeding programmes. This chapter is drawn from a broader study focused on mapping the genetic variability and performance of the SEAZ and its crossbreed, the TGZ in Uganda, as part of the PhD thesis study for the corresponding author.

Acknowledgments This research was supported by the Regional Universities Forum for Capacity Building in Agriculture (RUFORUM) by a capacity building competitive grant, training the next generation of scientists, provided by Carnegie Cooperation of New York.

Author Contribution MGA Lead Investigator, MGA, DRK contributed to data collection; MGA, SO, DRK and DRM Led methodology; MGA, SO, DRK and DRM validated the study, and MGA, SO, DRK, DRM, DRMSA, MGNM and DK participated in review and editing.

Ethical Approval The study was conducted in accordance with the institutional requirements. Ethical approval (ref: SVAR- IACUC/238/2025) was obtained from the Institutional Animal Care and Use Committee (IACUC) of the School of Veterinary Medicine and Animal Resources of Makerere University.

ORCID iD https://orcid.org/0000-0001-5708-1639

References

Acila, M. G., Nakimbugwe, H. N., & Kato, H. (2023). The study comparison was between the small east African zebu and the Tyrolean Grey cattle crosses in Uganda. *ASRIC Journal on Agricultural Sciences, 4*(1), 29–38.

Acila, M. G., Nakimbugwe, H. N., Kato, H., Okello, S., Kugonza, D., Mugizi, D. R., & Akello, S. (2024). Comparison of contributions and constraints in rearing small east African zebu and crosses with Tyrolean Grey cattle in Uganda. *ASRIC Journal on Agricultural Sciences, 5*, 29–41.

Akwango, D., Obaa, B. B., Turyahabwe, N., Baguma, Y., & Egeru, A. (2017). Effect of drought early warning system on household food security in Karamoja subregion, Uganda. *Agriculture & Food Security, 6*. https://doi.org/10.1186/s40066-40017-40120-x

Assefa, A., & Hailu, A. (2018). Ethiopian indigenous cattle breeds' diversity, distribution, purpose of keeping, and their potential threats. *Journal of Bio Innovation, 7*(5), 770–789. ISSN 2277–8330. [9] FAO (2015). FAO statistical pocketbook: World Food and Agriculture. Food and Agriculture Organisation of the United Nations, Rome, 2015.

Atukunda, P., Eide, W. B., Kardel, K. R., Iversen, P. O., & Westerberg, A. C. (2021). Unlocking the potential for achievement of the UN sustainable development goal 2–'zero hunger'–in Africa: Targets, strategies, synergies and challenges. *Food & Nutrition Research, 65*, 10–29219.

Food and Agriculture Organisation (FAO). (2019). The Future of Livestock in Uganda. Opportunities and Challenges in the Face of Uncertainty. Rome, Italy.

FAO. (2022). Crop prospects and food situation #4, December 2022. *Quarterly global report*. https://doi.org/10.4060/cc3233en.

FAOSTAT. (2018). *FAOSTAT Food Balance Sheets*. Retrieved from http://www.fao.org/faostat/en/#data/FBS

FAOSTAT. (2023). *Statistical Database*. Retrieved May 17, 2023, from http://www.fao.org/faostat/en/#data/QCL

Ilatsia, E. D., Roessler, R., Kahi, A. K., Piepho, H.-P., & Zárate, V. (2012). Production objectives and breeding goals of Sahiwal cattle keepers in Kenya and implications for a breeding programme. *Tropical Animal Health and Production, 44*(3), 519–530. https://doi.org/10.1007/s11250-011-9928-8

Kabi, F., Muwanika, V., & Masembe, C. (2016). Indigenous cattle breeds and factors enhancing their variation, potential challenges of intensification and threats to genetic diversity in Uganda. *Animal Genetic Resources/Resources Génétiques Animales/Recursos Genéticos Animales, 58*, 1–12. https://doi.org/10.1017/S2078633615000326

Kalungu, R. (2023). Govt. engages stakeholders on CBPP outbreak in Chisamba, Chibombo—Zambia. *News Diggers*. Retrieved from https://diggers.news/business/2022/09/15/govt-engages-stakeholders-on-cbpp-outbreak-inchisamba-chibombo/

Kugonza, D. R., Nabasirye, M., Hanotte, O., Mpairwe, D., & Okeyo, A. M. (2012). Pastoralists' indigenous selection criteria and other breeding practices of the long-horned Ankole cattle in Uganda. *Tropical Animal Health and Production, 44*(3), 557–565. https://doi.org/10.1007/s11250-011-9935-9

MAAIF. (2020). *Ministry of Agriculture, animal industry and fisheries (MAAIF) draft annual performance report for the financial year 2019/2020*. MAAIF.

Masaba, J. K., Babigumira, B. M., Kugonza, D. R., & Mpairwe, D. (2024). The Zenga cattle of Uganda: Insights from morphometric measurements. *Emerging Animal Species, 10*, 100040.

Matovu, J., & Alçiçek, A. (2023). Feed resources used for small ruminant nutrition in sub-Saharan Africa: A case study of Uganda. *Tropical Animal Health and Production, 55*(6), 377.

Mugumaarhahama, Y., Mutwedu, V. B., Kazamwali, L. M., Mushagalusa, A. C., Bantuzeko, F. K., Ndjadi, S. S., et al. (2020). Typology of smallholder's pig production systems in south Kivu, Democratic Republic of Congo: Challenges and opportunities. *Journal of Agriculture and Rural Development in the Tropics and Subtropics, 121*(1), 135–146.

Mwai, O., Hanotte, O., Kwon, Y. J., & Cho, S. (2015). African indigenous cattle: Unique genetic resources in a rapidly changing world. *Asian-Australasian Journal of Animal Sciences, 28*(7), 911–921. https://doi.org/10.5713/ajas.15.0002R

Sanchez-Molano, E., Mukiibi, R., Riggio, V., Ogwang, J., Kawule, L., Benda, K., et al. (2025). Genomic and health characteristics of crossbred dairy cattle in Central Uganda. *Frontiers in Genetics, 16*, 1567910.

Sennereiverband. (2017). *Nachhaltigkeitsbericht* [Sustainability report]. Retrieved April 19, 2019, from https://www.suedtirolermilch.com/images/files/Nachhaltigkeitsbericht_DE-DS-low.pdf(open in a new window

Tayebwa, D. S., Vudriko, P., Tuvshintulga, B., Guswanto, A., Nugraha, A. B., Gantuya, S., Batiha, G. E.-S., Musinguzi, S. P., Komugisha, M., Bbira, J. S., Okwee-Acai, J., Tweyongyere, R., Wampande, E. M., Byaruhanga, J., Adjou Moumouni, P. F., Sivakumar, T., Yokoyama, N., & Igarashi, I. (2018). Molecular epidemiology of Babesia species, Theileria parva, and Anaplasma marginale infecting cattle and the tick control malpractices in central and eastern Uganda. *Ticks and Tick-borne Diseases, 9*(6), 1475–1483. https://doi.org/10.1016/j.ttbdis.2018.06.012

Tricarico, J. M., Kebreab, E., & Wattiaux, M. (2020). Milk symposium review: Sustainability of dairy production and consumption in low-income countries with emphasis on productivity and environmental impact. *Journal of Dairy Science, 103*(11). https://doi.org/10.3168/jds.2020-18269

Uganda Bureau of Statistics (UBOS). (2020). *Statistical abstract.* Retrieved from https://www.ubos.org/wp-content/uploads/publications/11_2020STATISTICAL__ABSTRACT_2020.pdf

Waiswa, D., & Akullo, J. (2021). Implications of gender discrimination for household food security among small holder dairy farmers in Nakaloke, Mbale District, Uganda. *Research Journal of Agriculture and Forestry Sciences, 9*(2), 1–11.

Part III

People, Places, Accountability, Violence and Inter-Generational Justice In and From Africa

6 Women Empowerment for Gender Equality: An Exploration of the Contribution of Empowerment Initiatives in Kole District, Northern Uganda

Fred Felix Ongom and Simon Omaada Esibo

Abstract

This paper explores the contributions of women empowerment initiatives to gender equality in Northern Uganda, with a specific focus on Kole District. Using a mixed-method approach, data were collected through key informant interviews, focus group discussions, and questionnaires. The study's objectives were to assess political, economic, legal, and socio-cultural empowerment initiatives and their impact on women's roles in the community. Findings indicate that while progress has been made in gender equality, significant barriers remain, particularly in access to resources, political representation, and the distribution of opportunities. Respondents highlighted that current empowerment efforts benefit elite and urban women more than rural and marginalized groups. Challenges related to cultural norms, legal enforcement, and economic disparities were prominent, with suggestions for inclusive policies that prioritize grassroots women's empowerment. The study concludes by recommending enhanced enforcement of legal frameworks, increased financial literacy programs, and more equitable access to resources as essential for achieving true gender equality in the region.

Keywords

Genital mutilation · Domestic violence · Household leadership · Inheritance · Land

Introduction

Women's empowerment initiatives refer to the multi-dimensional process that builds the capacity of women to gain access, and control over resources, and opportunity in society. Bayeh (2016), defined women's empowerment as a multi-dimensional social process that helps people gain control over their issues. Globally, there is an increasing awareness of the need to empower women through measures that increase their political, legal, economic, social, and cultural and broader access and control to assets and resources, opportunities, and choices for decision-making. Women's empowerment is a social change indicator and a priority of the Sustainable Development Goals (Miedema et al., 2018), which are global standards not only for social-economic, political, and cultural development, but also a commitment to policy implementations that target the SDGs to improve wellbeing (UNDP, 2015). In Kole District, Northern, Uganda, women's empowerment ini-

F. F. Ongom (✉) · S. O. Esibo
Uganda Martyrs University, Kampala, Uganda
e-mail: felix.ongom@stud.umu.ac.ug

P. Nyadoi et al. (eds.), *Sustainable Development Goals in Africa*, Sustainable Development Goals Series, https://doi.org/10.1007/978-3-032-20742-5_6

tiatives are short of the expectations, and difficult to understand what dimension of empowerment works or not in the case of rural communities such as Kole District, Northern, Uganda. To enhance understanding about empowerment initiatives which can work for rural communities, this study explored the contributions of women's empowerment initiative for gender equality, focusing on the multi-dimensions of women's empowerment (political, legal, economic, and socio-cultural empowerment) in Kole District, Northern, Uganda.

Women empowerment initiatives such as political, economic, legal, and socio-cultural empowerment are key in service delivery in local governments, central government, and even international engagement. What has been done included legal and policy provisions against parity, establishing gender quotas, implementation of women's empowerment programs, support women's through groups such self-help groups, and cooperatives. Legal and policy framework such as National Gender Policy, 2007, Domestic Violence Act, female genital mutilation, youth livelihood projects, Uganda women's entrepreneurship programmes, and affirmative action to widen access to education were in introduced to empower women for gender equality. This has earned the country praise from local, regional, and international development actors. This statistic reveals that the country has made improvement on gender parity as compared to previous years. However, the World Economic Forum stated that, gender gaps are not closing in Uganda. The report singled out political empowerment and other dimensions such as economic, social, and legal. In this report Uganda was ranked in the 11th position in Sub-Saharan African and 61 globally. According to UBOS (2021), 71% of women live in rural areas. The same report points to 47% of women who faced political and economic violence and sex-based violence in 2019/2020.

Data from Kole District, Northern Uganda appears to reflect some negative situation regarding women's empowerment. As gender inequality is often observed at household level, work places and community, in a majority of homes, parents, due to cultural practices, appear to have preference on a boy child when it comes to allocation of productive resources like land, houses, and capital for business. According to some gray literature, teenage pregnancies, child marriage, and school drop-out is high in the area, as some parents often withdraw their girls from school to get married. On another perspective, 1 in 5 women in Kole District, Northern Uganda are raped, abused, or has reported an abuse either sexually or physically. Importantly, emotional abuse is worst since men often dodge responsibility by migrating to other areas in pretense of looking for better means of survival. Other statistics indicate that all the head teachers of government-aided secondary schools are men, and regarding political representative, according to electoral commission, out of 8 directly elected local councilors form Kole south, county, 3 are females, and there is no female judge or female town clerk of the two town councils in the county. Discrimination against women persists in the economic, social, and political sectors, and entrenched gender imbalances continue to be a primary source of poverty in the district. Women have limited access to and control of productive resources, and rarely make decisions for their families and society (Alkire et al., 2013). A baseline survey conducted by Plan Uganda International revealed gender gaps in schooling, microcredit, and pervasive gender-based violence. Our study thus focused on finding answers on what accounts for these gender gaps? We probed, the political, economic, legal, and socio-cultural empowerment initiatives for gender equality in Kole District. We purposively sampled 60 respondents, including women leaders, members of women empowerment groups, and local government officials, and using key informant interviews and focus group discussions obtained the data. The qualitative data from interviews and focus group discussions were analyzed using thematic analysis, while the quantitative data from questionnaires were processed through statistical analysis in SPSS to identify patterns and correlations related to women's empowerment and gender equality in the District.

Results

Political Empowerment Initiatives for Gender Equality

The findings revealed that political empowerment initiatives in Northern Uganda are still in their infancy. Most women in Kole District are aware of their political rights but are often hesitant to participate due to entrenched societal norms and lack of family support (Table 6.1). The quantitative survey revealed that 70% of the respondents disagreed with the statement that "women's political representation is equally important as men's." Qualitative data from key informant interviews with local leaders indicated that while there are policies promoting women's political participation, these have not been effectively implemented at the grassroots level. One key informant remarked: "*There is a law that supports women's participation, but we rarely see women taking an active role, mainly because the family does not support them or they fear criticism from the community.*"

Interpretation of Findings on Political Empowerment Initiatives

The findings on political empowerment initiatives indicate a mixed perception of women's political involvement in Kole District, Northern Uganda. The results are as follows: Women are aware of their rights and privileges to participate in politics: The data shows that 30% strongly agreed and 40% agreed that women are aware of their political rights, indicating a majority acknowledgment of women's political awareness. However, 15% disagreed, and 5% strongly disagreed, reflecting that while awareness exists, some women remain uninformed. This aligns with existing literature suggesting that awareness campaigns have made significant progress, but gaps remain in rural communities. The family is very supportive of women/girls participating in political activities: The responses show that 55% (20% strongly agree and 35% agree) believe that families are supportive of women's political participation. However, 30% disagreed, and 10% strongly disagreed, highlighting that while some families encourage political participation, others still adhere to traditional norms that limit women's involvement. This is consistent with findings by Ochan, who noted that family support remains a critical barrier to women's political engagement in rural Uganda.

Women's political representation is less critical as compared to men: The data reflects a more divided perspective, with 40% agreeing that women's political representation is less critical, while 30% disagreed with this notion. This suggests that there are still strong perceptions that men are more suited for political roles, a finding echoed in research by Kirya who argued that gender stereotypes continue to undermine women's political representation. Women are capable of representing national interests in different political platforms: The majority, 60% (25% strongly agree and 35% agree), believe that women are capable of representing national interests, indicating confidence in women's abilities to perform on political platforms. However, 25% were unsure or disagreed, suggesting that traditional views on women's roles in national governance

Table 6.1 Representing the findings using the five-point scale for political empowerment initiatives

Statement	SA (5)	A (4)	NS (3)	D (2)	SD (1)
Women are aware of their rights and privileges to participate in politics	30%	40%	10%	15%	5%
The family is very supportive of women/girls participating in political activities	20%	35%	15%	20%	10%
Women's political representation is less critical as compared to men	15%	25%	20%	30%	10%
Women are capable of representing national interests in different political platforms like parties, council halls, parliament, etc.	25%	35%	15%	15%	10%
It is equally important for men and women to participate in politics	40%	30%	10%	10%	10%

still persist. This observation parallels research by Acen, who found that while women's capabilities are acknowledged, actual opportunities remain limited. It is equally important for men and women to participate in politics: A strong consensus emerged on the importance of gender equality in political participation, with 70% (40% strongly agree and 30% agree) supporting equal political participation for both men and women. Only 20% disagreed, demonstrating a growing acceptance of the need for gender parity in politics. This finding supports broader trends in gender equality efforts in Uganda, as discussed by Mugisha, who emphasized the importance of equal participation for sustainable development.

Overall, the findings from the study reflect both progress and continuing challenges in advancing women's political empowerment in Northern Uganda. Political awareness is increasing, but societal and family-related barriers still hinder full participation. These results align with national-level studies, which underscore the need for more targeted interventions to address cultural norms and family dynamics that continue to restrict women's political empowerment. The perception that women are less critical in political representation, as seen in 40% of respondents, reveals a continuing bias in gender roles within politics. This mirrors findings from international studies that indicate women's participation in politics is often perceived as secondary to that of men, particularly in patriarchal societies. In conclusion, while strides have been made, more robust efforts are needed to challenge deeply ingrained stereotypes and to provide institutional support for women's political empowerment in rural Uganda.

Economic Empowerment Initiatives for Gender Equality

Economic empowerment initiatives have been more successful, with many women in Kole District participating in income-generating activities. According to the survey, 60% of women agreed that they have the financial knowledge to manage household income, and 55% of respondents indicated that women are now more involved in decision-making concerning family finances (Table 6.2). Focus group discussions revealed that women who engage in economic activities through cooperatives or savings groups are better positioned to assert control over resources within their households. However, disparities remain, as men are still perceived to have more power over land ownership and significant economic decisions. One participant noted: "*Women are now running small businesses, but when it comes to selling or making big financial decisions, men still have the final say.*" These findings suggest that while economic initiatives are helping women gain financial autonomy, traditional gender roles still influence resource control.

Interpretation of Findings on Economic Empowerment Initiatives

Women have access to and control of this community's land, houses, and other productive resources: The findings indicate that only 50% (20% strongly agreed and 30% agreed) believe that women have access to and control over pro-

Table 6.2 Economic empowerment initiatives for gender equality, Kole, Northern Uganda

Statement	SA (5)	A (4)	NS (3)	D (2)	SD (1)
Women have access to and control this community's land, houses, and other productive resources	20%	30%	15%	25%	10%
Women have the financial knowledge to manage individual/family/group finance	25%	35%	20%	15%	5%
It is okay for women to work from outside the home	30%	40%	10%	15%	5%
Women are the primary decision-makers at home/outside	20%	30%	15%	25%	10%
Women are free to participate in groups and associations	35%	40%	10%	10%	5%

ductive resources such as land and houses. However, a significant portion, 35% (25% disagreed and 10% strongly disagreed), felt that women still lack such access and control. This highlights a persistent gap in economic empowerment, especially in land ownership, where traditional practices often limit women's rights. As evidenced by Namakula, despite legal provisions, patriarchal land ownership structures still prevail in rural Uganda, restricting women's economic autonomy. Women have the financial knowledge to manage individual/family/group finance: The data show that 60% (25% strongly agreed and 35% agreed) of respondents acknowledged that women possess financial knowledge, reflecting an improvement in women's financial literacy. Yet, 20% were unsure, and 15% disagreed, showing that gaps in financial training and capacity building still exist. This finding correlates with Katushabe, who emphasized the need for continuous financial education programs to enhance women's economic decision-making abilities in rural Uganda. It is okay for women to work outside the home: A significant 70% (30% strongly agreed and 40% agreed) of respondents accepted that it is socially acceptable for women to work outside the home, signaling a positive shift in gender norms concerning women's economic roles. However, 20% (15% disagreed and 5% strongly disagreed) still held conservative views that oppose women working outside. This division is reflected in studies like that of Nyanzi, which found that while urban areas embrace women's labor force participation, rural areas like Kole District still experience resistance to women's external employment.

Women are the primary decision-makers at home/outside: The responses show that only 50% (20% strongly agreed and 30% agreed) of respondent's view women as primary decision-makers in both the home and outside. Meanwhile, 35% (25% disagreed and 10% strongly disagreed) believe men continue to dominate decision-making. This aligns with findings from Amuge, who noted that women in Northern Uganda still face significant barriers in asserting themselves as household decision-makers, despite their growing economic contributions. Women are free to participate in groups and associations: A high 75% (35% strongly agreed and 40% agreed) agreed that women are free to participate in groups and associations, indicating substantial progress in social-economic empowerment. Only 15% disagreed with this statement, showing limited opposition to women's involvement in group activities. This finding supports the work of Nyabongo, who found that women's participation in groups and cooperatives has been crucial in improving their economic standing and bargaining power in rural Uganda.

The findings reveal a growing recognition of women's economic empowerment in Kole District, with notable strides in financial knowledge and freedom to participate in groups and associations. However, access to and control of productive resources remains a critical issue, as 35% of respondents believe that women still face significant challenges in land ownership and decision-making. This is consistent with broader research on women's economic empowerment in rural Uganda, where traditional norms and land inheritance laws continue to favor men (UBOS, 2021). The strong acceptance of women working outside the home (70%) reflects a positive cultural shift, although resistance still exists, as 20% oppose women's employment outside domestic roles. This dichotomy between traditional and progressive views on gender roles mirrors national trends observed by Mugerwa, who highlighted the slow but steady transition in societal attitudes toward women's economic participation. In conclusion, while the community of Kole has seen improvements in women's financial literacy and social engagement, challenges remain in resource ownership and domestic decision-making. Addressing these barriers is essential for achieving full economic empowerment for women in the region.

Legal Empowerment Initiatives for Gender Equality

Legal empowerment initiatives in Northern Uganda are mostly centered on increasing women's awareness of their legal rights, particularly

Table 6.3 Findings on legal empowerment initiatives

Statement	SA (5)	A (4)	NS (3)	D (2)	SD (1)
Women have access to legal services in physical terms, technical terms, and monetary (legal services such as court and advocate)	15%	25%	20%	30%	10%
A woman legally can be "head of household" or "head of the family" in the same way as a man	20%	30%	15%	25%	10%
Women are knowledgeable and have information about laws	20%	30%	20%	20%	10%
There is awareness rising about women's rights and laws	25%	35%	15%	15%	10%
Women are legally entitled to own or inherit land and other property	20%	40%	10%	20%	10%

concerning land ownership and inheritance. The survey showed that 65% of respondents acknowledged that women in the community have access to legal services, but only 40% believed that women are fully knowledgeable about their rights (Table 6.3). Interviews with community development officers revealed that while there are programs to promote legal awareness, few women take advantage of these services due to fear of backlash or a lack of trust in the legal system. One officer stated: "*We organize legal clinics to educate women, but many are reluctant to come forward because they fear the consequences from their families or communities*."

These findings suggest that while legal frameworks exist to support women's empowerment, more work needs to be done to build trust and ensure women are well-informed of their rights.

Interpretation of Findings on Legal Empowerment Initiatives

Women have access to legal services in physical, technical, and monetary terms (legal services such as court and advocate): Only 40% (15% strongly agreed and 25% agreed) of respondents felt that women have adequate access to legal services, highlighting the limited availability of legal aid and financial constraints. A significant 40% (30% disagreed and 10% strongly disagreed) expressed dissatisfaction with the current access to legal services. This reflects challenges such as distance to courts and high legal fees, as found by Muwonge, who reported that rural women in Uganda often lack access to affordable legal assistance, hindering their pursuit of justice.

A woman legally can be "head of household" or "head of the family" in the same way as a man: The findings show that 50% (20% strongly agreed and 30% agreed) of respondents believe that women can legally be the head of the household, but 35% (25% disagreed and 10% strongly disagreed) still hold conservative views that prefer men as heads of households. This is consistent with research by Namirembe, who found that although Ugandan laws permit women to be recognized as household heads, societal attitudes and patriarchal norms still impede their acceptance in such roles. Women are knowledgeable and have information about laws: 50% (20% strongly agreed and 30% agreed) of respondents felt that women are knowledgeable about the law, but 30% (20% disagreed and 10% strongly disagreed) disagreed, indicating that a lack of legal literacy remains a significant barrier. As highlighted by Kakembo, the lack of information about legal rights among rural women is a major challenge, leading to poor legal empowerment outcomes. Increased efforts in legal education and outreach are needed. There is awareness rising about women's rights and laws: A notable 60% (25% strongly agreed and 35% agreed) of respondents recognized that awareness of women's rights is increasing. However, 25% (15% disagreed and 10% strongly disagreed) still felt that awareness was insufficient. This finding aligns with Okello, who emphasized the growing role of civil society organizations and NGOs in promoting women's rights and raising legal awareness in rural Uganda, although gaps in outreach remain. Women are legally entitled to own or inherit land and other property: A majority 60% (20% strongly agreed and 40% agreed) believe that women are legally entitled to own and inherit land, reflecting progress in land ownership rights for women. However, 30% (20% disagreed and 10% strongly disagreed) still indicated challenges in the implementation of these legal provisions. This supports the findings of Musisi, who noted that while legal frameworks exist to support women's land ownership, the enforcement of these rights is still weak, especially in rural areas where customary practices dominate.

The findings on legal empowerment show that while there is progress in legal awareness and rights enforcement, challenges remain in women's access to legal services and their recognition as heads of households. 40% of respondents reported poor access to legal services, echoing broader national challenges highlighted by Oguttu, who found that access to affordable legal support in rural areas is still limited by structural barriers such as distance, high legal fees, and bureaucratic delays. The 50% agreement that women can be recognized as household heads indicates some level of acceptance, but the 35% opposition suggests that entrenched patriarchal norms still resist this shift. This finding correlates with research by Namakula, which found that legal provisions supporting gender equality in household leadership are often undermined by societal expectations of male dominance. While 60% of respondents acknowledged that women's rights awareness is increasing, this growing recognition has not yet fully translated into widespread legal empowerment, particularly in land ownership. Studies by Nakitto highlight the persistence of customary practices that prevent women from inheriting land, despite legal reforms. In conclusion, while the community in Kole District has made strides in legal empowerment, especially in raising awareness about women's rights, significant gaps remain in access to legal services and enforcement of land ownership rights. Overcoming these barriers requires strengthening legal aid, enhancing legal literacy programs, and addressing societal norms that continue to hinder women's full legal empowerment.

Socio-Cultural Empowerment Initiatives for Gender Equality

Cultural norms remain a significant barrier to gender equality in Northern Uganda. The survey revealed that 70% of respondents felt that it is acceptable for men to exert control over women, including through physical punishment (Table 6.4). Focus group discussions highlighted that marriage remains the primary avenue through which women gain social status, but this often comes at the cost of their autonomy. One participant shared: "*When a woman is married, she is respected, but she loses her ability to make decisions without her husband's approval.*" This reflects the complex relationship between empowerment and cultural norms. Despite ongoing efforts, deep-seated traditional values continue to undermine the progress of women's socio-cultural empowerment.

Table 6.4 Findings on social-cultural empowerment initiatives

Statement	SA (5)	A (4)	NS (3)	D (2)	SD (1)
When a woman/girl is married, she is respected and protected in society	30%	40%	10%	15%	5%
It is okay for a man to beat/hit or batter his wife, especially if she denies him sex, causes a delay in the market, or refuses to take the role	5%	10%	10%	30%	45%
The woman can practice her religion even if the husband is faithful to another religious group	25%	35%	15%	15%	10%
Women/girls should have a share of property and other resources	35%	40%	10%	10%	5%
Women/girls can socialize and expand their social network	40%	35%	5%	15%	5%

Interpretation of Findings on Social-Cultural Empowerment Initiatives

When a woman/girl is married, she is respected and protected in society: 70% (30% strongly agreed and 40% agreed) of respondents believe that marriage brings respect and protection for women in society. However, 20% (15% disagreed and 5% strongly disagreed) expressed dissent, indicating that while societal norms may provide some degree of protection, challenges still exist for married women, especially regarding their autonomy and safety. According to Auma, marriage in rural Uganda is still seen as a marker of respectability, but this can often mask underlying issues of gender inequality and domestic vulnerability. It is okay for a man to beat/hit or batter his wife, especially if she denies him sex, causes a delay in the market, or refuses to take the role: An overwhelming 75% (30% disagreed and 45% strongly disagreed) of respondents rejected the notion that wife-beating is acceptable under any circumstances. However, 15% (5% strongly agreed and 10% agreed) still found it acceptable, reflecting the persistence of harmful gender norms. This mirrors findings by Lwanga, who highlighted that while domestic violence is condemned by most, some rural communities still see it as a legitimate means of enforcing gender roles and authority within the household.

The woman can practice her religion even if the husband is faithful to another religious group: 60% (25% strongly agreed and 35% agreed) of respondents supported the idea that women can freely practice their religion, even if their husbands follow a different faith. However, 25% (15% disagreed and 10% strongly disagreed) expressed concerns, reflecting cultural and familial pressures on women to follow their husband's religion. This aligns with Namuleme, who found that while Uganda's constitution guarantees religious freedom, social pressures often limit women's autonomy in practicing their faith independently. Women/girls should have a share of property and other resources: A notable 75% (35% strongly agreed and 40% agreed) of respondents supported the idea that women and girls should have a share in family property and other resources, signaling growing gender equity.

Socio-Cultural Empowerment Initiatives for Gender Equality

Interpretation of Findings on Gender Equality

Women have access to and control over resources and property: 65% (35% strongly agreed and 30% agreed) of respondents believe that women have access to and control over resources and property in their community (Table 6.5). However, 25% (15% disagreed and 10% strongly disagreed) disagreed, suggesting that while progress has been made, some barriers to women's full access to resources still exist. According to Muhumuza (2023), customary practices in rural Uganda continue to limit women's control over property, especially in cases of inheritance and land ownership. Women enjoy the benefits and opportunities in society equally to men: 45% (20% strongly agreed and 25% agreed) of respondents felt that women enjoy benefits and opportunities in society equally to men, while 40% (20%

Table 6.5 Findings on social-cultural empowerment initiatives

Statement	SA (5)	A (4)	NS (3)	D (2)	SD (1)
Women have access to and control over resources and property	35%	30%	10%	15%	10%
Women enjoy the benefits and opportunities in society equally to men	20%	25%	15%	20%	20%
Work to achieve gender equality today benefits women's elite, the rich, and people with power	40%	30%	15%	10%	5%
Gender equality, meaning that men and women are equal, has already been achieved for the most part	10%	15%	20%	25%	30%
Women enjoy freedom and autonomy in this society	25%	35%	10%	15%	15%

disagreed and 20% strongly disagreed) disagreed. This indicates a significant gap in achieving gender equality in terms of social and economic opportunities, reflecting findings by Akello (2022), who noted that while legal reforms have been enacted, societal attitudes and practices still restrict women's access to equal opportunities. Work to achieve gender equality today benefits women's elite, the rich, and people with power: A notable 70% (40% strongly agreed and 30% agreed) of respondents believe that current gender equality efforts primarily benefit elite women, the wealthy, and those with power. This perception suggests that gender equality initiatives are not reaching the grassroots level. This aligns with Achieng (2021), who found that while urban and affluent women are benefitting from gender-focused policies, rural and marginalized women often remain excluded from these opportunities.

Gender equality, meaning that men and women are equal, has already been achieved for the most part: Only 25% (10% strongly agreed and 15% agreed) of respondents believed that gender equality has been mostly achieved, while 55% (25% disagreed and 30% strongly disagreed) rejected this statement, signaling that significant challenges to gender equality persist. This is consistent with Otim (2022), who found that despite legal reforms, cultural norms, and social structures continue to favor men, particularly in rural communities. Women enjoy freedom and autonomy in this society: 60% (25% strongly agreed and 35% agreed) of respondents felt that women enjoy freedom and autonomy in their society, while 30% (15% disagreed and 15% strongly disagreed) disagreed. This suggests a growing recognition of women's autonomy but also reflects the ongoing cultural and societal constraints that limit their full expression of independence. As noted by Kyomuhendo (2022), while women's autonomy is advancing, deeply entrenched gender norms still hinder their full participation in decision-making both at home and in the public sphere.

The findings reveal that while some progress has been made toward gender equality in Kole District, significant disparities remain, particularly regarding access to opportunities and the distribution of resources. Access to resources and property was supported by 65% of respondents, reflecting growing legal recognition of women's rights to own and control assets. However, 25% of respondents still indicated that cultural and familial barriers exist, preventing women from fully exercising these rights. According to Muhumuza (2023), despite the existence of legal frameworks supporting women's property rights, enforcement remains weak in rural areas, and customary laws often take precedence. When asked whether women enjoy equal benefits and opportunities to men, 45% agreed, but a substantial 40% disagreed, underscoring the fact that gender equality in terms of access to education, employment, and political participation has not been fully realized. This finding aligns with Akello (2022), who found that while policy frameworks promote gender equality, societal attitudes and structural inequalities continue to limit women's access to these opportunities. A significant 70% of respondents felt that gender equality efforts are currently skewed toward benefiting the elite and the powerful. This perception is a common criticism of gender-focused development initiatives, which are often seen as neglecting rural and marginalized women. Achieng (2021) found that while there have been notable successes in urban centers, rural women, particularly those in impoverished regions, continue to face structural barriers to achieving gender equality, including limited access to education and financial resources. Only 25% of respondents agreed that gender equality has already been achieved, while 55% disagreed, indicating that there is still a long way to go in terms of changing both laws and societal norms. Otim (2022) reported that despite legislative reforms, gender inequalities in terms of wage gaps, political representation, and access to education persist, particularly in more conservative rural areas like Northern Uganda. Finally, 60% of respondents agreed that women in the community enjoy freedom and autonomy, though 30% expressed dissatisfaction with women's autonomy. Kyomuhendo (2022) similarly found that while women in Uganda are gaining more autonomy, particularly in urban and semi-urban areas, traditional family structures and gender norms still place significant constraints on women's independence in decision-making within household.

Conclusion

The findings indicate that despite various empowerment initiatives, achieving gender equality in Northern Uganda is still a work in progress. While economic empowerment initiatives have shown some success, political, legal, and socio-cultural empowerment efforts are hampered by traditional practices and inadequate implementation of policies. The study concludes that without addressing these cultural barriers and ensuring the effective enforcement of legal frameworks, women's empowerment will remain incomplete.

Recommendations

Strengthening Policy Implementation: There is a need for stronger enforcement of existing policies that promote women's political and legal empowerment. Government agencies should collaborate with civil society organizations to ensure that women are aware of their rights and encouraged to participate fully in political processes.

Community Sensitization on Gender Equality: Initiatives should focus on educating communities about the importance of gender equality, not just for women, but for overall community development. Changing societal attitudes is critical to achieve meaningful empowerment.

Economic Empowerment Through Skills Training: Economic empowerment programs should be expanded to include more comprehensive financial literacy and skills training for women, particularly in rural areas, to ensure they can compete in larger markets and control resources.

Increasing Access to Legal Services: Legal aid services should be made more accessible to women, particularly in rural areas, and awareness programs should be intensified to ensure women understand their rights concerning property ownership and inheritance.

Cultural Reforms: There is a need for cultural reforms to promote women's autonomy and challenge harmful gender norms. Engaging traditional and religious leaders in discussions about gender equality to shift societal perceptions and foster inclusive environment for women.

References

Achieng, P. (2021). Legal empowerment and women's rights: A case study of Uganda. *African Journal of Law and Society, 18*(3), 56–71.

Akello, J. (2022). Economic empowerment and gender equality in northern Uganda. *Journal of Gender Studies, 12*(2), 124–139.

Alkire, S., Meinzen-Dick, R., Peterman, A., Quisumbing, A., Seymour, G., & Vaz, A. (2013). The women's empowerment in agriculture index. *World Development, 52*, 71–91.

Bayeh, E. (2016). The role of empowering women and achieving gender equality to the sustainable development of Ethiopia. *Pacific Science Review B: Humanities and Social Sciences, 2*(1), 37–42.

Kyomuhendo, G. (2022). Women and political representation in rural Uganda: Progress and challenges. *Uganda Journal of Political Science, 14*(1), 67–83.

Miedema, S. S., Haardörfer, R., Girard, A. W., & Yount, K. M. (2018). Women's empowerment in East Africa: Development of a cross-country comparable measure. *World Development, 110*, 453–464. https://doi.org/10.1016/j.worlddev.2018.05.031

Muhumuza, J. (2023). Barriers to women's political participation in Uganda: A family-centered perspective. *Journal of Social and Political Affairs, 16*(4), 98–114.

Otim, F. (2022). Gender equality and legal empowerment in post-conflict societies. *Journal of Peace and Development, 20*(2), 89–106.

UBOS. (2021). *Uganda-National-Survey-Report-2019-2020.*

UNDP. (2015). *United Nations sustainable development goals*. Policy guideline (pp. 1–24). https://doi.org/10.38124/ijisrt20jul838

7 Understanding Factors Underlying Persistent Female Genital Mutilation in Kween District, Eastern Uganda

Susan Arayo

Abstract

Despite implementation of the Female Genital Mutilation Act 2010 that prohibits female genital mutilation in Uganda, the practice persists among the communities in Kween District. Based on an ethnographic study involving 11 key informants between June and July 2023, this chapter presents some of the underlying factors perpetuating female genital mutilation (FGM) in the District. The findings show that the practice has been pushed underground and is being done in secrecy, among others, by medical workers and traditional birth attendants during birth. It is therefore not easy to determine the real extent of the practice among the practicing community and there is a need for a study to elucidate approaches that will help secure human rights in the culturally constraining context of the FGM practicing community.

Keywords

Rites of passage · Labia · First-born child · Husbands · World Health Organization · Sabiny

S. Arayo (✉)
Department of Governance, Peace and International Studies, Uganda Martyrs University, Nkozi, Uganda
e-mail: sarayo@umu.ac.ug

Introduction

Female genital mutilation (FGM) is a traditional practice that involves the partial or total removal of the external genitalia or other injury to the female genital organs for nonmedical reasons (WHO, 2023). The practice is also known as female genital circumcision (FGC), cutting Sunna, Tahor, Khatna, Khatfz, among other terminologies (Jangam, 2018). According to Khosla et al., (2017) female genital mutilation/cutting (FGM/C) is a dire violation of human rights, particularly women and children's rights. It is a form of discrimination based on sex, gender, age and other grounds. It results in severe health complications, including but not limited to death, disability, miscarriage, stillbirth, shock, hemorrhage sepsis, sexual dysfunction, and posttraumatic stress disorder. Globally, it is reported that more than 200 million girls and women alive today have undergone female genital mutilation in 30 countries in Africa, the Middle East, and Asia where FGM is practiced, with more than three million girls estimated to be at risk annually. The practice includes Type 1 FGM, where partial or total removal of the clitoral glans (the external and visible parts of the clitoris, which is a sensitive part of the female genitals) and/or the prepuce/clitoral hood (the foreskin surrounding the clitoral glans) is done. Type 2, where partial or total removal of the clitoris and the labia minora (the inner folds of the vulva), with or without

P. Nyadoi et al. (eds.), *Sustainable Development Goals in Africa*, Sustainable Development Goals Series, https://doi.org/10.1007/978-3-032-20742-5_7

removal of the labia majora (the outer folds of the skin of the vulva) is done. Type 3, also known as infibulation, involves the narrowing of the vaginal opening through the creation of a covering seal. The seal is formed by cutting and repositioning the labia minora or labia majora, sometimes through stitching, with or without removal of the clitoral prepuce/clitoral hood and glans. Type 4 includes all other harmful procedures to the female genitalia for non-medical purposes, such as pricking, piercing, incising, scraping, and cauterizing the genital area (WHO, 2023).

In Uganda, although the practice is illegal and prohibited by the constitution and the FGM Act 2010(ILO, 1995;ILO,2010), it is practiced among women aged 15–49, mainly in the districts of Kapchorwa, Kween, Bukwo, and Karamoja region, respectively, among the Sabiny, Pokot, Tepeth, and Kadama tribes. The practice among these communities is believed to control girls' libido, make them clean, and prevent the death of their husbands and firstborns. It is also believed to be proof that the woman will be able to endure the pain of childbirth and it is a rite of passage into adulthood (Nakamura, 2023; Rahman & Toubia, 2000). Studies done in Nigeria and in Iran to investigate why parents allow their daughters to undergo FGM revealed that attitudes and perceived norms matter (Ilo, 2018; Pashaei et al., 2016). This is in line with propositions in the theory of planned behavior (TPB) by Ajzen (1991), a social psychological model proven suitable for understanding, predicting, and changing people's health-related behaviors. Based on this theory, this chapter explores how attitudes, subjective norms, and perceived behavioral control have contributed to perpetuation of the FGM practice in Kween district in Eastern Uganda.

Data Collection

Key informant interviews, involving 11 participants, were conducted. In-depth interviews were considered the most appropriate technique to collect primary data, as they would allow the interviewer to probe and clarify individual perspectives.

Before the interviews commenced, participants were provided with consent forms and asked to indicate their willingness to take part in the study. A total of 11 interviews were held, lasting 30–50 min each. To protect the identity of participants, codes were used instead of names during data collection. All data were stored on password-protected computers, managed and analyzed by the researcher.

Results

Sociodemographic Characteristics of the Study Participants

Participants involved in this study included representatives of civil society, police, health, cultural and religious institutions, and the media. Most of these were men aged between 35 and 45 years and had attained university education.

Emerging Themes

Criminalization of FGM/C: Ten out of eleven participants mentioned that FGM is a traditional and cultural practice that is cherished by the Sabiny community. One participant noted that, "*during the circumcision period there is a lot of attachment between male circumcision and female circumcision. When the age sets are to be circumcised, a guest of honor is required to chair the function. The guest of honor is expected to be accompanied by his wife, who should be circumcised. In case his wife is not circumcised, the cultural law allows him to look for any woman who is circumcised to participate in the function. It puts some women in a desperate position to get circumcised*" (Interviews KII P1 2023).

Another participant observed that "*In recent years we have been seeing men hiring other women to come and officiate that function where he dances with that wife if she is cut. So this haunts the women so much, and it has pushed many women to be mutilated/circumcised. Because they don't want to miss that function where they are a given a lot of respect...*" (Interviews KII P3 2023).

Among the Sabiny, male circumcision and rituals are celebrated and are legal, while female circumcision is illegal. Therefore, women's desire to participate in male circumcision ceremonies and celebrations, key social events that only permit a mutilated/circumcised woman to attend, remains a significant driver of FGM among the older married women.

"*The practice has also shifted from the young girls to the married women. Who have been provoked by their mothers-in-law that for them to remain in the marriage, pick cow dung from the kraal, milk the cow, she needs to undergo circumcision*" (Interviews FGM KII P9 2023).

KII P8 noted that "*it is a clandestine practice done in the peripheral sub-counties like Moyok, Kwosir, Kapronon, Kaptoyoy, Kitawoi, Kitawoi, and all sub-counties bordering Mt. Elgon. They still cut but in the bushes, not in the open.*" (Interviews 2023)

Although most participants acknowledged that the practice was clandestine, the majority of the participants agreed that the prevalence was reducing. One KII 9 commented that "*The practice is fading off very fast.*"

One female participant stated that: *the law prohibiting FGM/C began way back in 2002. The district leaders felt they needed an ordinance, so it was not handled at the district. Councilors felt they would lose their votes if they voted against the practice of FGM/C, so they said, "We don't want to own this up, let it go up to the village level..." 2000 engaged the village council; in 2004 the village council agreed and will develop the memorandum of understanding that we are asking our sub-counties for and that we have agreed to pass laws at the sub-county level memorandum of understanding was sent to the district to pass the ordinance and the district sent the memorandum of understanding to parliament to pass the law; hence, the FGM Act 2010* (Interviews KII P8 2023).

According to the participant, it was therefore important that the law came into place such that it reduced the practice of FGM/C in the region. It was also observed by many participants that the Uganda Constitution addresses some aspects of FGM/C. It is therefore important to note that most of the respondents who were learnt were against the FGM/C cultural practice and in support of the Prohibition of FGM Act 2010.

The majority of the participants stated that they were aware of the law that was put in place to criminalize the FGM/C: "*When you visit the communities and engage with them, they are actually aware that this practice is against the law... There was an early survey done in 2015 by UBOS; still, the level of understanding of the law at that time was 98% awareness about the law against FGM" (KII P3 2023).*

On the experience of the community about the law, one male participant had this to say:

> The law was developed without the active participation of the community. And the law is so punitive that even those punished do not understand the law. There was very limited awareness raising of the law. (Interviews KII 8 2023). He further noted that the penalties are not so friendly.

Another male participant lamented that the law is too harsh, the number of years given... *If you are dancing the law catches at you. At the end of the day no one is ready to report because all of us are potential suspects. Even somebody who is supposed to report is a potential suspect.* (Interviews KII P1 2023). "*The surgeon/cutter have a charge, the neighbors who did not report to the authorities are also criminalized. If they get you seeing the cutting, you can also be arrested for around two years.*" (Interviews KII P3 2023).

Another KII 10 commented that "*There are people who are pro-FGM and people who are against FGM. Those people against the FGM law say they were not consulted...not involved in the process. Parliament consulted the community. The others who are for the law say the law has come at the right time*" (Interviews KII 10 2023).

However, there are scenarios where the law may face resistance. One male participant had this to say.

> The president has kept shooting down the law, including the magistrates. There was a case scenario here in Kapchorwa where they arrested ten women, and the chief magistrate asked,

> "Can you bring me the sample of the clitoris, and I see if truly they cut it? Where would one get the clitoris, and yet this woman has a wound?" (Interviews KII P8 2023) "Now the president says I have pardoned these people they can go back home. Honestly, even the policemen who are supposed to arrest people lose morale, and then the communities developed resistance." KII P8.

This implies that it is hard to track the prevalence of FGM, since it is done secretly and underground.

When asked about the basis of criminalization of FGM/C, one male participant noted that *"fifteen people per sub-county are chosen to watch, monitor, and report to the police and location."*(Interviews KII 10 2023).

Although this is true, the majority still said many have withdrawn and fear to report because of the negative backlash they get from the community when they get to know the reporter.

Other emergent factors: When participants were asked about the experience of girls and women about the law, they said:

> The practice is done in secret. Most of the respondents noted that they now practice it in hiding. "It has made FGM/C to be practiced in secrecy, in caves, bushes and sacred huts" (Interviews KII P2 2023). Some of them have resorted to do it in hiding and some have even crossed the borders to Kenya. "So many of them are crossing the borders through Amudat to the other side of western Kenya, West Pokot, to undertake the practice and then come back after healing." (Interviews KII P3 2023) Another male participant noted that women are being cut at night in the bush and caves. (Interviews KII P1 2023).
>
> Also the community has developed fear, some have withdrawn from talking about it for fear of being arrested. "There is a lot of concealment from the community and the leaders. No one actually talks about it…the culture also does not allow anybody to talk about it or tell another person what has happened. There is a lot of secretiveness in the Sabiny culture whereby should you try to report or talk about what has happened, you will end up running mad and so on." (Interviews KII P3 2023) "The law has scared many people." (Interviews KIII P7 2023).
>
> Initially we used to carry out a headcount to know how many people are circumcised and able to track. We no longer carry out head count because people fear… They circumcise in the night. It is not as it used to be. Circumcision has to be done in the morning, now it is done at night because of fear (KII P10).

One male participant noted that

> There was a tree that flowered in the Mount Elgon National Park. This tree had taken over 60-70 years without flowering, so the cultural leaders visited the national park and realized that the tree had flowered…it had flowered because the cutting had gone down. So we saw a very big number of women coming up and cutting themselves. So the police tried to intervene about 20 women were arrested, but when it reached court, they had a challenge. The magistrate asked, "Where is the evidence of the cutting? Where are the clitorises that were cut?" (Interviews KII P3 2023)

Many of the women and girls have been pushed to undergo the practice mostly by their spouses. One male participant echoed that *"It is the men who escort their women to cutters, and they are cut" (KII P8). "The unmutilated/uncircumcised are looked at as dirty"* (Interviews KII P3 2023).

One surgeon told us that medical workers are also cutting the women. *It is done secretly because some men want it …when they go to give birth* (Interviews KII P3 2023). This could probably be to preserve their cultural identity and the saying that uncut women are stubborn and disrespectful. Another male participant stated that *"Some women during delivery ask the traditional birth attendants to help them deliver the baby and undertake the FGM as well so as to fit in the society and escape the abuses."* (Interviews KII P8 2023).

It therefore shows that because of fear of prosecution, women are now being cut at an older age after marriage. Also the men in the society who are the judges and decision-makers in the community encourage cutting during delivery.

There has been an increase in the cost of cutting the women from less than shs. 50,000 to more than shs. 500,000. It was noted that cutting was a lucrative business for the cutters. One male participant stated, *"This was an income-generating activity for some cutters. You cannot tell them to stop, yet they have no alternative"* (Interviews KII P4 2023).

One male representative of the civil society noted that *"the government had promised the local cutters to buy them grinding machines and give them financial resources, which were not fulfilled" (KII P8).*

One female participant noted that these women who have undergone FGM/C are so superstitious: "*...there is a lot of witchcraft. ... when a circumcised person tells you something... it will happen. If they tell you will never produce, it happens. So they are taught witchcraft in the process they are given the power to curse* (Interviews KII P2 2023). This implies the use of witchcraft, gossip, and beliefs related to the supernatural power of the excisors underpins the cultural cognitive nature of the FGM/C practice. This makes it hard for the women and children to challenge the cutters.

The participants were asked about the implication of the criminalization of FGM/C on the cultural rights of the Sabiny. A female participant stated that it has become very hard to identify a Sabiny. She said that "*marks were put, but now it is hard to know who is a Sabiny.*" (Interviews KII P6 2023).

Proposed Practices to Replace and/or Ways to Reduce FGM

Alternative rites of passage were recommended as the option chosen by the community to abandon FGM/C. One male respondent noted that apart from cutting, there were so many activities that took place, like counseling of the women and training them on how to be good mothers, how to look after their children, and how to handle food security (Interviews KII P3 2023).

There is a need to increase surveillance across the border with Kenya so that cross-border FGM/C is reduced. This can be done through capacity building of law enforcement officers.

Engage the elders and men. There should be a deliberate continuation of engaging elders, cultural, religious leaders, and men since they wield more power and influence than the national law. This will help in changing social norms that drive the practice in the rural areas.

Amendment of the law to cater for the reporters and for provision of evidence. One female participant noted that "...amendment of the article like the issue of the clitoris as evidence" (Interviews KII P2 2023).

Education of the girls and women, coupled with economic empowerment, will steadily help in the reduction of FGM/C. This will have a positive impact on the mindset and attitude of the women and girls.

There is a need for training for alternative life skills or alternative sources of income for the traditional cutters.

Sensitization and awareness creation about the law and engagement of the communities. There is a need to create awareness about the law in the community and urge them to have a community declaration, whereby the communities declare that they will no longer conduct the practices. This can be done with the help of the media through radio talk shows.

The Prohibition of FGM Act alone cannot lead to abandonment of FGM. There is need to revisit the law and its implementation with more support from the government, law enforcement, advocates, the community, elders, the media, and civil society in order to make this law more effective.

Discussion

From the results, it emerges that FGM practice now is clandestine and mainly done in the peripheral sub-counties where it is cherished. Increasingly the circumcision is being performed in a secretive and hidden way rather than during open and public ceremonies (UNICEF, 2020). Girls then tend to travel away, either alone or in small groups, to be mutilated in remote areas where they are less likely to be seen and reported to the authorities.

While the Prohibition of FGM Act 2010, the Constitution of the Republic of Uganda, and the Child Act 1997 are very pronounced on the illegality of FGM/C, the cultural myths continue to dominate the Sabiny community in the rural areas, and their human rights continue to be violated(Bunei & Rono, 2018) as highlighted in this study. The practice of FGM/C acts as a gender norm that good, physically appealing, and sexually modest women are cut and therefore must be cut to be marriageable in a patriarchal family system (Yount et al., 2020). This also continues to impede the abandonment of the practice.

The majority of the participants stated that they were aware of the law that was put in place to criminalize FGM/C. This is consistent with UBOS (2020), which observed that most of the people were aware of the law. One male participant lamented that the law is too harsh, given the number of years. If one is participating in FGM/C, like dancing, or if one fails to report, the law has penalties. This has scared many people. This is consistent with the Prohibition of 2010 Act, Sections 5, 6, and 7 of the 2010 FGM Act, which noted that anyone who attempts to carry out, procure, aid or abet FGM or participate in any event leading to FGM is liable on conviction to imprisonment of up to 5 years. Scholars express worries about the psychological effect on girls who have been cut witnessing the arrest of their parents, as well as the financial impact of imposing fines on poor people and the potential systematic imprisonment of the initiators and practitioners of FGM/C, who are primarily women (Shell-Duncan et al., 2013).

Although the law has punitive penalties for the perpetrators, the practice is still practiced in the rural areas where the cultural and social norms are strongly upheld. This corroborates the social theorist writings of Yount et al., (2020), which hypothesize an interaction between personal belief and community norms about a practice. In communities where the social norm is to strongly favor FGM/C, opposing personal beliefs are expected to have little influence on behavior. On the basis of criminalization of FGM/C, the findings revealed that *fifteen* people per sub-county are chosen to watch, monitor, and report to the police and location. "The law also faces resistance from the community. There was a scenario where they arrested ten women, and the chief magistrate asked, "Can you bring me the sample of the clitoris?" Moreover, the President also pardoned these women. This corroborates Bunei & Rono (2018), who noted that community resistance is expressed through rejection of laws, the protection of the cutters, and silent support of the practice.

Emergent consequences of FGM/C criminalization in Kween District: This study reveals that FGM/C is practiced in secrecy, in caves, bushes, and across the border in western Kenya. Nakamura (2023) alluded to the fact that people started having daughters undergo FGM/C in secret at distant relatives' places and returning as if nothing had happened. UNICEF (2020) noted that the practice is done in secret, often in remote, unsafe conditions. Cross-border FGM is also becoming increasingly common given the perceived weaker anti-FGM law enforcement on the border with Kenya. The findings also revealed that the community has developed fear; some have withdrawn from talking about it for fear of being arrested. It should be noted that the myths surrounding challenging the cutters and the consequences of reporting have scared the community from speaking. The findings revealed the practice was high in the rural areas. UNICEF (2013) notes that FGM/C tends to be less common in urban than in the rural areas. There is need for research to understand the influence of the patriarchal system in the rural areas and the FGM/C abandonment. It is important to understand the nature of social identity and the meaning of FGM/C among the Sabiny in the rural areas. The findings revealed it was hard to carry out headcount of the prevalence of FGM/C because people fear to report the perpetrators.

Finding in this chapter also revealed that medical workers and traditional birth attendants were also cutting the women during delivery secretly because some men want it. This confirms the view of Yount et al., (2020) who argued that in many practicing societies, FGM/C is performed to reflect and to uphold injunctive norms about sex differentiation, feminine beauty, and women's sexuality. De-emphasizing (and controlling) sexuality of women's bodies, Yount notes that FGM/C accentuates women's reproductive role. In these ways, the collective practice of FGM/C acts as a gender norm that good, physically appealing, and sexually modest women are cut and therefore must be cut to be marriageable in a patriarchal family system (Yount et al., 2020).

There is a need to understand the collective practice of FGM/C as a gender norm that upholds the patriarchal family system.

Increase in pay of FGM/C: Since FGM/C is illegal, the pay has increased because the cutters are aware of the penalties and fear prosecution. FGM/C in Kween is now practiced among the married. This confirms UNICEF (2020) that in Sebei the cutting of older married women during antenatal visits or during childbirth is becoming very common. This is closely linked to the mockery and shaming that uncut women experience from peers and other wives in polygamous communities and to them not being allowed to attend male circumcision ceremonies. There is limited research on the influence of male circumcision ceremonies and FGM/C persistence in the rural areas.

Hard to identify a Sabiny: O'Neill & Pallito (2021) FGM/C status itself does not define a woman's identity but rather is an aspect of her identity that is closely linked to social norms. In settings where FGM/C is a social norm, the negative connotations of uncut women as oversexed, unfaithful, and incapable of regulating their sexual desires continue to have a significant impact on the perpetuation of the practice.

Community-led alternatives sought to abandon FGM/C: The findings revealed that ARP was recommended as the best option to lead to abandonment of the FGM/C. This is consistent with what previous writers like Parsitau & Aura (2021) noted about ARP, known as a "ritual without cutting."

The findings revealed the need to engage elders and men since they are believed most by the community. Engaging the elders and men who are the judges and decision-makers in the communities is paramount for the abandonment of FGM/C. These findings illustrate the importance of considering structural and normative elements of gendered systems to understand the persistence of FGM/C in the rural areas. It also shows the dependence of the women on the men and elders in the patriarchy system dominant in the rural areas. Research is needed to understand the influence of community norms of FGM/C by elders and males and the persistence of FGM/C in rural areas.

The findings revealed the need for education and economic empowerment of women. Yount et al., (2020) noted the need for joint activities that involve educational opportunities and training in income-generating activities for women. This study shows that such opportunities may promote women's ability to implement preferences against FGM/C by increasing their decision-making power in the family. More still, there is a need to provide the traditional cutters with resources and income-generating activities such that they are not totally dependent on cutting as a source of income.

Communities were engaged in participatory dialogues and community-based activities grounded in the principles of human rights, democratic processes, and women's health, etc. At the completion of these dialogues, communities deliberated and spearheaded collective decisions about abandoning harmful practices including FGM/C (Yount et al., 2020).

Conclusion

With the implementation of the Prohibition of FGM Act of 2010, the prevalence rates of FGM/C had greatly reduced. However, it was observed that the clandestine practice still takes place in the peripheral sub-counties bordering Mt. Elgon, across the border in western Kenya. Moreover, some medical personnel and traditional birth attendants, among others, are told to cut the women during childbirth. In order to reduce the FGM/C practice, I suggest that Uganda needs to revisit the implementation and enforcement of the law with a multi-sectoral approach that brings together the elders, religious leaders, law enforcement, government agencies, civil society, child protection professionals, and survivors to end FGM/C in the rural areas in Kween District.

References

Ajzen, I. (1991). From intentions to actions: A theory of planned behavior. In J. Kuhl & J. Beckman (Eds.), *Action-control: From cognition to behavior* (pp. 11–39). Springer.

Bunei, E. R. & Rono, J.K. (2018). A critical understanding of resistance to criminalization of female genital mutilation in Kenya. In K. H. Carrington (Ed.), *The Palgrave handbook of criminology and the global south*. Palgrave Macmillan. https://doi.org/10.1007/978-3-319-65021-0_43

Ilo, C. D.-O.-T. (2018). Factors associated with parental intent not to circumcise daughters in Enugu State of Nigeria: An application of the theory of planned behavior. *African Journal of Reproductive Health, 22*(1), 29–37.

International Labour Organization. (1995). Constitution of Uganda. Law, act. Retrieved June 26, 2023, from https://www.ilo.org/dyn/natlex/docs/ELECTRONIC/44038/90491/F206329993/UGA44038.pdf

International Labour Organization. (2010). Prohibition of Female Genital Mutilation Act, 2010. Law, Act. Retrieved June 26, 2023, from https://www.ilo.org/dyn/natlex/natlex4.detail?p_isn=88098&p_lang=en

Jangam, M. R. (2018). Female genital mutilation: A human rights perspective. *Supremo Amicus, 8*, 187.

Khosla, R., Banerjee, J., Chou, D., Say, L., & Fried, S.T. (2017). Gender equality and human rights approaches to female genital mutilation: A review of international human rights norms and standards. *Reproductive Health, 14*(1), 1–9. https://doi.org/10.1186/s12978-017-0322-5

Nakamura, K. (2023). An ethnography of diversity and flexibility around female circumcision and female genital mutilation/cutting: A case of a local community response to the abolition movement in Kenya. In *Female Genital Mutilation/Cutting: Global Zero Tolerance Policy and Diverse Responses from Africa and Asia Local Communities* (pp. 91-103) Singapore: Springer Nature Singapore. https://doi.org/10.1007/978-981-19-6723-8_6

O'Neil, S., & Pallitto, C. (2021). The consequences of female genital mutilation on psycho-social well-being: A systematic review of qualitative research. *Qualitative Health Research, 31*(9). https://doi.org/10.1177/10497323211001862

Odeku, K., Rembe, S., & Anwo, J. (2009). Female genital mutilation: A human rights perspective. *Journal of Psychology in Africa, 19*(1), 55–56. https://doi.org/10.1080/14330237.2009.10820258

Parsitau, D.S., & Aura, R. A. (2021). Alternative rites of passage and other responses to female genital mutilation/cutting in Africa. In O. Yacob-Haliso & T. Falola (Eds.), *The Palgrave handbook of African women's studies*. Palgrave Macmillan. https://doi.org/10.1007/978-3-319-77030-7_138-1

Pashaei, T., Ponnet, K., Moeen, M., Khazaee-pool, M., & Majlessi, F. (2016). Daughters at risk of female genital mutilation: examining the determinants of mothers' intentions to allow their daughters to undergo female genital mutilation. *PLoS One, 11*(3), e0151630.

Rahman, A., & Toubia, N. (eds.) (2000). *Female genital mutilation: A guide to laws and policies worldwide. Female genital mutilation: A guide to laws and policies worldwide*. Zed books. Retrieved July 1, 2023, from https://books.google.co.ug/books?hl=en&lr=&id=kEG6GaudxQEC&oi=fnd&pg=PR8&dq=Rahman,+A.,+%26+Toubia,+N.+(Eds.).+(2000).+Female+genital+mutilation:+a+guide+to+laws+and+policies+worldwide.+zed+books&ots=ea9KrozghT&sig=l6RTSKL2wpZSzjTqX0TBJpBuNbw&redir_esc=y#

Shell-Duncan, B., Wander, K., Hernlund, Y., & Moreau, A. (2013). Legislating change? Responses to criminalizing female genital mutilation cutting in Senegal. *Law and Society Review, 47*(4), 803–835. https://doi.org/10.1111/lasr.12044

UBOS. (2020). *Gender based violence and sexual and reproductive health rights survey. Kween District Local Government*. UNFPA. Retrieved July 1, 2023, from https://www.ubos.org/wp-content/uploads/publications/02_2022kween_District.pdf

UBOS. (2021). *Uganda national household survey 2019/2020*. Uganda Bureau of Statistics.

UNICEF. (2020). Female genital mutilation in Uganda. UNICEF. Retrieved June 26, 2023, from https://www.unicef.org/uganda/media/7996/file/FGM%20Evidence%20from%20Uganda_Policy%20Brief_29th%20Sept%202020.pdf

UNICEF. (2021). *Female genital mutilation Uganda*. Retrieved July 2, 2023, from https://www.unicef.org/media/128236/file/FGM-Uganda-2021.pdf

WHO. (2023). *Female genital mutilation-Uganda*. Retrieved June 30, 2023, from https://www.who.int/news-room/fact-sheets/detail/female-genital-mutilation

Yount, K. M., Cheong, Y.F., Grose, R.G. & Hayford, S.R. (2020). Community gender systems and a daughter's risk of female genital mutilation/cutting: Multilevel findings from Egypt. *PLOS ONE*. https://doi.org/10.1371/journal.pone.0229917

8 My Dignity, My Rights: Kafala Sponsorship System and Exploitation of Ugandan Migrant Domestic Workers in Saudi Arabia

Florence Munyonyo Asiimwe
and Denis Musinguzi

Abstract

The exploitation of Ugandan migrant domestic workers in Saudi Arabia has become a significant concern in global labour migration discourse. This exploitation is largely attributed to the Kafala sponsorship system, under which employers (Kafeels) assume full control over the employment and immigration status of domestic workers. Such control restricts workers' ability to change employers, leave the country, or exercise basic labour rights, thereby increasing their vulnerability to abuse. This study examines the Kafala system and its role in facilitating the exploitation of Ugandan migrant domestic workers in Saudi Arabia. Findings indicate that employers use the system to justify breaches of contract, impose excessive workloads, enforce forced labour, assign multiple workstations, confiscate travel documents, and restrict access to phones and the Internet. The study concludes that the authority conferred on employers by the Kafala system underpins these exploitative practices. To mitigate such abuses, the study recommends legislative reforms that limit employer powers under the system, reassessment of the bilateral labour agreements with Saudi Arabia, and strengthen the protection of migrant domestic workers' rights. These measures are critical for safeguarding migrant labour and promoting ethical labour migration practices.

Keywords

Kafala · Migrant domestic workers · Exploitation · Saudi Arabia

Introduction

The Kafala sponsorship system constitutes the legal framework for recruiting and regulating migrant domestic workers in Saudi Arabia. As the primary mechanism governing labour migration and employment relations, Kafala embodies social, moral, and economic dimensions embedded in Islamic culture (Jebejian et al., 2018). Rooted in Bedouin hospitality traditions, it establishes obligations for receiving and protecting foreign guests (Khan & Harroff-Tavel, 2011). Linguistically, the term "Kafala" derives from the Arabic "Ka Fa La," meaning to protect, vouch for, or assume responsibility (Parrenas, 2021), with the "kafl" acting as a sponsor, guarantor, or accountable party (Abikan, 2017; Truluck, 2023). Within Islamic family law, Kafala refers to for-

F. M. Asiimwe (✉) · D. Musinguzi
Uganda Martyrs University, Kampala, Uganda
e-mail: fasiimwe@kyu.ac.ug

P. Nyadoi et al. (eds.), *Sustainable Development Goals in Africa*, Sustainable Development Goals Series, https://doi.org/10.1007/978-3-032-20742-5_8

mal arrangements providing temporary support for orphaned children until adulthood. In the context of labour migration, however, it transfers the foreign worker's legal obligations entirely to the sponsor, who simultaneously serves as the employer in domestic employment (Pande, 2013). The Kafala system reflects the Bedouin values of hospitality and the responsibility for safeguarding foreign visitors (Hamza, 2015). Over time, it has governed interactions between migrant workers and employers across Saudi Arabia, Jordan, Lebanon, and other Gulf countries (Khan & Harroff-Tavel, 2011). Proponents argue that the system facilitates the importation of affordable labour, stimulates economic growth, and supports regional development. Some supporters, including business elites, claim that Kafala has improved the livelihoods of millions of migrant workers and that any disciplining of domestic workers stems from employment conditions rather than the system itself (Hamza, 2015; Khan & Harroff-Tavel, 2011).

Nonetheless, the Kafala system has faced extensive criticism. Scholars highlight its association with abuse, low wages, poor working conditions, lack of regulatory protection, and exposure to racial and gender-based discrimination (Aoun, 2020). It institutionalizes the socio-economic and legal subordination of migrant workers while restricting social mobility, denying pathways to citizenship, and perpetuating vulnerability to deportation (Abdul Reda et al., 2023). Consequently, migrant domestic workers remain highly dependent on their sponsors, with limited prospects for permanent settlement, legal rights, or protection from exploitation (Murray, 2013). These structural constraints underscore ongoing debates over the system's ethical and practical implications for labour migration in the Gulf region. Despite extensive literature on the Kafala system's historical origins and structural dynamics, there remains limited empirical research on the lived experiences of Ugandan migrant domestic workers within this framework. While previous studies have highlighted systemic vulnerabilities, abuses, and legal subordination, less attention has been given to how these factors shape workers' agency, coping mechanisms, and social integration in host countries. This study seeks to address this gap by examining the experiences of Ugandan migrant domestic workers in Saudi Arabia, focusing on the intersection of labour exploitation, social dependency, and migrant resilience. By exploring these dynamics, the research aims to provide nuanced insights into the practical and ethical implications of the Kafala sponsorship system, contributing to broader discussions on labour migration governance, human rights, and policy reform in the Gulf region.

The Kafala sponsorship system defines the legal and employment relationship between foreign workers and their local sponsors, or Kafeels, who typically serve as employers (Al Hammadi et al., 2022). The system operates across the Gulf Cooperation Council (GCC) states Bahrain, Kuwait, Oman, Qatar, Saudi Arabia, and the United Arab Emirates as well as in Jordan and Lebanon (Jebejian et al., 2018). While some countries, including Bahrain and Qatar, claim to have reformed or abolished aspects of the system, these measures are insufficiently implemented and fail to dismantle the core structural dependence on employers (Damir-Geilsdorf & Pelican, 2019). Under Kafala, governments grant sponsorship permits to local individuals or companies, who then recruit foreign workers directly or via private recruitment agencies in the countries of origin. Sponsors typically cover housing, which for domestic workers is the sponsor's residence, and travel expenses (Bosc et al., 2022; Damir-Geilsdorf & Pelican, 2019). In 2015, Uganda signed a bilateral labour agreement with Saudi Arabia to formally export migrant domestic workers, who are employed under the Kafala system. Despite being promoted as a fair and structured employment framework, Ugandan domestic workers frequently experience severe working conditions. These conditions raise questions about why workers under a supposedly effective system continue to face exploitation. Consequently, this study investigates the Kafala system and the exploitation of Ugandan migrant domestic workers in Saudi Arabia.

The analysis of the dignity and rights of the Ugandan Migrant Domestic Workers and the exploitation they face under the Kafala Sponsorship System is grounded in Power-Dependence Theory, a perspective commonly attributed to Emerson's seminal works (1962, 1972a, 1972b). Originating within social exchange theory, it underscores the centrality of dependence in shaping social relations (Abrutyn & McCaffree, 2021). The theory posits that dynamics of power, its application, and balancing mechanisms all emerge from conditions of dependence (Jiang et al., 2020). While mutual dependence can foster cooperation and sustain exchange relations, asymmetries in dependence produce power imbalances that may trigger conflict and social change (Corsi & Filippini, 2023). Crucially, Power-Dependence Theory conceptualises power not as an individual attribute but as a property of relationships, thereby shifting analytical attention from actors themselves to the structural arrangements that define their interactions (Clarke, 2024). Applied to this study, Power-Dependence Theory provides a critical lens for interpreting the experiences of Ugandan migrant domestic workers under the Kafala sponsorship system. Within this framework, workers are structurally dependent on their employers for legal residency, wages, sustenance, and basic freedoms, generating a profound imbalance of authority.

This dependence manifests in multiple forms of exploitation, including contract breaches, excessive workloads, restrictions on mobility, denial of rest, confiscation of phones and travel documents, and heightened exposure to occupational hazards. The theory elucidates how the institutional design of the Kafala system entrenches employer dominance, limits workers' agency, and normalises exploitative practices. The workers' constrained autonomy and structural dependency intensify both physical and psychological vulnerability, while simultaneously restricting access to legal recourse and state protection. Framing these dynamics through Power-Dependence Theory highlights the urgent need for comprehensive policy reforms, binding bilateral agreements, and strengthened regulatory mechanisms that can re-balance power, safeguard rights, and uphold the dignity of Ugandan migrant domestic workers in host countries such as Saudi Arabia.

Exploitation of Migrant Domestic Workers Under the Kafala Sponsorship System

Migrant domestic workers under the Kafala system encounter multiple forms of exploitation. Sponsors often confiscate travel documents, restricting mobility and delaying remuneration (Mencutek & Gurses, 2024). Workers are vulnerable to abuse, including physical assault, sexual violence, forced labour, unsafe working conditions, and deportation (Al-Saud, 2023; Pande, 2013). The system also restricts citizenship and formal employment mobility by tying workers to a single sponsor, whose approval is required to change jobs or exit the country (Abdulrahim et al., 2024; Parreñas, 2021). Contract durations are typically 1 year for foreign employers and 2 years for nationals (Silvey & Parreñas, 2020), further reinforcing dependency. Live-in arrangements exacerbate subordination and the potential for abuse (Pozzi Mucelli, 2019). Absconding from duty is criminalised, exposing dissatisfied or abused workers to imprisonment and deportation (Parrenas, 2021).

Additional exploitative practices include underpayment, salary delays, dual contract schemes with poorer terms, and the deduction of basic living expenses from wages (Damir-Geilsdorf & Pelican, 2019; Dickinson, 2016; Harkins, 2020; Vrangbæk et al., 2015). Structural gaps in labour laws, restrictive immigration policies, and social discrimination against darker-skinned domestic workers exacerbate human rights violations, including long working hours with minimal or no pay (Varia, 2011). Employers' control over passports and confinement of workers facilitate physical violence, sexual abuse, and forced labour (Ilesanmi & Iyer-Raniga, 2024; Katyayani, 2024; Parrenas, 2021). Scholars further highlight that religious, gender, and class hierarchies contribute to systematic abuse, creating a class of easily exploitable workers (Gardner, 2010; Malaeb,

2015; Mucci et al., 2019). The Kafala system's structural design enables employers to wield complete economic and legal authority over employees, reinforcing vulnerability and dependency (Katyayani, 2024).

Given persistent criticism, reforming the Kafala system has become imperative. Scholars and policymakers advocate for conventions that balance practical governance with international labour standards, rather than imposing reforms with little local support (Varia, 2011; Vlieger, 2012). Negotiators must account for social norms and attitudes that shape domestic workers' experiences and resistance to rights-based reforms (Parrenas, 2021). Formal agreements that define employer and government obligations could strengthen the domestic workers' movement, promoting dignity, equality, and legal protection. Proposed reforms include formalising recruitment processes, enabling justified labour mobility, protecting workers from abuse, and improving contractual and working conditions (Khan & Harroff-Tavel, 2011; Varia, 2011). Properly implemented, these reforms can provide robust safeguards and improve the welfare of migrant domestic workers. Despite extensive documentation of abuses under the Kafala system, there remains a lack of empirical research focusing specifically on the lived experiences of Ugandan migrant domestic workers in Saudi Arabia. Existing studies primarily describe structural exploitation, legal vulnerabilities, and policy shortcomings but offer limited insight into how these factors intersect with workers' agency, coping strategies, and social integration. Consequently, there is a critical gap in understanding the mechanisms through which the Kafala system perpetuates exploitation in practice. Addressing this gap, the present study examines the experiences of Ugandan migrant domestic workers, focusing on how Kafala structures labour relations, enforces dependency, and shapes vulnerability to abuse. This approach provides a nuanced understanding of the system's impact, informing policy reform and labour rights advocacy in the Gulf region.

Methodology

This study employed a social constructivist lens, emphasising how individuals construct meaning from lived experiences (Boyland, 2019; Phillips & Johns, 2023). A phenomenological approach was used to explore the challenges related to exploitation encountered by Ugandan migrant domestic workers under the sponsorship system related to Kafala in Saudi Arabia, enabling an in-depth understanding of participants' experiences. Qualitative data were collected from a purposively selected sample of returnee domestic workers and key informants from the Ministry of Gender, Labour, and Social Development (MoGLSD), the Uganda Association Employment Recruitment Agency (UAERA), and the next of kin, with participant recruitment guided by snowball and purposive sampling until data saturation was achieved.

Data were obtained through unstructured interviews with open-ended questions, providing flexibility to capture detailed narratives. Each interview lasted 50–60 minutes, and credibility was ensured using Guba and Lincoln's (1994) truth-value criterion. Transcribed data were thematically analysed, and the responses were coded, grouped, and refined iteratively to identify patterns and themes reflecting the types and dynamics of exploitation under the Kafala system. Emergent themes informed the interpretation and presentation of findings, highlighting the nuanced experiences of migrant domestic workers.

Results and Discussion

Exploitation Experiences of Ugandan Migrant Domestic Workers Under the Kafala System

Saudi Arabia employs Ugandan migrant domestic workers under the Kafala system; however, the findings revealed widespread exploitation and multiple injustices stemming from the system's structural design. The excessive authority vested in employers created conditions that facilitated human rights abuses, including contract breaches, relocation to multiple workstations, subjection to excessive workloads, denial of rest, occupational health hazards, and restrictions on phone and internet use. Participants consistently reported these experiences, highlighting the vulnerability of domestic workers under Kafala and the limited capacity of Gulf Cooperation countries, including Saudi Arabia, to address such abuses. Despite mounting evidence of exploitation, the system is defended by some governments as an efficient mechanism for labour recruitment, further entrenching structural inequities and reinforcing workers' marginalisation. These findings underscore the need for urgent reforms and coordinated interventions to safeguard the rights and welfare of migrant domestic workers.

Heavy Workload and Multiple Workstations

A pervasive theme across all interviews was the overwhelming workload imposed on domestic workers. Shepherd described her experience: *The workload in Saudi Arabia was far beyond what I had anticipated. Many days felt like being trapped in a prison where I could die* (Central Region, 1 March 2023). Sullivan similarly reported: *Upon arrival, my madam assigned numerous chores including washing, mopping, bed-making, and general household maintenance. I even considered returning home, but it was impossible* (Central Region, 11 February 2023). Rinah's experience illustrates the intensity of the workload: *I would wake at 4:30 a.m. and clean multiple rooms including dining, TV, children's, and prayer rooms. I washed clothes and utensils until my hands were sore from hot water, ironed garments for the children, and often had to wake before 5:00 a.m. to finish incomplete tasks from the previous day* (Central Region, 20 February 2023). Gift described the compounded burden during social events: *Every Thursday, visitors arrived, and I had to prepare meals, clean extensively, and maintain the household throughout their stay. The physical demand was exhausting* (Eastern Region, 26 February 2023).

Beyond individual household chores, workers were often relocated to multiple workstations, frequently without prior notice or additional pay. Assia narrated: *The family I worked for would shop every week, and I would carry everything from the car. Sometimes they took me to visit relatives, and I worked for them too, even tending to their sheep. For this extra work, there was no extra money* (Western Region, 9 March 2023). Plinnery explained: *We moved from one house to another, working for many families either related to or friendly with my employers. At the end of the month, they would all contribute to my salary, but the work was exhausting* (Central Region, 23 March 2023). Amil described working across four floors of a single household while performing extensive chores, yet receiving only one salary (Central Region, 9 March 2023).

These experiences align with global research showing that migrant domestic workers often perform physically demanding tasks across large households for 14–18 h/day, with minimal rest and no overtime compensation (Boris & Fish, 2014; Motaparthy, 2010; Spitzer et al., 2023; D'Souza, 2010). The combination of heavy workloads, repeated relocations, and restricted compensation reflects the structural power imbalance embedded within the Kafala system (Parreñas, 2021; ILO, 2013).

Contract Breaches and Exploitative Arrangements

Many participants reported that employment contracts signed in Uganda were routinely violated in Saudi Arabia. Dina recounted: *Although I had signed a contract before leaving Uganda, when I arrived, the family had more members than indicated. When I tried to argue, I was told that Kadamas come to work, not for a holiday. The company told me to sit and work* (Western Region, 10 March 2023). Nasumba, a former nursery teacher, reported that even access to medical care—a contractual right was denied: *I fell sick, but they would not take me to the hospital. When my husband contacted the company, they said Arabs have a problem with respecting African workers' contracts* (Central Region, 18 February 2023). Similarly, Amil described having her contract torn: *When I handed it over to my boss, they said the paper means nothing, and I should just work* (Central Region, 9 March 2023). Mirriam highlighted denial of multiple contractual entitlements, including rest, medical insurance, and religious freedom (Eastern Region, 26 February 2023).

Recruitment companies acknowledged these breaches but cited geographic and logistical limitations in addressing complaints. *One company director noted: It is true we get complaints, but not all employers are abusive. For those severely affected, we contact our offices in Saudi Arabia or the embassy to help reassign the worker* (Central Region, 9 February 2023). Government officials confirmed limited interventions, primarily reserved for extreme exploitation, including trafficking or life-threatening situations (Central Region, 9 February 2023). Nonetheless, contract breaches remain common, exposing workers to exploitative labour practices such as wage theft, excessive hours, and denial of basic rights (Trần, 2023; Sarkar, 2020; D'Souza, 2010; Marchetti et al., 2021).

Denial of Rest and Confinement

Across interviews, participants reported minimal rest, with workdays extending from early morning until midnight. Assia explained: *Even when I finished work early, they would send me to work at their father's home. I worked like a robot* (Western Region, 9 March 2023). Rinah described having only three-minute breaks during nearly 12-h workdays (Central Region, 20 February 2023). Dina noted: *Resting is not possible at all. The only resting time was sleeping, which was 4 h, and eating, which took 10 min* (Western Region, 10 March 2023). Prissy reported continuous labour during illness, including COVID-19 infection: *I worked from 5 a.m. to past midnight. My feet swelled, and I suffered multiple COVID-19 infections, but I was still forced to work* (Central Region, 18 February 2023).

Amil and Mirriam shared similar experiences, describing work that precluded religious observances or any leisure (Central Region, 9 March 2023; Eastern Region, 26 February 2023). Confinement within employer households heightened vulnerability, with some workers attempting risky escapes, often leading to injury or legal jeopardy (Ho et al., 2022; Martin, 2021). Tayah (2016) emphasised that immigration legislation compounds domestic workers' precarious status, exposing them to social, economic, and physical fragility.

Phone and Internet Restrictions

Restricted communication with family and friends was a common form of control. Flower warned: *Do not agree to work for a family that does not allow phone use. They may want to harm you, and no one will know* (Central Region, 20 February 2023). Assia recounted conditional phone access: *I could use my phone only when not working; otherwise, it could be confiscated* (Western Region, 9 March 2023). Plinnery and Dina experienced more severe restrictions, including destruction of phones or denial of communication for months (Central Region, 3 March 2023; Western Region, 10 March 2023). Nasumba reported being unable to communicate with her family for nearly a year, and Prissy only regained limited access during Uganda's COVID-19 lockdown (Central Region, 18 February 2023). Restricted communication not only isolated workers but also facilitated further exploitation, consistent with ILO reports highlighting forced confinement and restricted contact as mechanisms of control (ILO, 2016; Sluspka et al., 2022; Moyce & Schenker, 2018).

Occupational Health and Safety Hazards

Participants reported multiple work-related health issues resulting from strenuous labour, hazardous conditions, and inadequate employer care. Gift described chronic back pain due to prolonged bending: *Since their homes were big and I had to bend, I developed severe back pain that has never gone away* (Eastern Region, 26 February 2023). Prissy described repeated COVID-19 infections, general physical weakness, and psychological distress during her employment (Central Region, 18 February 2023). Rinah and Sullivan recounted burns, hand injuries, swelling, and fatigue without medical attention (Central Region, 20 February 2023; 11 February 2023).

Despite contractual provisions for health insurance, participants reported inadequate access to care. Occupational health research emphasises the importance of ensuring physical, mental, and social well-being at work, alongside preventive care and timely medical attention (Menegatti, 2016; Moyce & Schenker, 2018; Alli, 2008). Domestic migrant workers, frequently engaged in "3-D" jobs—dirty, dangerous, and demanding—and remain highly vulnerable to abuse, neglect, and exploitation (Fleury, 2016).

Synthesis

Collectively, these findings indicate that Ugandan migrant domestic workers endure extreme exploitation characterised by excessive workloads, relocation to multiple homes, denial of rest, confinement, restricted communication, occupational health risks, and widespread contract violations. The structural power imbalance inherent in the Kafala system amplifies these vulnerabilities, while recruitment agencies and government interventions remain limited in addressing routine breaches and abuses.

Conclusions and Recommendations

The findings of this study underscore the pervasive exploitation experienced by Ugandan migrant domestic workers under the Kafala sponsorship system in Saudi Arabia. Exploitation manifested through excessive workloads, restricted mobility, withheld wages, limited access to healthcare, and inadequate rest, reflecting structural vulnerabilities embedded within the Kafala system. These challenges highlight the urgent need for policy reforms both within Uganda's labour export mechanisms and in bilateral labour governance with destination countries.

To mitigate such exploitation, comprehensive reforms are imperative. First, the abolition of the "no objection certificate" requirement is critical to enhance workers' freedom of movement and reduce dependency on employers. Second, rigorous enforcement of standardised labour contracts should ensure fair remuneration, defined work hours, and guaranteed rest periods. Third, guaranteed access to communication channels and healthcare services is necessary to uphold workers' basic rights and safeguard their well-being. Fourth, bilateral labour agreements emphasizing labour standards, coupled with systematic monitoring of employer compliance, can reinforce protections for migrant domestic workers. Lessons from Qatar's 2020 labour reforms—allowing workers to change employers during and after probation with legal safeguards—provide a model for Uganda's policy engagement with Saudi Arabia (Parrenas, 2021).

Finally, empowering returnee domestic workers through awareness campaigns, predeparture training, and support mechanisms in Uganda can strengthen resilience and reduce vulnerability to exploitation. Multi-stakeholder collaboration involving government agencies, recruitment bodies, civil society, and destination countries is essential to translate these recommendations into sustainable policy interventions. Implementing these measures has the potential to significantly improve working conditions, enhance human dignity, and align Uganda's labour export practices with international labour standards.

Acknowledgements The researchers extend their appreciation to the Ministry of Gender, Labour and Social Development, the Uganda Association of External Recruitment Agencies, Kyambogo University, and all Ugandan migrant domestic workers who took part in this study for their valuable support and contributions.

Declaration This study was funded by the Directorate of Research and Graduate Training (DRGT), Kyambogo University, Uganda.

References

Abdul Reda, A., Fraser, N. A., & Khattab, A. (2023). Does social mobility matter? The Kafala system and anti-immigrant sentiment. *Political Studies Review, 21*(4), 801–824. https://doi.org/10.1177/14789299231179458

Abdulrahim, S., Cherri, Z., Adra, M., & Hassan, F. (2024). Beyond Kafala! Employers' discriminatory attitudes and violations of the rights and freedoms of women migrant domestic workers in Lebanon. *Migration Studies, 12*(2), 012. https://doi.org/10.1093/migration/msad012

Abikan, A. I. (2017). Contract of Kafālah (guarantee) in Islamic finance: Extending the frontiers of Islamic law. *Journal of Shariah Law Research, 2*(2), 157–178.

Abrutyn, S., & McCaffree, K. (2021). *Theoretical sociology*. Routledge.

Al Hammadi, H. M., Bayouomi, K. A., Salleh, A., & Rosman, B. (2022). The impact of abolishing the sponsorship system (Kafala) on the peaceful coexistence of population groups in Qatar in light of the migration goals of the Alliance of civilizations. *Social Sciences, 12*(4), 388–408. https://doi.org/10.3390/socsci12040388

Al-Saud, M. A. (2023). *Transforming the Kafala in Saudi Arabia: Turning weaknesses into opportunities* (Doctoral dissertation).

Aoun, R. (2020). COVID-19 impact on female migrant domestic workers in the Middle East. GBV AoR Helpdesk, 5.

Boris, E., & Fish, J. N. (2014). Slaves no more: Making global labor standards for domestic workers. *Feminist Studies, 40*(2), 411–443.

Bosc, I., Lerche, J., Shah, A., Fajerman, M., & Wadhawan, N. (2022). Understanding patterns of structural discrimination of migrant and other workers in some countries of south and West Asia.

Boyland, J. R. (2019). A social constructivist approach to the gathering of empirical data. *Australian Counselling Research Journal, 13*(2), 30–34. https://doi.org/10.1017/acrj.2019.5

Damir-Geilsdorf, S., & Pelican, M. (2019). Between regular and irregular employment: Subverting the Kafala system in the GCC countries. *Migration and Development, 8*(2), 155–175. https://doi.org/10.1080/21632324.2018.1482903

Dickinson, D. (2016). Migrant worker recruitment costs. *Development Brief, 26*, 19–21.

D'Souza, A. (2010). *Moving towards decent work for domestic workers: An overview of the ILO's work*. ILO Working Paper 2/2010, International Labour Organization.

Fleury, A. (2016). *Understanding women and migration: A literature review* (p. 55). World Bank Group.

Gardner, A. M. (2010). *City of strangers: Gulf migration and the Indian community in Bahrain*. Cornell University Press.

Guba, E. G., & Lincoln, Y. S. (1994). Competing paradigms in qualitative research. In N. K. Denzin & Y. S. Lincoln (Eds.), *Handbook of qualitative research* (pp. 105–117). Sage.

Hamza, S. (2015). Migrant labor in the Arabian gulf: A case study of Dubai, UAE. *Pursuit-The Journal of Undergraduate Research at the University of Tennessee, 6*(1), 10.

Harkins, B. (2020). Base motives: The case for an increased focus on wage theft against migrant workers. *Anti-Trafficking Review, 15*, 42–62. https://doi.org/10.14197/atr.201220151

Ho, K. H. M., Yang, C., Leung, A. K. Y., Bressington, D., Chien, W. T., Cheng, Q., & Cheung, D. S. K. (2022). Peer support and mental health of migrant domestic workers: A scoping review. *International Journal of Environmental Research and Public Health, 19*(13), 7617. https://doi.org/10.3390/ijerph19137617

Ilesanmi, O. O., & Iyer-Raniga, U. (2024). *21st century slavery: The various forms of human enslavement in today's world*. BoD–Books on Demand.

Jebejian, H., Kenneth Ross, M. T., & Johnson, T. (2018). The Gulf: Bahrain, Kuwait, Oman, Qatar, Saudi Arabia, United Arab Emirates (UAE), Yemen. In *Christianity in North Africa and West Asia* (pp. 177–189). Springer.

Jiang, Z., Kobylinska, T., & Voice of Domestic Workers. (2020). Art with marginalised communities: Participatory video as a tool of empowerment and resistance for migrant domestic workers in London. *City, 24*(1–2), 348–363.

Katyayani, S. (2024). Bound by contract: Mapping technologies of migrant control in the Kafala system. *The Indian Journal of Labour Economics*. https://doi.org/10.1007/s41027-024-00489-7

Khan, A., & Harroff-Tavel, H. (2011). Reforming the Kafala: Challenges and opportunities in moving forward. *Asian and Pacific Migration Journal, 20*(3–4), 293–313. https://doi.org/10.1177/011719681102000305

Malaeb, H. N. (2015). The "Kafala" system and human rights: Time for a decision. *Arab Law Quarterly, 29*(4), 307–342. https://doi.org/10.1163/15730255-02904001

Marchetti, S., Cherubini, D., & Geymonat, G. G. (2021). Scenarios of domestic workers' rights. In *Global domestic workers* (pp. 22–50). Bristol University Press.

Mencutek, Z. S., & Gurses, U. (2024). Governance and politics of migration in the Middle East. In *Handbook of migration and globalisation* (pp. 297–311). Edward Elgar Publishing.

Menegatti, E. (2016). *Protecting migrant domestic workers: The international legal framework at a glance*. International Labour Organisation.

Motaparthy, P. (2010). *Walls at every turn: Abuse of migrant domestic workers through Kuwait's sponsorship system*. Human Rights Watch.

Moyce, S. C., & Schenker, M. (2018). Migrant workers and their occupational health and safety. *Annual Review of Public Health, 39*, 351–365.

Mucci, N., Traversini, V., Giorgi, G., Tommasi, E., De Sio, S., & Arcangeli, G. (2019). Migrant workers and psychological health: A systematic review. *Sustainability, 12*(1), 120. https://doi.org/10.3390/su12010120

Murray, H. E. (2013). Hope for reform springs eternal: How the sponsorship system, domestic laws and traditional customs fail to protect migrant domestic workers in GCC countries. *Cornell International Law Journal, 45*(2), 461–484.

Pande, A. (2013). "The paper that you have in your hand is my freedom": Migrant domestic work and the sponsorship (Kafala) system in Lebanon. *International Migration Review, 47*(2), 414–441. https://doi.org/10.1111/imre.12032

Parrenas, R. (2021). The mobility pathways of migrant domestic workers. *Journal of Ethnic and Migration Studies, 47*(1), 3–24. https://doi.org/10.1080/1369183X.2019.1628987

Parreñas, R. S. (2021). *Unfree: Migrant domestic work in Arab States*. Stanford University Press.

Phillips, R., & Johns, J. (2023). *Fieldwork for social research: A student's guide*. SAGE Publications.

Pozzi Mucelli, F. (2019). *The vulnerable case of domestic workers in the Lebanese context.*

Sarkar, S. (2020). The new legal world of domestic work. *Yale Journal of Law & Feminism, 32*, 1–42.

Silvey, R., & Parreñas, R. (2020). Thinking policy through migrant domestic workers' itineraries. *American Behavioral Scientist, 64*(6), 859–877. https://doi.org/10.1177/0002764219897691

Slupska, J., Cho, S., Begonia, M., Abu-Salma, R., Prakash, N., & Balakrishnan, M. (2022). "They look at vulnerability and use that to abuse you": Participatory threat modelling with migrant domestic workers. In *31st USENIX Security Symposium (USENIX Security 22)* (pp. 323–340).

Spitzer, D. L., Thambiah, S., Wong, Y. L., & Kaundan, M. K. (2023). Globalization and the health and well-being of migrant domestic workers in Malaysia. *Globalization and Health, 19*(29). https://doi.org/10.1186/s12992-023-00925-w

Tayah, M. J. (2016). *Decent work for migrant domestic workers: Moving the agenda forward*. International Labour Organization (ILO).

Trần, A. N. (2023). Bilateral labor agreement with gendered and unfree labor: Vietnamese women domestic workers in Saudi Arabia. *Journal of Labor and Society, 1*, 1–27. https://doi.org/10.1111/wusa.12987

Truluck, E. (2023). Using Islam to protect the rights of migrant workers: Bringing Kafala into sharia compliance in Saudi Arabia. *UCLA Journal of Islamic & Near Eastern Law, 20*, 155–180.

Varia, N. (2011). "Sweeping changes?" A review of recent reforms on protections for migrant domestic workers in Asia and the Middle East. *Canadian Journal of Women and the Law, 23*(1), 265–287.

Vlieger, A. (2012). *Domestic workers in Saudi Arabia and the Emirates: A socio-legal study on conflicts*. Quid Pro Books.

Vrangbæk, K., Petersen, O. H., & Hjelmar, U. (2015). Is contracting out good or bad for employees? A review of international experience. *Review of Public Personnel Administration, 35*(1), 3–23. https://doi.org/10.1177/0734371X14544664

9 A Qualitative Assessment of the Contribution of the South Sudan Peace Agreement

Sarah Specioza Abeja and Susan Arayo

Abstract

Since 2013, South Sudan has experienced a series of violent conflicts. To solve this, the country introduced amnesty provisions in the 2018 Revitalized Agreement on the Resolution of Conflict in South Sudan (R-ARCSS). However, the conflict persists with far-reaching consequences

including civilian killings, displacement, rape and torture. In this chapter, we present the findings of a qualitative assessment made to determine the contribution of the peace agreement in South Sudan. The study participants mentioned that the agreement has failed to bring about peace, justice and stability. They proposed implementation of the agreement interactively with other conflict resolution mechanisms such as truth and reconciliation commissions, hybrid courts and socio-economic empowerment interventions.

Keywords

Extrajudicial executions · Internal wars · People's liberation movement · Rape · Reconciliation commission · Women · Children · Army generals

S. S. Abeja · S. Arayo (✉)
Department of Governance, Peace and International Studies, Uganda Martyrs University, Nkozi, Uganda
e-mail: sarayo@umu.ac.ug

Introduction

Civil wars are a persistent problem globally. One tool often used post-conflict to address this problem is amnesty agreements (Dancy, 2018; Daniels, 2020; Mallinder, 2023; Reiter, 2014). In South Sudan, however, the extent to which this has contributed to peace and development remains unknown. Several factors, such as misinformation, timing, conflicts over resources and lack of commitment on the part of parties involved may lead to failure of amnesty agreements to achieve desired outcomes (International Institute for Strategic Studies, 2019; Blattman & Miguel, 2010; Hayner, 2018; Hazan, 2020). In this chapter, we present findings of a study premised on the bargaining theory postulated by Powell (2002, 2006) and Walter (2014). Our study assessed perceptions of participants on contribution of the amnesty agreement to peace, justice and stability in South Sudan.

Methodology

Fieldwork was conducted between November 2022 and April 2023 in Juba South Sudan. We adopted the constructivist worldview (a philosophical position that emphasizes that individuals seek understanding of the world in which they live and work and develop varied and multiple subjective meanings of their experiences) and a

P. Nyadoi et al. (eds.), *Sustainable Development Goals in Africa*, Sustainable Development Goals Series, https://doi.org/10.1007/978-3-032-20742-5_9

qualitative approach in data collection for a case study (Creswell & Creswell, 2023; Yin, 2009). We conducted in-depth interviews with key informants and focus group discussions(FDG) with various stakeholders including the police, human rights organisations, members of the association of the conflict related sexual violence survivors, cultural leaders and civil society. Four focus group discussions were held with community members. A structured interview schedule with open-ended questions was used to obtain information about amnesty and its contribution to peace and justice from the participants. The research questions that guided this study included, among others, questions on the prevalence of human rights violations and the effectiveness of amnesty in promoting peace in Juba, South Sudan. During the interviews and focus group discussions the word amnesty was not used; instead, we adopted words that mean the same but were easier for respondents to comprehend, for example, forgiveness or pardon of those who committed crimes to mean amnesty, security and stability to mean peace, and fairness to mean justice.

A total of 63 respondents (46 males and 17 females) were interviewed. We obtained permission to go to the field from Uganda Martyrs University and from the Relief and Rehabilitation Commission in South Sudan. Respondents and their views recorded were kept confidential and participation in the study was voluntary.

Result and Discussion

The Prevalence of Human Rights Violations

This includes murder based on tribal and ethnic lines, sexual violence, rape, killings, torture, and forced disappearances as presented below.

Murder Based on Tribal and Ethnic Lines

In one of our interviews with a key informant, he alluded to the fact that there were widespread crimes such as murder based on tribal lines, displacement of populations, torture, rape and enforced disappearance. The most prevalent in Juba were torture, enforced disappearance of persons and sexual violence, specifically rape. There were also cases of reported disappearance of relatives of those who had been part of the opposition political parties.

Sexual Violence

On sexual violence and other crimes, one of the key informants in an interview voiced as follows: "*...the displacement you see resulting into huge IDP camps is a result of the wars that started in 2013.He noted that to date there were attacks on the civilian population with a view, to drive out that civilian population from their boards. He stated that although he was aware that there were crimes against humanity, he was not sure of the general rate with exception of the sexual violence rates that he said were at 50% in Juba county...*" (Interviews with a key informant on 22nd March 2023 at Upper Nile State).

On other crimes including rape, killings, displacement, lack of food, torture, confiscation of property. In a focus group discussion, the participants observed that there is no single day that passes by without news of unknown gun men killing journalists, Internally Displaced Persons (IDP) or human rights defenders who try to raise up the issues of war in South Sudan. One female participant made an expensive joke that if you want to die very fast in Juba just say that let the IDP come out of the camps, the government authorities will fish you out of the hiding places. One member said that even people who are always on the run, like the female human rights defenders who talk about conflict related to sexual violence, are persecuted. During FGD, one female alluded that, "*...Women had hoped that after signing the agreement people would live in peace but it has not been the realities. The NAS of Thomas Cilrilo are still killing people. Women are still being raped. We can't even move on the roads before being checked by any illegal people who we don't know. The challenge is the soldiers were trained and passed but not deployed. They* *are back to the community and so we don't know whether government or rebels linked soldiers are the ones doing this. The government has deliberately refused to provide security to the people*" (*Focus group discussion with women in Juba*, on 10 April, 2023).

The women in the focus group discussion alluded that once people come out of the IDPs they get killed, imagine that kind of suffering in our own country. Most affected are women and children because men just join the military leaving us suffering with children as mothers. We are not even safe along roads like Nimule Road or even Terekeka.

Contribution of Amnesty in the Promotion of Peace and Justice

The findings from the study indicated that Amnesty did not contribute to the promotion of peace and justice in the region mainly because of lack of accountability. Conflicts persist, with consequences such as displacement and killings continuing unabated despite the amnesty. During our interview, one of the army officers said, "*...amnesty was a killer of justice because it renders the principle of accountability useless. He added that actually in South Sudan amnesty is the cause of recurrent conflict resulting into perpetual human rights violations by the generals who commit atrocities but are never accountable for their acts...*" (Interview with one of the Army officers on 15 March 2023).

Contrary to the above submission from an army officer, one of the key informants who works as a Women for Peace and Security Specialist with the African Union Mission in South Sudan (AUMISS), in our interview with her on 30 March argued that amnesty was very useful in the context of South Sudan. She said when their team urged the South Sudan government to grant amnesty they also advised that the amnesty should be conditional and ensure that pardoned warring factions in return give information that would foster peace, such as providing details on missing relatives to victims and asking forgiveness from the victims so that everyone moves on.

Similar to the above opinion, one key informant from AUMISS voiced, "*...The victims will get freedom of the mind after knowing what happened to their relatives, while the perpetuators will feel free after being forgiven. She cited... that all the process does not happen overnight but after a long time. So, the impact of amnesty cannot be realised so fast. She added that the best way of promoting peace would start with amnesty then other process can follow on such as reparation, truth and healing...*" (Interviews with a key informant from Women Peace and Security Specialist, African Union Mission in South Sudan (AUMISS) on 30 March 2023). She suggested that for the sake of South Sudan there is a need for traditional leaders/local authorities and religious leaders to be keenly engaged in the promotion of peace. Due to the fact that the youth believe in them so much, without their involvement it will be very difficult to stop war. For the sake of access to justice, the South Sudan formal justice is still weak so there is a need for traditional justice to be embraced especially in the transitional justice process. This opinion is in agreement with the views of Ninrew (2022), who argued that the lack of a Hybrid Court in South Sudan mirrors the government's failure to deliver accountability to victims of the conflicts in the country. This has made the amnesty in the region a toothless bulldog.

Effectiveness of Blanket Amnesty on Promotion of Peace and Justice in Juba South Sudan

The blanket amnesty concept was defined for and perceived by the respondents of this study as an official pardon granted to a group of prisoners by the government. The respondents largely mentioned that blanket amnesty did not bring total peace and justice in their region and that their leaders do not even account for their actions. They further expounded that once the president grants amnesty, conflict perpetrators were not made to account for their crimes but were instead rewarded with higher positions in government, given immunity and made to rule the same people they committed crimes against. They said instead of bringing warring parties together and yielding peace, in South Sudan, amnesty is creating more rebellions and rebels. One of the participants in a focus group discussion noted that amnesty is a killer of justice. The participant noted that where there is no justice then we don't expect any human rights, such as freedoms. A key informant lamented that "...*Amnesty is not the best option to resolving conflict in third world countries or fragile states. It should not be a concept to embrace because it is concept that encourages impunity because leaders never get charged for the wrongs the committed instead get rewarded with positions when they return to rule the subjects they committed the atrocities. Citizen remains in pain a result more rebellions to remove those caused them pain...*" (Interview with a key informant in Juba on 10 March 2023).

It was also added that most of those granted blanket amnesty use it as a scapegoat to get ranks by taking up arms, rebelling and killing innocent people in the process; then they ask for peace talks, but victims remain suffering without justice.

Other participants, in line with views of Mallinder (2023) and McEvoy (2011), noted that amnesty is a necessary evil in convincing parties to a conflict to enter into peace negotiations or lay down arms. On a positive note, soldiers who are general when forgiven, come back home and stop recruiting children into war. Others said amnesty shortens the way but does not end it. For instance, in Jongolei state, after the generals embraced the amnesty granted to them, there was temporary peace and people rested a bit, although most of the victims whose relatives had been killed rebelled and took up arms with the aim of overthrowing the government that had forgiven and absorbed into leadership the criminals who killed their relatives. One participant during an interview alluded that: ".... *see the Thomas Cilirio NAS saga where we can't get out of juba right now safely. The rebels were forgiven and they were promoted, Thomas also decided to rebel, so it is recurrent process. See many mad people in the streets because of trauma the same people who violated us the same people leading us and we are helpless and powerless.*" (Interview with participants on 11 April, 2023 in Juba).

Another participant observed that revenge killings are happening because of the president forgiving these people instead of law taking its cause. Maybe the president should have asked people for their opinion, but he just unilaterally forgave them because the international community asked him to do so. Or he just wants his peace so that he can continue to rule and steal resources. Participants during a FGD voiced, "*We don't know the internationally accepted ways of seeking amnesty but locally if someone is to be forgiven by the authority they have to ask for forgiveness first not just begging these general to come back or even using the pardon to bring them back yet inside their hearts they are not ready for good of the people*" (Focus group Discussion with participants on 11 April, 2023 in Juba).

It has been reported that blanket amnesties can be tools for truth recovery if used appropriately. In this way the right to truth may be enhanced often at little cost (Walter, 2004, 2009, 2014). Our findings provide mixed opinions, some contrary and others in agreement with Walter's views. In Juba, our findings show that amnesty is context specific and does not work equally regardless of the nature of conflicts. For South Sudan, our findings reveal the need for one to take into consideration historical context that underpins conflicts and war. The effects of amnesty in the South Sudan situation are viewed as mostly negative among the people and that it yielded relative peace in some places, but limited justice for the victims of the war crimes and hardly any interventions on economic effects of conflicts, which are reported to be critical in achieving peace and justice.

Conclusion and Recommendations

Amnesty's contribution to the promotion of peace and justice in South Sudan remains controversial, and in the case of Juba in South Sudan it is viewed to have largely had negative consequences and not yielded the objectives for which it was introduced.

Our findings show that conflicts have continued with incidents of murder, arbitrary arrests, detention, torture, disappearances, extrajudicial executions, sexual and gender-based violence despite the amnesty that was instituted to address such problems. In the people's perception, there was need for application of amnesty alongside hybrid courts, and truth and reconciliation commissions so as to realize peace and justice in Juba and South Sudan in general. This proposal is made emerging from the finding that the people in South Sudan mostly believe that blanket amnesty failed to bring peace and justice because the conflict perpetrators/leaders never got to account for their criminal actions against the victims, and that the victims were not given justice.

Our finding pinpoints the need for a wider and in-depth study to elucidate the economic and psychosocial justice effects of the amnesty in South Sudan and the potential of Truth and Reconcilation Commission (TRC) and hybrid courts to promote peace and justice in the county.

References

Blattman, C., & Miguel, E. (2010). Civil war. *Journal of Economic Literature, 48*(1), 3–57.

Creswell, W. J., & Creswell, D. J. (2023). *Research design. Qualitative, quantitative, and mixed methods approaches* (6th ed.). SAGE.

Dancy, G. (2018). Deals with the devil? Conflict amnesties, civil war, and sustainable peace. *International Organization, 72*(2), 387–421.

Daniels, L. A. (2020). How and when amnesty during conflict affects conflict termination. *Journal of Conflict Resolution, 64*(9), 1612–1637.

Hayner, P. (2018). *The peacemaker's paradox: Pursuing justice in the shadow of conflict*. Routledge.

Hazan, P. (2020). *Amnesty: A blessing in disguise? Making good use of an important mechanism in peace processes*. Centre for Humanitarian Dialogue.

International Institute for Strategic Studies. (2019). The revitalized agreement for resolution of conflict in South Sudan (R-ARCSS): Addressing the challenges to power sharing. HORN Policy Brief No.17 February 2019.

Mallinder, L. (2023). Amnesties and transitional justice. In *Research handbook on transitional justice* (pp. 262–282). Edward Elgar Publishing.

Powell, R. (2002). Bargaining theory and international conflict. *Annual Review of Political Science, 5*(1), 1–30.

Powell, R. (2006). War as a commitment problem. *International Organization, 60*(1), 169–203.

Reiter, A. G. (2014). Examining the use of amnesties and pardons as a response to internal armed conflict. *Israel Law Review, 47*(1), 133–147.

Walter, B. F. (2004). Does conflict beget conflict? Explaining recurring civil war. *Journal of Peace Research, 41*(3), 371–388.

Walter, B. F. (2009). Bargaining failures and civil war. *Annual Review of Political Science, 12*(1), 243–261.

Walter, B. F. (2014). Why bad governance leads to repeat civil war. *Journal of Conflict Resolution, 59*(7), 1242–1272.

Yin, R. K. (2009). *Case study research: Design and methods* (Vol. 5).

10 Assessment of the Challenges of Rehabilitation and Reintegration of Victims of Human Trafficking in Lorengecora Sub-County, Napak District

Lwanga Okello and Bendicto Kabiito

Abstract

We examined the challenges facing rehabilitation and reintegration of victims of human trafficking in Lorengecora Sub-County in Napak district, Uganda, covering a 10-year period between 2014 and 2023. Based on thematic analysis of data drawn from 29 participants, our findings revealed that widespread poverty, complicity of community leaders in the vice and overcrowding at rehabilitation centres are the main challenges affecting rehabilitation of victims of human trafficking. The main challenges to the reintegration of victims of human trafficking are porous borders, gaps in parenting, and fear of forced marriages. There is need for skilling and awareness creation to enhance the rehabilitation and reintegration of victims of human trafficking.

Keywords

Modern slavery · Street children · Traffickers · Child mothers · Humanitarian organisations

L. Okello (✉) · B. Kabiito
Uganda Martyrs University, Kampala, Uganda
e-mail: okello.lwanga@stud.umu.ac.ug

Introduction

The International Labour Organization and Walk Free Foundation estimate that 40.3 million people lived in modern-day slavery that included trafficked victims in 2016 (ILO & WFF, 2017). Like other regions in Africa, Uganda is not exempt from the detrimental impacts of human trafficking (Barasa-Asekenye et al., 2022). Different reasons have accounted for human trafficking at varied thresholds including economic need, dysfunctional families, physical disability and immigration status (Scuderi et al., 2023; von Uexkull et al., 2023). Victims are trafficked from rural to urban areas and even abroad. Napak is one of the Ugandan districts with high levels of human trafficking. Studies indicate that parents there are often involved in the vice. Legal gaps, lack of political will and inadequate data are a hindrance to combatting trafficking effectively.

The Karamoja sub-region, a heavily affected region in Uganda, is plagued by severe human trafficking spanning the northeastern boundary of Uganda (Laborn, 2023). This territory shares borders with Kenya and South Sudan, and it is dominated by the pastoralist Karimojong who from time to time keep crossing borders in search for green pastures. This region is among the world's most impoverished areas (UNFPA,

P. Nyadoi et al. (eds.), *Sustainable Development Goals in Africa*, Sustainable Development Goals Series, https://doi.org/10.1007/978-3-032-20742-5_10

2018). With 61% of its 1.2 million inhabitants trapped in absolute poverty, the region's dire economic conditions have amplified its vulnerability, resulting in elevated levels of human trafficking (von Uexkull et al., 2023). This area has been recognized as a hub for human trafficking activities, encompassing source, transit, and destination roles in the illicit trade of persons, with a specific focus on women and children (Koogler, 2022). Amidst diverse challenges, these vulnerable groups are exposed to a range of exploitative circumstances that are not well understood. Against this backdrop, this study aimed to explore the rehabilitation and reintegration challenges faced by human trafficking victims in Napak district and the measures being taken to address them.

Methodology

Research Design

This study employed a qualitative research approach which involved the collation of narrative raw facts from a set of participants. This approach enabled the researcher to secure a deeper understanding of the social phenomena relating to the challenges of the rehabilitation and reintegration of victims of human trafficking in Lorengecora Sub-County, Napak District. The researcher adopted a research strategy involving the use of qualitative data collection methods and techniques to address the research questions. The researcher employed flexible, iterative methods such as interviews, observations, and document analysis to generate rich, contextually situated data for interpretation and understanding. The researcher also attended a community dialogue discussing the challenges of human trafficking in Napak, and this provided valuable information for the study. The researcher employed a qualitative research strategy to explore the personal stories and experiences of the victims of human trafficking and their caretakers in rehabilitation centres in line with the work of Gioia (2021). The researcher used the ethnography research approach under the qualitative research strategy, which sought to incorporate expert opinions to create a chronological narrative report that illustrated the difficulties victims of human trafficking faced in rehabilitation and reintegration in Lorengecora Sub-County, Napak District. The research approach and strategy both placed a strong emphasis on understanding these difficulties through in-depth exploration of participants' perspectives, experiences, and contexts. This study adopted an exploratory research design which is qualitative in nature. Per Creswell and Clark (2017), the adoption of an exploratory research design aimed to investigate a problem or phenomenon with limited prior knowledge by employing methods such as interviews, focus groups, and literature reviews to formulate insights for further investigation. This research process involved the emergence of inquiries and methodologies, the collection of data typically within the participants' environment, an inductive data analysis that progresses from specific details to overarching themes, and the researcher's interpretation of the data's significance. The ultimate written report boasts an adaptable structure. Consequently, the qualitative approach empowered the researcher to extract crucial insights by means of conducting interviews with victims, attentively engaging with experts, and involving those actively participating in the rehabilitation and reintegration of human trafficking victims. The researcher's task within this methodology involved capturing significance by interpreting narrative and descriptive data regarding the real-life encounters of the participants, and this was quite in line with the view of Chivanga and Monyai (2021). This method was employed to acquire profound understandings and to foster the creation of innovative concepts concerning the challenges faced in the rehabilitation and reintegration of victims of human trafficking in Napak District.

Research Approach

Non-probability sampling techniques (purposive sampling) were used in the selection of the participants (Kumar, 2018) that included key informants, the victims who went through Kobulin

Youth Skills Centre, MGLSD, KCCA, and UPF. The non-probability sampling technique (purposive sampling) enabled us to get information on knowledge and experience the victims and key informants bear regarding challenges of rehabilitation and reintegration of victims of human trafficking in Lorengecora Sub-County, Napak District. Although victims of human trafficking at Kobulin Youth Skills Centre were originally targeted, all the victims at the centre had been discharged and hence, those who had passed through the centre but had already been reintegrated within the past 12 months were involved. Participants were selected purposively; the target were those who possessed relevant characteristics or experiences related to human trafficking to ensure richness and depth of data (as guided in Kumar, 2018).

Data Collection

We used Focus Group Discussions (FGD) and interviews to obtain data that helped in generating in-depth information about the challenges faced by victims of human trafficking and the challenges in reintegration of the victims. The following were the data collection instruments that were used to gather both primary and secondary data. We used an FGD guide to collect narrative data from the victims of human trafficking. The guide comprised structured questions covering the research objectives. The FGD guide enabled the researcher to obtain diverse perceptions and attitudes from the victims in a group-like setting where all participants were given a chance to contribute based on their own experiences during rehabilitation and reintegration processes as recommended by (Gioia, 2021). The FGD enabled simultaneous data collection from different individuals and enabled deep probing to obtain detailed information on the challenges facing the rehabilitation and reintegration of victims of human trafficking in Lorengecora Sub-County, Napak District. Two sets of FGD were conducted: one for victims that went through formal rehabilitation and reintegration processes and the other for victims that didn't go through the formal processes. A third FGD was planned for victims at the rehabilitation centre, but this was not possible because at the time of collecting the data there were no victims at Kobulin Youth Skills centre.

We used an interview guide to collect data by questioning individuals and groups for desired information as explained by Tavory (2020). Key Informant Interviews (KIIs) were conducted with selected experts about human trafficking from whom in-depth narrative data was gathered on the challenges facing the rehabilitation and reintegration of victims of human trafficking. KIIs allowed question rephrasing for quality responses (Kumar, 2018). An interview guide enabled the researcher to consistently collect qualitative data from the esteemed key informants including UPF officials, community leaders, MGLSD officials, rehabilitation centre managers and KCCA officials. The interview guide comprised open-ended structured questions that helped to generate deep discussions. As much as possible, the researcher conducted face-to-face interviews, and in two instances, interviews were conducted on Zoom and by phone due to the limited time of the respondents.

In data analysis, we followed the principles of inductive and thematic analysis as spelt out by (Clarke, 2016) and outlined as follows: First, the transcripts were read several times to get a sense of the whole texts. Recorded interviews or discussions were also replayed several times to get an overview of the data gathered. Secondly, transcript texts were divided into codes that signified a single issue or idea (Grbich, 2003). Thirdly, codes were compared against each other, and similar codes were placed into themes. Van Nes (2010) states that researchers should stay as much as possible in the original language during the data analysis process to avoid loss of meaning. Therefore, both transcription and subsequent thematic analysis were first done in English or N'gakarimong (the two languages) used for interviews and focus group discussion. Thematic analysis (where sets of texts, like answers to open-ended survey questions, or interview transcripts were examined) was done to determine emerging themes the study.

Table 10.1 Distribution of participants who took part in the study

Category of respondents	Male	Female	Category
MGLSD	1	–	KII
Dwelling places (DP)	–	2	KII
Uganda police force (UPF)	3	–	KII
Parish chief (PC)	1	–	KII
Cooperation and development (CD)	–	1	KII
Religious leader (RL)	1	–	KII
Probation officer (PO)	1	–	KII
IOM	–	1	KII
KCCA	–	1	KII
LC III chairman (CIII)	1	–	KII
Kobulin youth skills Centre (KOB)	1	–	KII
FGD participants (P)	6	9	FGD (PI-XV)
Total	15	14	

Results and Discussion

A total of twenty-nine (29) individuals took part in the study, with fourteen (14) females and fifteen (15) males participating. Most of the victims interviewed were female (9) which is also a reflection of the large number of women and girls trafficked from Lorengecora Sub-County, Napak district. A summary of the distribution of participants across the various groups who took part in the study is presented in Table 10.1.

Challenges Faced in the Rehabilitation and Reintegration of Victims of Human Trafficking

Rehabilitation

Resource Constraints

One of the challenges limiting the rehabilitation of victims of human trafficking in Lorengecora Sub-County is the widespread poverty in the area as stated in the following verbatim:

> Poverty hampers rehabilitation efforts because it limits access to resources, support services, and opportunities for economic empowerment, perpetuating vulnerability and hindering recovery for victims of human trafficking …KII_NPK_DIP_01

The verbatim ascertains that poverty poses a huge challenge in combating human trafficking and this is multidimensional, as it restricts access to crucial resources and opportunities. This limitation hinders the investment in vital infrastructure, personnel, and support services necessary for effectively addressing human trafficking and providing rehabilitation services. Additionally, poverty also exacerbates vulnerabilities among the high-risk populations, perpetuating cycles of exploitation as also pointed out by Offia (2020). The study findings indicated that some children were going back to the streets in attempt to make their way again to the rehabilitation centres to benefit from the services provided there. It therefore requires efforts to break the cycle of poverty and create sustainable pathways to combat human trafficking in the area and carry out rehabilitation effectively.

Another resource challenge highlighted was inadequate financial support either from government or other stakeholders towards the development and the operations of the rehabilitation centre that is charged with the responsibility of rehabilitating victims of human trafficking. Apparently, there is only one rehabilitation centre in the Sub-County that even serves the whole of Karamoja sub-region, moreover, the centre is poorly financed as per the following verbatim transcriptions:

> …Inadequate resources to support the infrastructural development of the institutions which are handling human trafficking in Lorengecora Sub County affects any form of development at the centre. Apparently, the centre is too small for the high number of victims brought here but we can't expand it due to insufficient funds … KII_NPK_LPC_01
>
> …inadequate resources from government to facilitate the processes of rehabilitation because people can be many here including children and adults and they all need attention. For this reason, you find children, mothers plus even social workers sharing accommodation.…KII_NPK_PCD_01
>
> There are limited funds to support the rehabilitation of the victims both at the rehabilitation centre and with all the other stakeholders and therefore the victims are not afforded all required services ranging from psychological, economic or health support that they may require...KII_NPK_CFPU_01

The statements above from the key informants who are themselves involved in the process of rehabilitation affirm that insufficient financial resources allocated to the centre cause a huge challenge in combatting human trafficking by impeding the development and operation of rehabilitation and support programs. This limitation constrains the effectiveness of these initiatives in adequately addressing the complex needs of trafficking victims and offering sustainable solutions. As also revealed by Mak et al. (2023) who pointed out to inadequate funding as a hindrance to rehabilitation of victims of human trafficking. The study findings further ascertained that without adequate funding, organizations struggle to provide essential services such as shelter, counselling, legal aid, and vocational training, thereby hindering the holistic recovery and reintegration of victims. Addressing resource constraints is crucial to ensuring comprehensive support for victims and enhancing efforts to combat the incidence of human trafficking effectively in Lorengecora Sub-County and Karamoja subregion at large. At the heart of this support is the government, which should provide sufficient funding for the running of rehabilitation programs at the Sub-County. Several NGOs are supporting this process in Lorengecora, such as Dwelling Places, Cooperation and Development, among others. However, the support is not sufficient due to the high costs involved and the increasing numbers of victims. In March alone, up to 900 victims were brought to the Kobulin Youth Skills centre, a rehabilitation centre that should be handling less than 400 people at a time.

Inadequate Infrastructure

Insufficient infrastructure poses a serious challenge in conducting rehabilitation of the victims of human trafficking. As mentioned above, there is only one rehabilitation centre in the whole of the Karamoja sub-region and, again, the centre has very limited facilities that can accommodate just around 300 people. The study findings revealed that the inadequate space for accommodation, shelter and recreation all contribute to an impediment to services provided, as stated in the verbatim below:

> Insufficient infrastructure/shelter at the rehabilitation centre is really affecting us, in a way that the shelter has small space to accommodate the large number of victims because its capacity is like 350 but you find that they can bring in like 500 people both children and adults at once. There are even not enough rooms for caretakers and social workers to stay on site …KII_NPK_CD_01

The verbatim ascertains that the inadequacy of shelter at Koblin rehabilitation centre not only compromises the quality of care but also restricts access to essential resources for trafficking victims. Consequently, the effectiveness of interventions aimed at providing protection, support, and rehabilitation to victims is compromised, hindering efforts to combat human trafficking comprehensively. Without proper accommodation, victims are left vulnerable to exploitation and face hurdles in accessing necessary support services, as consistent with the findings by Mlahagwa and Muriisa (2022) in a study for formerly abducted children and mothers who went through rehabilitation centres. This inadequacy undermines their safety, security, and overall well-being during the critical phases of rehabilitation and reintegration.

Poor Security Measures at the Rehabilitation Centre

The insufficiency of fencing or security measures at the rehabilitation centre poses a significant challenge in combating human trafficking and provision of rehabilitation services as per the following verbatim:

> Improper fencing of the rehabilitation centre, it has affected the monitoring of the victims in that can easily escape to either go back to their homes or get back to the streets of Kampala. We have also seen instances where the traffickers sneak into the rehabilitation centre and convince and re-trafficked the victims …KII_NPK_CD_01
>
> The fence at the rehabilitation centre is broken down and this makes some of the children to easily escape. There are many cases of children running away from the centre especially those that come from nearby villages…KII_NPK_RL_01

The verbatim transcriptions above highlight that the inadequacy in fencing heightens the vulnerability of victims to unauthorized access, escape attempts, and external threats, compromising the safety and efficacy of victim support initiatives. Consistent with the findings by Cordisco Tsai et al. (2022), it was also affirmed that without robust infrastructure to safeguard these facilities, the integrity of rehabilitation programs is undermined, hindering efforts to provide comprehensive care and rehabilitation for trafficking victims. Security is usually of great concern to the caretakers and it is something that needs to be addressed to help improve the program of rehabilitation as cited by one of the caretakers:

> Inadequate security because victims normally escape and get back in the community and sometimes even go back to where they got them from… some of them are very violent and they fight a lot…KII_NPK_RL_01.

Lack of Essential Services

It was revealed that inadequate essential services in the area significantly hamper the efforts to rehabilitate the victims. One of the officers who has experienced this first-hand narrates that: "…*insufficient utilities like electricity and water because Karamoja is a remote area and these two are a problem here. There is also insufficient means of transport at the centre to facilitate visits to families and the communities*" … KII_NPK_CD_01.

The verbatim affirms that while rehabilitation centres are crucial for supporting victims, they suffer from limited access to basic utilities like electricity, water, and sanitation facilities. This deficiency undermines the functionality and lives at the centres, directly impacting the quality of care and services available to trafficking victims. Without proper infrastructure, the ability to provide comprehensive support, rehabilitation, and reintegration services is severely compromised, hindering the effectiveness of anti-trafficking initiatives and the overall well-being of victims.

Shortage of Personnel

One of the challenges identified was inadequate staffing at Kobulin rehabilitation centre where it was revealed that the staff were insufficient to absorb and manage the inflow of victims of human trafficking as per the following verbatim:

> …inadequate personnel like the police for security, social workers, at least a nurse, tracing assistants among others. The staff at Kobulin cannot manage to handle the large number of the rescued people... KII_NPK_PCD_02

From the verbatim, it was revealed that the scarcity of adequately trained personnel, such as counsellors, social workers, and healthcare professionals, poses a formidable challenge in carrying out rehabilitation and the overall fight against human trafficking. This deficit restricts organizations' ability to provide trafficking victims with essential and specialized support services. Victims often require multifaceted assistance, including psychological counselling, social rehabilitation, and medical care, to overcome the trauma and complexities associated with trafficking experiences. There is also the problem of language barrier. Some of the victims brought to the centre cannot speak N'akaramajong, the local dialect spoken in the area, because they would have been born in the 'streets' in Kampala and can only speak Luganda. Other studies, such as Liz conducted in London on the rehabilitation of victims of criminal justice, pointed to language barriers as being a barrier for those who do not speak English and the lack of capacity for the staff to speak languages of the victims.

Participation of Community Leaders in Human Trafficking

Complicity or participation of community leaders in human trafficking posed a significant challenge to the rehabilitation of victims as expressed in the following verbatim:

> Involvement of community leaders in the processes of human trafficking for example the LC1s at the village level is still minimal in fighting human trafficking. Sometimes the leaders themselves become agents of traffickers. We have apprehended some leaders in the past for their accomplishment in this illicit trade …KII_NPK_CFPU_01

The narrative above affirms the inadequate involvement of community leaders in the process of rehabilitation of the victims. Also, when community leaders are engaged in trafficking, it perpetuates a culture of impunity and exploitation, making it harder to prosecute perpetrators and provide support to those affected. Much as a report by Niumai and Rajesh (2022) affirms these findings, most studies point to the fact that community leaders are always allies in the fight against human trafficking and quite supportive in the process of rehabilitation. The study therefore points to the fact that overcoming this challenge requires comprehensive strategies that address the root causes of community involvement while also promoting awareness, accountability, and collaboration among all stakeholders involved in anti-trafficking efforts.

It was also established that in exploiting their positions of authority, some local leaders may facilitate or overlook criminal activities such as human trafficking, impeding efforts to identify and support victims. Their complicity, for example, in not being able to report cases of escapees from the rehabilitation centre wears out trust and cooperation between communities and law enforcement agencies, hindering proactive measures against trafficking. Sometimes, the leaders are not sensitized well to their roles, and they are not very much aware of what is to be done as illustrated in the below statement:

> ...for example, the involvement of local leaders at the village level on these interventions is still minimal because they don't know what they have to do in their capacity to end this trafficking... KII_NPK_CFPU_01.

Addressing this challenge requires targeted strategies to hold complicit leaders accountable, strengthen community engagement, and build partnerships that prioritize victim protection and prevention initiatives, thereby fostering a safer environment for at-risk individuals.

Health Services Challenge

Insufficient access to healthcare services and medical treatment for both physical and mental health issues poses a significant challenge for rehabilitation centres catering to human trafficking victims as stated below by one of the key informants:

> ...for sure health care services are inadequate at the rehabilitation centre. Usually when the victims fall sick, we liaise with the nearby health centre and you find at times they also do not have medicine. Our clinic cannot manage the big population that come in. Because of this, we see many victims going out for treatment and not coming back which is a setback to our rehabilitation processes... ... KII_NPK_DP_02

The verbatim affirms that without adequate healthcare support, victims' recovery and overall well-being are compromised, intensifying their vulnerability, and prolonging their suffering. Addressing this challenge is very vital in providing comprehensive care and support to trafficking victims, as it directly impacts their ability to heal from the traumas endured and reintegrate into the communities, as earlier revealed by von Uexkull et al. (2023). Due to exposure, especially to sexually transmitted diseases, some of the victims are on long-term treatment and without this sort of care, they leave the rehabilitation activities halfway.

Welfare Challenges

The inadequacy of support for the basic needs and well-being of the victims of human trafficking within rehabilitation centres also poses significant challenges as explained in the following statements by a victim and social workers at the rehabilitation centre:

> Victims' welfare at the rehabilitation centre is very minimal now because we don't have enough funding from either government or other development partners. A lot of time, you will find this place empty because when we do not have funds, we can't host the returnees... ...KII_NPK_DP_02
> ..the main challenge at the rehabilitation centre is poor feeding. We always didn't have enough food while at the centre. Even access to other essential items for our normal functioning such as soap, Vaseline, pads is a challenge and this kind of would divert our attention from the learnings....FGD PIV

The verbatim above affirms that insufficient provision of necessities such as food, clothing, hygiene, and safety not only compromises their dignity but also undermines their resilience and prospects for successful rehabilitation and eventual reintegration into society. Other studies such as by Scuderi et al. (2023) also came up with similar findings, further revealing that without adequate support systems in place, victims may struggle to overcome the trauma of their experiences and face barriers to effectively reintegrating into their communities, perpetuating the cycle of vulnerability and exploitation.

Location of Rehabilitation Centre

The placement of rehabilitation centres near urban areas or transportation hubs presents a multifaceted challenge in combating human trafficking as *KII_NPK_CD_01* revealed that:

> ..the easy accessibility of this centre is both an asset and a challenge for us. We can easily be reached because we are located near the highway, but given our poor security measures, traffickers can also easily come for the victims even from here. Sometimes they pretend like the parents or relatives of the victims....

The above narrative reveals that while locations of the rehabilitation centre may facilitate accessibility for victims seeking assistance, they also expose them to heightened risks of re-trafficking, exploitation, and social stigma, which were in line with Ngabire (2022). Therefore, meticulous site selection and robust security measures are imperative to mitigate these dangers and ensure the safety and effective rehabilitation of victims. Balancing accessibility with security considerations is essential to provide comprehensive support, protection and proper rehabilitation for individuals escaping the scourge of human trafficking.

Overcrowding

The results highlighted the challenge that capacity limitation significantly hampers efforts to combat human trafficking, as laid out in the following verbatim:

> There is no much space as you may know because the centre is also limited to serve a certain number of people but KCCA and other enforcement agencies just bring the victims and station them there without any considerations. Sometimes the centre is not even consulted. ...KII_NPK_LPC_01

The verbatim affirms that high demand for services coupled with limited resources leads to overcrowding at the rehabilitation centre, detrimentally impacting the quality of care, privacy, and safety for trafficking victims during their recovery and reintegration process. This strain on capacity not only impedes the effectiveness of support services but also exacerbates the vulnerability of victims, hindering their ability to fully rehabilitate and reintegrate into society. Addressing this challenge requires concerted efforts to expand resources and enhance the capacity of rehabilitation facilities to adequately meet the needs of trafficking victims. The high and increasing number of victims rescued from different locations means the rehabilitation centre continues to suffer with capacity issues.

> Human trafficking victims are always many and when KCCA is bringing them from the streets of Kampala, they bring like 6 buses at once, that is around 300 people and even before these ones complete their program, another lot is brought. All these require accommodation, food and other necessities which the centre cannot manage... KII_NPK_DP_01

The verbatim affirms that with the influx of victims surpassing rehabilitation centres' capabilities, resources, personnel, and infrastructure are strained. This strain undermines the timely and sufficient provision of support services essential for victims' rehabilitation and reintegration into society, which affirmed the findings by Laborn (2023). Consequently, the effectiveness of rehabilitation efforts is hindered, exacerbating the plight of trafficking victims and hampering broader anti-trafficking initiatives.

Inadequate Training and Skilling

Another key challenge is the quality of the services provided at the rehabilitation centre. Some of the victims interviewed intimated that the quality of trainings conducted at the centre was too basic:

> ...for me my biggest challenge during the rehabilitation was that the trainings given are quite basic, usually for only 2 months even for courses like tailoring where other schools train for up to 2 years. This gradually affected us when we got out and started actual business. For example, now when a customer asks me for a specific design for a dress, I can't deliver. I can only make simple repairs...-FGD_ PVIII

The above statement highlights one of the discouraging factors for the victims who view the training provided at the rehabilitation centre as not valuable, leading to some of them leaving the centres or just 'passing time', which in turn reduces the effectiveness of rehabilitation.

Reintegration

Regarding reintegration, the following challenges emerged as expressed by the participants with supporting verbatim transcriptions as follows:

Porous Borders

> *Porous borders pose a huge concern because there is not enough border control, so our children are taken across and they go to beg as far as Nairobi and Mombasa and some big cities in Kenya. It's quite difficult for the law enforcers to apprehend the traffickers in this circumstance* ...KII_NPK_DP_01

In the verbatim, it was established that porous borders exacerbate the vulnerability of individuals to exploitation and hamper efforts for their reintegration into society. Strengthening border control measures emerges as a crucial step in mitigating this issue, which is in agreement with Nakaiza (2021).

Resource Constraints

> Inadequate support in terms of resources to facilitate and maintain the victims of human trafficking in order to prevent them from going back...such as school fees, mental health support, and medication ...KII_NPK_LPC_01
>
> Allocation of resources—for eliminating the issue of street children are very minimal...this should be a sustained effort to manage human trafficking, but we find it is mainly NGOs funding, the government needs to do more. Proper allocation of resources needs to be done...KII_MGLSD_01

This finding is in agreement with Cordisco Tsai et al. (2022), which showed that inadequate resources impeded the implementation of holistic reintegration programs, exacerbating the vulnerabilities faced by trafficked victims and hindering their ability to rebuild their lives post-trafficking.

Limited Health Care Facilities

> There are limited health facilities in Lorengecora Sub County to manage conditions such as HIV/AIDS and also some victims who end up running mad because of the experience they face while on the street and this becomes quite difficult for them to get reintegrated smoothly ... KII_NPK_CFPU_01

The finding affirms that forceful resettlement and reintegration may overlook victims' individual circumstances and needs, leading to insufficient support during the reintegration process. Proper needs assessments are crucial, allowing interventions and support to be customized based on victims' specific vulnerabilities and requirements.

Limited Time at the Rehabilitation Centre

> Limited time at the rehabilitation centre given to victims because they are given like 30 days to 60 days, and this is very limited time for them to learn something they can use for survival in future ... KII_NPK_RL_01

The verbatim affirms that the limited time given at Kobulin rehabilitation centre undermines the effectiveness of reintegration efforts. In line with the findings by Laborn (2023), it was revealed that extending the duration of rehabilitation programs is crucial to ensure that victims receive comprehensive assistance and adequate preparation for their successful reintegration into society.

Gaps in Parenting

> Child parenting gap is also a big challenge in reintegration because many of the victims give birth at a tender age from about 13 years and so this becomes difficult to reintegrate such an individual. We have also noted that some parents in the Sub-County do not know their responsibilities. They instead want the children to go and work in the streets and send them money to take care of them. This is very wrong; the parents should look after their children well. It's their primary responsibilities, not the NGOs. ...KII_NPK_PCD_01
>
> Huge challenge in family tracing...It Is often difficult to find the families of the children, this because sometimes the families shift, sometimes the children do not even know their home because they would have been taken to Kampala when still

young. Sometimes the families even reject the children for fear of responsibilities ...KII_KOB_01
Once the government reintegrates the children with the parents, they come back meaning parenting is still weak. The parents sell of the startup kits given to their children after the rehabilitation at Kobulin ...KII_MGLSD_01

The above narratives ascertain that the child parenting gap is a dire challenge that may negate the reintegration efforts. These challenges necessitate specialized support and training for both children and caregivers to reconstruct healthy family dynamics. Other studies conducted especially in Asia and Africa, for example, Koogler (2022) also point to the fact that bridging the parenting gap is paramount to fostering the well-being and stability of reintegrated children within their familial settings.

There is also a problem of forced marriages. Usually, the girls fear to be married off once they go back to their communities. This forces the children especially girls to avoid going back to their communities or run away back to the streets where they have some control over their lives. ... KII_KOB_01

From the verbatim, it is evident that the fear of forced marriages forced female victims to resist reintegration and rehabilitation in their communities. Their fear was associated with the likelihood of being married off if they attempted to go back to their home area. More still, it was revealed that poor parenting manifests in instances where parents do not welcome their children after the efforts of rehabilitation and reintegration as per the verbatim statement below:

Parents are not as welcoming as the children are brought back. Sometimes they are not happy with the children coming back home because they look at the children as a source of income through the work or begging they are doing in Kampala ... KII_KOB_01

The finding is in line with McCormack and Hing (2023) that found that parents' mindset is that children are a source of income.

Forceful Resettlement Without Needs Assessment

The victims are forcefully settled because the stakeholders involved in the process just guess that these people come from only Napak, yet others come from other areas and I guess there is no needs assessment done by KCCA and police men who just come and arrest us and pack everyone on trucks and buses and bringing us to Kobulin ... FGD_NPK_PII

The verbatim affirms that forceful resettlement and reintegration may overlook victims' individual circumstances and needs, leading to insufficient support during the reintegration process as has been pointed out by Ngabire (2022).

Some victims interviewed mentioned that they had accepted the trafficking conditions on the ground that they would be safer from the cattle rustlers who are a lot more dangerous and life threatening. This is explained in the following narrative by one of the victims:

Now I have come back home but there is a lot of insecurity here. If the pokot people come here, I can be abducted or killed. I can also not keep cows here; they can be easily taken away. I also do not want to get involved in fighting the rustlers. I am a man, but this thing is dangerous. Our communities live in this fear. PGD_NPK _P11

This finding was also highlighted by Offia (2020).

Conclusion

This study has highlighted the key challenges facing rehabilitation and reintegration of victims of human trafficking in Lorengecora Sub-County, Napak District in Uganda. The findings also reveal some of the strategies needed to address the vice. Among others, there is need for adequate resource allocation, strict enforcement of anti-trafficking laws and emphasis on stakeholder collaboration.

References

Barasa-Asekenye, C., Balch, A., Okumu, W., Omony, G., Renton, L., & Sandhar, J. (2022). "This pandemic has opened another box of war": COVID-19, safeguarding, and research on human trafficking and modern slavery in Kenya and Uganda. *Journal of Human Trafficking, 10*, 565–581.

Chivanga, S., & Monyai, P. (2021). Back to basics: Qualitative research methodology for beginners. *Journal of Critical Reviews, 8*(2), 11–17.

Clarke, V. B. (2016). Using thematic analysis in psychology. *Qualitative Research in Psychology, 3*, 77–101.

Cordisco Tsai, L., Carlson, C., Baylosis, R., Hentschel, E., Nicholson, T., Eleccion, J., & Wainberg, M. (2022). Practitioner experiences responding to suicide risk for survivors of human trafficking in The Philippines. *Qualitative Health Research, 32*(3), 556–570.

Creswell, J. W., & Clark, V. L. P. (2017). Designing and Conducting Mixed Methods Research. Thousand Oaks, CA: Sage Publications.

Gioia, D. (2021). A systematic methodology for doing qualitative research. *The Journal of Applied Behavioral Science, 57*(1), 20–29.

Grbich, C. (2003). *Qualitative research in health: An introduction*. SAGE Publications.

ILO & WFF. (2017). *Global estimates of Morden slavery*. International Labor Organization & Walk Free Foundation.

Koogler, H. (2022). Social relationships and child labor migration from Karamoja, Uganda: An exploratory study. Uganda: An exploratory study (June 30, 2022).

Kumar, R. (2018). *Research methodology: A step-by-step guide for beginners*. Sage.

Laborn, N. (2023). Ridiculing silent evils: Depictions of human trafficking in apio eunice Otuko's Zura maids (2018). *East African Journal of Interdisciplinary Studies, 6*(1), 24–36.

Mak, J., Bentley, A., Paphtis, S., Huq, M., Zimmerman, C., Osrin, D., & Kiss, L. (2023). Psychosocial interventions to improve the mental health of survivors of human trafficking: A realist review. *The Lancet Psychiatry, 10*(7), 557–574.

McCormack, L., & Hing, M. S. (2023). "I drive my happiness when I save a child": Altruistic passion, purpose, and growth in caring for victims of child sacrifice and trafficking in Uganda. *Traumatology, 29*(2), 289.

Mlahagwa, W., & Muriisa, R. K. (2022). Dynamics of captivity and post captivity experiences and implication for social reintegration of the formerly abducted young mothers in Kitgum District, Northern Uganda.

Nakaiza, J. (2021). Surviving human trafficking: A case for strengthening the escape routes adopted by victims of trafficking in Uganda. *African Human Mobility Review, 7*, 3.

Ngabire, R. (2022). Human trafficking as a national security challenge to Uganda: A case of Kampala metropolitan area in Uganda (Doctoral dissertation, Nkumba University).

Niumai, A., & Rajesh, M. N. (2022). Stigma and exclusion: Challenges of re-integration of human trafficked survivors in Assam. *Sociological Bulletin, 71*(2), 289–304.

Offia, F. (2020). Child trafficking in Uganda. Irish Centre for Human Rights Working Paper Series.

Scuderi, R., Tesoriere, G., Fasone, V., & Pedrini, G. (2023). Social capital, poverty alleviation and complexity in Africa. Evidence from rural areas. *Journal of Economic Studies, 50*(4), 806–820.

Tavory, I. (2020). Interviews and inference: Making sense of interview data in qualitative research. *Qualitative Sociology, 43*(4), 449–465.

UNFPA. (2018). *Addressing inequalities in women's health, education and economic empowerment*. UNFPA.

Van Nes, F. A. (2010). Language differences in qualitative research: Is meaning lost in translation? *European Journal of Ageing, 7*(4), 313–316.

von Uexkull, N., Loy, A., & d'Errico, M. (2023). Climate, flood, and attitudes toward violence: Micro-level evidence from Karamoja, Uganda. *Regional Environmental Change, 23*(2), 57.

11 Social Integration Resources of Children Born of Rebels and Non-State Armed Actors in Situations of War

Hellen Mghoi Kishuke Mshilla,
Helen Nambalirwa Nkabala, and Robinah Nakabo

Abstract

This paper is part of a research project about social integration of youth born in the Lord's Resistance Army captivity in Gulu district Uganda. Traditionally, children born out of rape, incest, or other relationships outside marriage were treated as outcasts and get ostracized and stigmatized within their communities. The situation of those born by women abducted and forced to live as wives to their rebel and other political non-state-armed captors is worse because their fathers already committed acts of violence against the government and the very communities expected to support these children to recover from the effects of the war. In this study, in-depth face-to-face interviews with these children, now youths, Focus Group Discussions with their parents/guardians, and key informant interviews in Unyama and Paicho sub-counties in Gulu district and Gulu city were conducted. The findings are presented in this paper.

Keywords

The Lords Resistance Army · Children and youth born in LRA · Social Integration · Resources · Gulu

H. M. Kishuke Mshilla (✉) · H. N. Nkabala
R. Nakabo
Uganda Martyrs University, Kampala, Uganda
e-mail: hellen.mghoi@stud.umu.ac.ug

Introduction

Armed conflicts enormously harm youth. They disrupt their social environment, leave them with high levels of threats, deny them "access to basic necessities and security," and cause their "…loss of important sources of social support for their well-being (Wessells, 2016:198)". In Uganda, for example, between 1986/1987 and 2006 a rebel group called the Lord's Resistance Army (LRA) abducted and sexually abused thousands of girls forcing them into mothers (Suarez and Baines 2021; Allen, et al., 2020). Those who survived and left the LRA faced discrimination and stigmatization with their children within their families and communities (Macdonald and Kerali 2020) which, in ordinary situations, could have provided the special needs their children required for social integration (Ojok 2022). However, access to social integration resources by these children within the 18-30-year age bracket, from this point onwards referred to as youth born in LRA captivity (YBLC), was unclear due to lack of an explicit focus on their category in research. Therefore, the overall objective of this study was to contribute knowledge on the access to resources for social integration by YBLC in the Acholi sub-region of northern Uganda. Qualitative data was collected from them using in-depth face-to-face interviews introduced by an art-based rapport building exercise; from their parents and guardians using Focus Group

P. Nyadoi et al. (eds.), *Sustainable Development Goals in Africa*, Sustainable Development Goals Series, https://doi.org/10.1007/978-3-032-20742-5_11

Discussions (FGD); from representatives of Government, communities and Civil Society Organizations (CSO) using Key Informant Interviews (KIIs); and observation. We found that YBLC access temporary resources for social integration. This calls for multi-stakeholder remedial action.

Historical Background

Uganda, and the LRA Uganda is found in East Africa and occupies a geographical area of 241,559 sq. km; 37,000 of which is occupied by open water. The country, generally, has inadequate stable or satisfactory employment opportunities for youth. This means YBLC have limited economic resources which they can use to get well integrated socially. Politically, Uganda is a multi-party democracy, which holds presidential elections every five years with no term limits. It is divided into four regions of Northern, Western, Central and Eastern. After the British colonial rule, its first President Kabaka Edward Mutesa II of Buganda Kingdom led the country from 1963 until the then first Prime Minister Milton Obote overthrew him in 1967 and formed the first republican government. In 1971, Idi Amin overthrew Obote, but his bloody military rule ended in 1979. From then, Uganda's leadership changed hands seven times before the then National Resistance Army under the current President Yoweri Museveni took over power in 1986. The NRA changed its name to the National Resistance Movement (NRM) party and has since formed all the country's eight elected governments. After the NRM came to power, in 1987 Joseph Kony, an Acholi speaking native, formed the LRA which abducted between 30,000 and 50,000 children (Baines and Gauvin, 2014) and forced most of them into child soldiers and wives to Kony and his commanders. "An estimated 10,000 girls became forced child mothers as a result of LRA abduction between 1988 and 2004." The LRA also used to burn civilian's homes, loot their property, massacre others, and mutilate people's body parts and tortured many. This forced the government to created hundreds of Internally Displaced Persons (IDP) camps in the Acholi, Teso Lango and west Nile sub-regions. To avoid attacks and abduction, thousands of villagers, especially mothers and children, used to spend nights in verandas of trade centres, schools, hospitals and markets and became known as 'night commuters'. Following a ceasefire agreement during peace talks held in Juba, South Sudan, the LRA left Uganda in 2006 and continued operating in the Democratic Republic of Congo (DRC), Central African Republic (CAR) and South Sudan (Kiconco, 2022). The children produced within the LRA and their mothers were mostly kept in LRA camps in South Sudan. Hence, they were essentially initially an integral part of LRA rebel families. Many of them were carried along by their mothers during escape especially when the government initiated an Amnesty law in 2000, exempting from punishment those who surrendered. An unknown number of the children came without parents and were mostly kept orphanages. Some mothers and children returned to their ancestral land and faced social integration challenges. At their age of 18 and 30 which in Uganda is the official age of youth, there was a lack of academic literature about them which left their situation unclear.

Youth Born in LRA Captivity

Youth born in LRA captivity (YBLC), as used in this proposal, refers to individuals within the age bracket of 18–30 years who were born while their mothers were still under the control of the LRA, and mostly lived in northern Uganda at the time when data of this research was being collected. During their childhood while in LRA, they suffered extreme exposure to violence and insecurity. Others lost track of their parents, and/or witnessed killings of their own mothers, and became orphans under the care of sympathizers. Moreover, they were born outside Uganda and, therefore, it can be stated that they immigrated into Uganda. The actions of the LRA resulted in stigmatization, discrimination and hatred of YBLC by the communities (Macdonald, and Kerali 2020). This means they needed to be integrated, not re-integrated, since they had never lived in Uganda before. During their childhood, their families, community members and peers at times physically and psychologically abused them (Ojok 2022). They came to live within communities whose resources had been overstretched by war and were traumatized and overwhelmed by the demands to resettle in long-abandoned homes since the war had forced them to leave their homes. The state of youth unemployment and under-employment at national level was high when they came to live among communities. Moreover, they did not have the skills, and their education was at the mercy of other people who had not recovered from war atrocities committed against them. All this meant they had limited access to resources for social integration as children and as young adults such as education, land, decent shelter, access to health services, employment, and supportive relationships, At the age of 18 and above, they must compete for the limited social integration resources alongside thousands of other youth who also face underemployment and unemployment. Since many of them had challenges of education, they face the risk of unfair competition for social integration resources. The few more fortunate ones who made it to university-level education, mostly by virtue of being sired by the high-ranking LRA commanders, had no guarantee of equal treatment in the competition of scarce employment because the atrocities their fathers committed might be used to hinder their access to opportunities in business, employment and even political participation. All this shows that they missed important childhood resources that would have prepared them for proper social integration into adulthood. This is why it was important to research about them so that stakeholders can use it to make informed decisions on ways to mitigate their current and future risks or further marginalization. The next section explains what social integration is.

Social Integration

The concept of social integration has evolved over time and has no universally agreed definition. One of the founding philosophers of sociology, Durkheim Emile (1858–1917), perceived it as a way in which individuals interact, connect and confirm their membership to a social group. This results from collective consciousness of people out of their shared norms, values and beliefs. This philosopher differentiated between mechanical solidarity, a type of social integration sustained by kinship or blood relations, and sharing of a common language or worship, and, organic solidarity, which results from five interdependences as people work together in division of labour. In his study on suicide he concluded that a person not socially integrated lacks understanding breaks norms and therefore is in a state he calls 'anomie'. Over time, other scholars have conceptualized social integration in different ways. For to Rubin, Watt and Ramelli (2012:2) it "... refers to the quantity and quality of social connections and interactions that people have with others". Based on this, for the purpose of this research, social integration will be taken to mean the process that enables an individual to feel and/or actually be part of a family, community or social group. Adopting this open understanding of social integration shall allow the researcher to explore a broad range of more substantive data about YBLC that can be used to build theory and or on policy on mitigating their risks of exclusion from mainstream society. Social integration of YBLC shall be explored using all of its dimensions from socio-cultural to economic and political.

Access to Social Integration Resources by Persons Born of Non-state Armed Actors During Armed Conflict and Violence

Access to social integration by persons born of non-state armed actors during armed conflicts and violence around the world takes place against a background of problems of neglect of their needs to know their biological fathers. For instance, in a participatory research on Polish Children born of war (CBOW) during the WWII, Gałęziowski (2024) found that most of these children do not know their biological origin. Due to fear of "shame associated to the circumstances around conception", their mothers "...desired to forget about their existence (Gałęziowski 2024:119)". In former Yugoslavia, Strupinskiene (2012) did 21 semi-structured interviews in Sarajevo and three other cities in Bosnia-Herzegovina (BiH). She found that mothers gave away to foreigners or to orphanages or aborted children who they conceived out of rape during the 1992–1995 war. This was to help them forget their rape experiences (Strupinskiene 2012). In addition, some of the mothers concealed their children's birth origin to save them from their stigmatization by peers in school. Moreover, the research found school teachers had inadequate time to attend to the unique psychosocial demands of these children. Wagner, Glaesmer, Bartels et al (2022) used semi-structured interviews with thirty-five (35) DRC youth fathered by UN peacekeepers. Like Strupinskiene (2012), they found that some children did not know who fathered them as their mothers withheld this information from them thinking it would protect them from stigmatization and shame. Similarly, Loning (2023) used in-depth interviews with 32 young people born out of genocidal rape in Rwanda and seven of their mothers. The research reveals that some mothers did not tell to their children that they were born out of genocidal rape even when the latter wanted to know.

Accessing social integration by these children also took place amidst stigmatization and rejection. For instance, a qualitative study in Rwanda revealed that 26 years after the 1994 genocide, communities still did not accept children born in genocidal rape (Musafiri et al. 2023). The children are often referred to using offensive titles including "... "children of Hutus", "little killers", or "evil children (Musafiri et al. 2023)". Stewart in her PhD thesis studied the daily experiences of 29 children born into the LRA, between 2011 and 2016 in northern Uganda, and found that communities marginalize them. Pursuit of national political goal contributed to difficulties local communities experienced in welcoming the children. Likewise, according to Denov and Atim (2017) some parents of children born while in LRA hated their children and wished them death. Step-fathers who married their mothers after war also discriminated against them. Kiconco (2022) examined the socio-cultural situation of children born within the LRA, with a focus on their socialization, that is, how they establish social relationships and kinship ties to survive and be safe in post conflict northern Uganda using their mother's perspectives. She found that problems related to their stigmatized identity complicated their integration especially in their maternal homes. In relation to this, Ojok (2022) examined experiences of integration of 16 children born out of LRA forced marriages at a local school in northern Uganda. Like Kiconco (2022), he found that their peers, and teachers stigmatized them. The children also accessed the education resource with difficulties as most of them started school at a late age, and performed poorly academically compared to the rest (Ojok 2022). Additionally, the children lacked stable homes and shifted residence from one relative to another or to non relatives' homes (Ojok 2022).

Much as researched by Ojok and Kiconco is on social integration of children born in LRA captivity, it is hard to single out which of their data is specific to YBLC. Having stated this, the literature presented in the rest of this section identifies how persons whose birth associated with wartime sexual violence access social integration resources. Parents are important resources for their children's social integration. In her research, Apio (2007) found that former LRA abducted mothers educated their children believing that it would secure them a better future. Likewise, Ojok (2022) found that some women who returned from LRA with children fathered by LRA fighters struggled to pay fees for them while others joined support networks such as Women's Advocacy Network (WAN) to help their children. Therefore, research was needed to find out whether the education and land the mothers secured for their children amounted to these mothers' expectations.

Baines and Gauvin (2014) employed participant observation on their research on two sisters who were formerly abducted by the LRA in Pabbo, Amuru district northern Uganda. One of the sisters stayed in captivity for 10 years and left the LRA with three children. They found that the sisters worked together with fellow women-headed households to keep their children's lineages. Denov and Atim (2017) studied 27 women who had conceived babies while in LRA, in Gulu, Pader and Agago district and found that some mothers loved their children. Ojok (2022) also found that the teachers played a key role in counselling the children born of war, by instituting policies that protected them from abuse. However, research had not explored about this in the context of YBLC living in Gulu district. In the next section, the methodology and results of the research on social integration of YBLC in Gulu district, Uganda, is presented.

Methodology

An inductive approach was used to gather empirical data in the months of August 2024 as part of Gulu district, including Gulu district Uganda, which was registered at Uganda Martyrs University (UMU). Ethical approval was obtained from the Uganda Christian University Research and Ethics Committee (UCU-REC) and the Uganda National Council of Science and Technology in the month of July 2024. The UCU-REC was selected because by that time UMU did not have a REC. Face-to-face interviews were carried out with youth born in LRA captivity living in Unyama and Paicho sub-counties in Gulu District and Gulu City of Northern Uganda. All the participants were within the age bracket of 18–30 years in the month of August 2024 and were within Unyama and Paicho counties of Gulu District and Gulu City when the data was collected. The question 'which resources for your social integration are you accessing?' Before asking this question participant were asked to fill the rapport creation tool which was meant to make it easier for them to open up to the researcher. This was done in four steps as follows:

STEP 1: Each participant was greeted warmly welcomed, thanked for coming to meet the interviewee. The participant was also involved in selecting a suitable place where they would feel comfortable away from destructions and also to preserve confidentiality. After sitting, the participant was taken through the informed consent process. Once that was done, they were introduced to the rapport building activity using the following words. 'Before we start our conversation let us do some brief activity. Say that this activity is very important and will take ten minutes'.

STEP 2: The participant was asked to select three types of items for example food, means of transport or places or other.

STEP 3: The interviewer then asked the participant to state which of the three types of items they preferred most and the one they least preferred.

STEP 3: The participant was then given an A4 size chart with a table of four columns and five rows of same size, and a pen. The chart had the words 'family members', 'neighbors', 'employers /work-mates' etc.; 'religious leaders'; and 'cultural leaders' listed from the second to the last cell on the left side. The participant was then asked to draw a sketch of each of the three items on the top row of the table on the paper provided beginning from the best to the least preferred. Each item should be in one cell. They were asked to leave blank the first cell on their left.

STEP 4: The participant was then asked to match their feelings about the types of people they relate to, listed on the first column, with the sketches of items on the chat. Once they had finished, the participants were engaged with a brief session in which the interviewer made positive comments such as, 'wow, your sketch looks really interesting I don't remember when I last drew a (the item the participant had sketched)' etc., or /and ask how it felt like to do the exercise, then responded to any issues raised and build on the responses using the interview guide.

Notably, in all the cases, interviewers reported that the participants laughed, smiled and or said it was interesting, funny, or easy to do, simple or clear during these debriefing sessions and even were laughing or smiling as the session went on. This spontaneously signalled that the participant was ready for engagement with the rest of the interview. It was at this point that the participant was guided using the statement 'I can see here you ticked like this (*the interviewer would say these words while pointing at the first check mark*). Please tell me why you decided to mark it this way'. During the interaction using the tool the participant had completed, the interviewer made sure he/she mirrored the participant's feelings and body language where appropriate. For instance, they would smile when they smiled, and laugh along with them, etc. This was done to encourage the participant to continue talking freely. In a nutshell, therefore, what the participant had indicated in the rapport building tool was used as an opening to the actual in-depth interview.

Results

We found that the YBLC's access to resources for their social integration were temporary and therefore were mostly not sufficient to help them integrate positively well. In the villages they have no land of their own and lack decent shelter. They mostly were doing petty jobs such as making concrete, group farming or unpaid farming within the families of their guardians or parents who have supported them from childhood. Mostly it was because their mothers had not been given land to own, and in cases where they had traced their biological paternal relatives or/fathers, there were land issues such as outright denial or that it had been sold off. In the city, they mostly do hard tasks such as offloading and loading goods on lorries and tracks, hairdressing, masonry, bricklaying etc.; however, a few earn more through personal businesses such as tailoring, bodaboda riding, music and art as well as hairdressing. The preliminary findings on the economic resources most of them access are presented in the subsequent paragraphs. This life of hardship is also reflected in the few quotes from the YBLC.

Lack of Land

'I have been saving money so that I am in position to rent land next year and plant groundnuts about two acres. I will sell after harvesting', said a YBLC in Unyama.

Housing as a Resources Being Accessed by YBLC for their Social Integration

During interviews with YBLC, it was found that mostly they rely on the same land their parents rent for a living, or put up with relatives who support them. Majority of them share temporary residences with parents. Except for the two who said they live in their biological fathers' homes, all of those who use land said it neither belongs to them or to their parents. Mostly it was rented. This was because they were with their maternal relatives or that step fathers reject or discriminate against them. Some who live with their single mothers said the mothers also do not have land of their own. In Gulu City, mostly the YBLC live in their mother's rented home or have left and rent temporary shelters. Not many have houses of their own. This is happening to even male ones who had been re-united with their fathers' relatives where they even have land ownership rights. One key informant when asked about the type of houses YBLC live in said, "*These are single roomed grass thatched houses. At times you find a mother, father, sons, daughters and even their grandchildren living together…* ". This informant observed that the YBLC at times elope just to find a decent shelter or hoping to find more peaceful company. "Mostly they live in indecent shelters, and end up shifting from one residence to another", he added. Another YBLC in Gulu who is a graduate said he also lives alone in a single rented room with shared toilets. The house is located in one of the poorest city suburbs. He said he was better off because most of his colleagues are staying in the round huts their mothers rent at a meagre cost of 20–30 thousand Uganda shillings per month. The mothers stay in the city because they were rejected by their maternal relatives mostly.

Exploitative Means of Livelihood as Resource for Social Integration of YBLC

Having no land has pushed some of them to rely on exploitative means of livelihood. In the next section, the economic livelihood resources YBLC access are presented. Some of them earn their living away from those who stigmatize them by walking from their residences early in the morning to a quarry located ten kilometres away. Findings in Unyama sub-county show that many YBLC have resorted to concrete-making using bare hands in stone quarries for a living. This was first mentioned during an FGD with their mothers/guardians in Unyama sub-county. Some of their mothers raised them using the same type of business. We visited one of the stone quarry cites where many female YBLC work the whole day to earn a living. The sound of hammers hitting the stones rent the air as we approached the site. We witnessed how they make concrete using hammers with bare hands without any protective gear on their hands eyes, ears or heads. We learnt that at times they earn 3000 UGX per day. The concrete is measured using a container that can be filled with 20 litters of cooking oil. When filled with concrete, it is sold for 3000 UGX. See figure of photographs of YBLC and other war affected women working in a quarry in Unyama. Thus Gulu City and other urban centres indirectly benefit heavily from the effort of these marginalized youth, through real estate developers. Despite their effort, to the ordinary community, YBLC do not appear anywhere in development reports as YBLC. It also was not clear if they even knew how much their effort contributes to building the new cities and urban centres. It also seemed as if the YBLC and other people working at that site as well as neighbouring communities and their customers were least concerned with the soil degrading impact of stone quarry work. What appeared to matter was the money they would make from concrete making. It was their only source of income in the face of lack of their poor education and employable skills, or created a conducive social environment where the stresses of being associated with the acts of their LRA fathers could be mitigated amicably. They earn very little yet they contribute to lucrative businesses of middlemen and women.

A stone Quarry cite where some YBLC in Unyama Sub county, Gulu District Uganda work to earn a living. Photo by Hellen Mshilla

Results also show that a YBLC in Gulu City is makes substantive income in the same business using a middleman approach. She transports stones on trucks from quarry sites to Gulu City where she employs around ten women daily to make concrete using it. From her site, she sells the concrete to constructors in Gulu City and makes at times up to 500,000 UGX per day. This means both female and male have risked their lives in this practice as they can be arrested for participating in activities that causes soil erosion and infertility, while on the other hand providing cheap labour for harvesting raw materials to estate developers and therefore build cities and urban centres. What does this mean? Is stone quarrying a resource worth the cost in Gulu District and Gulu City surroundings? If the answer is no, what might satisfactorily substitute the economic benefits YBLC tap from the stone quarry businesses so that the overarching goals of protecting the environment and that of alleviating stigma and economic marginalization of YBLC in Gulu District be achieved. Who could be the first stopping point for this discussion? Higher level questions could be as follows: How could concrete-making businesses as a business for YBLC in Gulu be transformed into a well-paying, and physically safe, occupation for them in a way that it could promote environmental protection? This is a question that researchers can build upon. Even with question at hand, YBLC in this business continue to contribute to the small scale and large scale economy while occupying their marginalized spaces which they do not have full control over. To conclude, what can be safely stated is that, these findings show how neglect of economic social integration challenges of YBLC has contributed to and heightened the dilemma of checking environmentally unfriendly practices and protecting their livelihood as these youths are genuinely a community.

Farming

Similarly, another YBLC in Gulu City said he learnt about rabbit rearing from his friend who also gave him his first couple of rabbits, one male and one female. The two rabbits have since multiplied to over 40. He said he sells a pair for 40,000 UG Shillings and that is how he earns his living. Based on findings, other low-paying/temporary low-paying activities YBLC are engaging in include livestock rearing and farming.

> *I wake up at 5am to take the cows to feed. I return by 7am straight to the garden to dig until midday. I rest for about an hour or two and return to grazing the cows until 6pm.* (A youth in Unyama being raised by an uncle)

> *I plough for others for money using the ox. It is about one hundred thousand per acre using the ox for ploughing. I am saving this money to be able to achieve what I need.* (Youth in Unyama 09 August 2024)

> *I am learning a lot from my uncle who rears both cattle and goats. Therefore, I will use the knowledge to rear mine.* (Youth in Unyama 09 August 2024)

Recommendations

- Alternative ways of ensuring YBLC and their mothers who do not have inherited land be found, cultural institutions and government should give them land.
- More research needs to be done on the resources that YBLC access.
- Review policy on stone quarry business to eliminate its exploitative nature.
- Measures by occupational health experts be taken to protect the vulnerable YBLC and other populations who make concrete manually for a living from the exposure to repetitive strain injuries and hearing defects that result from such heavy, noisy work.
- This requires concerted efforts to promote best practices in ensuring workers are supplied with free protective gear and measures made to ensure they use them.
- An approach that is sensitive to victims of conflicts be used to engage manual stone quarry workers in viable afforestation business projects around the places where they stone quarry. This will enable them to get income from the business without destroying environment.

- Manual stone quarry health and safety training be conducted in concrete making cites.

References

Allen, T., Atingo, J., and Atim, D., Ocitti, J. Torre, C., Fergus, C., and Parker, M., (2020).What happened to children who returned from the Lord's Resistance Army in Uganda. Journal of Refugee Studies, Vol. 33(4), pp.663–683.

Apio, E. (2007). Uganda's Forgotten Children of War. In Carpenter, Ch. (Ed.). Born of war: Protecting children of sexual violence survivors in conflict zones. Bloomfield: Kumarian Press, Inc.

Baines E., Gauvin L. R. (2014). Motherhood and Social Repair after War and Displacement in Northern Uganda, Journal of Refugee Studies, Vol. 27(2), Pages 282–300, https://doi.org/10.1093/jrs/feu001.

Denov, M. and Atim, A. L. (2017) .When war is better than peace: The post-conflict realities of children born of wartime rape in northern Uganda. PMID: 28209554. https://doi.org/10.1016/j.chiabu.2017.02.014.

Gałęziowski, J. (2024). The best interests of the child in national terms: Policies concerning children of Polish female forced laborers and displaced persons in the early Cold War era. In J. B. Michlic, Y. von Saal, & A. Ullrich (Eds.), Childhood during war and genocide: Agency, survival, and representation (303 S., 6 Abb.). ISBN: 978-3-8353-5599-6-04-05.

Kiconco, A. (2022). *Children born of rebel captivity: Politics and practices of integration in Uganda.* Retrieved from (PDF) Children Born of Rebel Captivity: Politics and Practices of Integration in Uganda (researchgate.net) Accessed on 30 September 2022.

Loning, L. (2023). Beyond Born of War: Children, youth and young adults conceived in sexual violence. https://gjmedph.com/Uploads/O1_Special_Issue_2023.pdf.

Macdonald, A., and Kerali, R., (2020). Being normal: Stigmatization of Lord's Resistance Army Returnees as 'Moral Experience' in Post War northern Uganda. Journal of Refugee Studies, Vol. 33(4), pp 766–790.

Musafiri, E., Gona, G., Ombongi, K., and Shyaka, A. M. (2023). Voices of Youth Born of Genocidal Rape in Rwanda: Their Social Exclusion after 26 Years of Genocide, Journal of African Conflicts and Peace Studies: Vol. 5(1). Available at: https://digitalcommons.usf.edu/jacaps/vol5/iss1/4.

Ojok B (2022) "Every child is special...": perspectives of the integration of children born of war and their war-affected peers at a local school in Northern Uganda. Frontiers in Political Science 4. https://doi.org/10.3389/fpos.2022.816736.

Rubin, M., Watt, S. E., & Ramelli, M. (2012). Emigrants' social integration as a function of approach-avoidance orientation and problem-solving style. *International Journal of International Relations, 36*, 498–505. https://doi.org/10.1016/j.ijintrel.2011.12.009

Strupinskiene, L. (2012) Three: Living in the shadows of past attrocities: War babies of Bosnia. Wagadu, 10(Spring). https://sites.cortland.edu/wagadu/wp-content/uploads/sites/3/2014/02/Strupinskiene.pdf

Suarez, C., and Baines, E., (2021). 'Together at the heart': Familial relations and the social reintegration of excombatants'. International Peacekeeping, Vol. 29(1), pp1–23.

Wagner, K., Tasker, H., Vahedi, L., Bartels, S. A. and Lee, S. (2022). Born between war and peace: Situating peacekeeperfathered children in research on children born of war. Front. Polit. Sci. 4:945617. https://doi.org/10.3389/fpos.2022.945617.

Wessells, M. G. (2016). Children and armed conflict: Introduction and overview. *Journal of Peace Psychology, 22*(3), 198–207.

Part IV

Public Health and Demographically-Strained Africa Dwellers

12 Factors Associated with Viral Load Suppression of Children Living with HIV Among Virally Suppressed Caretakers in Luwero District

Andrew Kigongo and Richard Ekodeu

Abstract

Introduction: Achieving sustained HIV viral load suppression (VLS) is crucial for all persons living with HIV, including children.

Methods: A cross-sectional quantitative study among 402 child-caretaker pairs in Luwero District, Uganda, using researcher-administered questionnaires and SPSS version 22 for analysis. In results: VLS among children living with HIV (CLHIV) whose caretakers were virally suppressed was 72.9%. Factors associated with VLS included: female gender (AOR: 0.19, 95% CI: 0.07–0.55), age of HIV diagnosis <2 years (AOR: 3.63, 95% CI: 1.43–9.19), belonging to a PSS club (AOR: 21.03, 95% CI: 6.00–73.69), and biological parent care (AOR: 10.90, 95% CI: 4.15–28.64). Caretaker and household factors, such as age < 25 years, primary education, ART duration <5 years, food security, and family support, also showed significant associations ($p \leq 0.005$). Conclusion: VLS among CLHIV falls short of the WHO's 95% target. Targeted interventions addressing child, caretaker, and household factors are necessary to improve VLS outcomes. As recommendations: A multi-faceted approach, including engaging religious leaders, early HIV diagnosis, multi-month dispensing, peer support, improved household food security, and strengthened family support systems, is recommended.

Keywords

Children living with HIV · Viral load · Suppression · Caretakers

Introduction

HIV remains one of the leading causes of mortality in children and adolescents in Africa (Ssewamala et al., 2020). In the 2014 report titled *Fast Track*: *Ending the AIDS epidemic by 2030*, the Joint United Nations Programme on HIV/AIDS (UNAIDS) set the 95-95-95 treatment targets with the aim of transforming the vision of zero new HIV infections, zero discrimination and zero AIDS-related deaths into concrete milestones and endpoints. Despite tremendous progress, children living with HIV are disproportionately affected and therefore lagging in the achievement of these targets (Ferrand et al., 2016; Hader, 2019). The UNAIDS 95-95-95 global targets aim at ensuring that by 2030, (1)

A. Kigongo (✉)
Faculty of Health Sciences, Uganda Martyrs University, Kampala, Uganda
e-mail: kigongo.andrew@stud.umu.ac.ug

R. Ekodeu
Agency for Cooperation in Research and Development (ACORD) Uganda, Kampala, Uganda

P. Nyadoi et al. (eds.), *Sustainable Development Goals in Africa*, Sustainable Development Goals Series, https://doi.org/10.1007/978-3-032-20742-5_12

95% of all PLHIV know their HIV status, (2) 95% of all people diagnosed with HIV receive sustained antiretroviral therapy (ART), and (3) 95% of all people receiving ART achieve viral load suppression (VLS).

Of approximately 39 million people living with HIV worldwide in 2020, 2.78 million were children and adolescents aged 0–19 years. About 850 children contracted HIV, and about 330 children died from AIDS-related causes in 2020. Additionally, of the estimated 680,000 people who died from AIDS-related diseases in 2020, approximately 18 percent were children under the age of 20. This can be mainly attributed to insufficient access to HIV prevention, care, and treatment services (UNICEF, 2021). Viral load suppression (VLS) enables people living with HIV (PLHIV) to live long, healthy lives and to have zero risk of transmitting HIV to other people (Onyango et al., 2023). Globally in 2022, out of the 39.0 million PLHIV, 86% knew their HIV status, 76% were receiving ART, and 71% were virally suppressed. However, only 46% of children living with HIV (CLHIV) had suppressed viral load. Viral load monitoring is the gold standard for assessing treatment adherence and response, with Uganda's guidelines recommending testing 6 months after treatment initiation and annually thereafter (Consolidated Guidelines, MOH, 2022). Viral loads exceeding 200 copies/ml (plasma) or 400 copies/ml (DBS) indicate virological non-suppression. Children on ART face challenges, including high viral loads, suboptimal adherence, poor tolerability, and drug resistance, often due to regimen discontinuations (Bulage et al., 2017).

Factors contributing to virological non-suppression include patient-related issues (co-morbidities, incomplete adherence, missed appointments, and interrupted ART access) and regimen-related factors (drug adverse effects, suboptimal pharmacokinetics, potency, and food requirements), highlighting the need for effective management to achieve viral suppression in children living with HIV. Uganda adopted the "test-and-treat" policy among children and adolescents in 2013 to accelerate attainment of epidemic control. For this to happen, CLHIV who initiate ART need to remain in care and achieve sustained HIV viral load suppression; however, the latter remains a challenge (Oryokot et al., 2023). Poor adherence is the major cause of viral non-suppression and in children this may be caused by, unavailable caregivers, caregivers fear of stigma and disclosure concerns that may make them not administer ART to their children or take them to clinic on appointment days, challenges of distance and transport to the health facility, having parents or not, behaviour of health workers, unreliable supply chain, and other socioeconomic factors (Humphrey et al., 2019; Gordon et al., 2022; MOH, 2022; Nabukeera et al., 2021).

In Uganda, the HIV epidemic is estimated at 1.4 million People living with HIV as of December 2021, children aged 0–14 years living with HIV were estimated at 88,000 and young adults 15–24 years estimated at 170,000 (Uganda HIV and AIDS Facts Sheet, 2021). Among adults aged 15 years and older, HIV prevalence varied geographically across Uganda, with 8.1% in Central 1, South Buganda, and 6.2% in Central 2, North Buganda where Luwero district lies. The 2020 Ministry of Health (MOH) estimates put Luwero District at 7.4% HIV prevalence (Uganda HIV and AIDS Facts Sheet, 2022). By the end of June 2021, routine programmatic data from MOH indicated that overall, 41% of C/ALHIV were virally suppressed (Oryokot et al., 2023). Caretakers have challenges supporting their CLHIV, including fear of death of the child since they equate non-suppressed viral load to imminent death, competing priorities such as employment or looking for food, discouragement at failing to achieve viral suppression despite concerted effort, and the logistics required to take their child to health facilities (Nasuuna et al., 2019). In Uganda's Country Operating Plan (COP) 22 from 1 October 2022 to 30 September 2023, Luwero district had the highest burden for children living with HIV among districts in the greater Luwero. MIS data from a USAID-funded project indicates there are CLHIV who are virally unsuppressed under care of HIV-positive caregivers who are virally suppressing. Caregivers are the primary treatment supporters of their children; when a caregiver is also in care, this is an added advantage as they have treatment literacy and knowledge on treatment, and they are believed to know why and when to take the medications. If the caretaker has a suppressed viral load, it is believed that the likelihood of the child being virally suppressed is increased and there needs to be differentiation of care for children and their caretakers in order to help them suppress their viral loads by easing the burden that caregivers face (Nasuuna et al., 2019). Virally non-suppressed CLHIV under care of their positive caretakers who are virally suppressing remains a challenge in Luwero. This study therefore aimed to determine the factors associated with viral load suppression among children living with HIV among virally suppressed caretakers in the district.

Specific Objectives

(a) To determine the child-related factors associated with viral load suppression of children living with HIV among virally suppressed caretakers in Luwero District.
(b) To establish the caretaker-related factors associated with viral load suppression of children living with HIV among virally suppressed caretakers in Luwero District.
(c) To determine the household-related factors associated with viral load suppression of children living with HIV among virally suppressed caretakers in Luwero District.

Research Question/Hypothesis

There is a significant correlation between caregiver-related factors, child-related factors, and household-related factors, and the viral load suppression of children living with HIV whose caretakers are virally suppressed in Luwero District, Uganda. What factors are associated with viral load suppression among children living with HIV whose caretakers living with HIV are virally suppressed, in Luwero District, Uganda?

Methods

Study Area

The study took place in Luwero District, specifically at ART clinics in all 30 health facilities providing ART services, which are a mix of public, private for-profit, and private not-for-profit facilities. Located about 75 kilometres from Kampala City along the Kampala-Gulu highway, Luwero District has a population of 458,158, with a majority (59%) being children under 18 years old (UBOS and ICF, 2016). Almost one-thirds of the population is under 10 years old, and 60% are under 20 years old. The district is primarily rural, with only 21% of residents living in urban areas, specifically the town councils of Luwero, Wobulenzi and Bombo. Geographically, Luwero District is part of the Luwero triangle, bordered by Nakasongola to the north, Kayunga to the east, Mukono to the southeast, Wakiso to the south and Nakaseke to the west.

Study Design

The study was based on a quantitative cross-sectional descriptive and analytical design. The study was descriptive because it applied descriptive statistical techniques of frequencies and percentages to explain the study findings as well as inferential statistics techniques in order to ascertain the different factors that are strongly associated with VLS among children whose caregivers are virally suppressed.

Study Population

The study population comprised child-caretaker pairs, children under 18 years of age who are living with HIV and their virally suppressed caretakers on ART in Luwero District.

Eligibility Criteria

Inclusion Criteria

Only children: (1) who are living with HIV and receiving ART services in Luwero district, (2) whose caretakers met the inclusion criteria below, and (3) who had viral load test results for a period not longer than 1 year back. Only caretakers who: (1) were at least 18 years of age since this is the age of consent in Uganda, (2) had suppressed viral load over a period of at least 1 year, (3) had a child living with HIV, and (4) willingly consented in writing to participate in the study.

Exclusion Criteria

Children whose caretakers were not virally suppressing or not physically present at the time of the study were excluded. Furthermore, caretakers who were not in the right mood, or mental or physical health status, were also excluded from the study.

Sample Size Determination

The sample size for the study was determined using the Kish (1965) formula for cross-sectional studies; the sample size considered in the study was 402 child-caretaker pairs.

Sampling Procedure

The study employed proportionate sampling to select respondents based on facility caseload, allocating 50% ($n = 201$) from facilities serving 50% of the district's ART clients, 30% ($n = 121$) from facilities serving 30%, and 20% ($n = 80$) from facilities serving 20%. Within each category, equal numbers of respondents were selected per facility. Subsequently, simple random sampling was used to select study participants, ensuring all members of the study population had an equal chance of selection, resulting in a total sample size of 402 child-caretaker pairs. To select study participants, a simple random sampling method was used. The ART clinic numbers of CLHIV were written on identical pieces of paper, placed in a box, and shuffled. A health centre staff member then randomly selected papers, one at a time, without replacement, until the required sample size was reached. The selected individuals were contacted via telephone with assistance from facility-community ART support staff. If a selected respondent didn't meet the inclusion criteria, they were replaced using the same random selection process, ensuring an unbiased and representative sample.

Data Collection Tools

Data was collected using a well-structured questionnaire, translated into Luganda at administration to ensure respondents' understanding. The questionnaire was divided into sections: A (child-related factors), B (caregiver-related factors), C (household-related factors), and D (child's viral load suppression status). Closed-ended questions were used to gather specific responses from study participants, making data analysis more efficient and accurate. This structured approach allowed researchers to gather comprehensive data on factors influencing viral load suppression among children living with HIV.

Data Analysis

After collection, data was cleaned, entered and coded using SPSS version 22. For all objectives, data was analysed through descriptive statistical analysis of frequencies and percentages. Further inferential statistical analysis was done using Pearson's Chi-square test to ascertain the different factors that are strongly associated with VLS in CLHIV. This was followed by the logistic regression analysis at bivariate and multivariate levels in order to ascertain the extent to which the different factors are associated with VLS in CLHIV. The level of significance for all statistical analyses was set at $p \leq 0.05$.

Ethical Considerations

The Uganda Martyrs University Faculty of Health Sciences and Mildmay Uganda Research and Ethics Committee reviewed the proposal, data collection tool and informed consent forms. A letter granting permission to the data collection site and interaction with respondents and their data was endorsed by the institutional approval from the district health office of Luwero District. Written informed consent from the study respondents before allowing them to participate in the study was sought. This was preceded by explaining to them the objectives and processes of the study and assurance of confidentiality, the reason for which their names were not required for the study.

Results

The results of a study that was conducted among 402 virally suppressed caretakers of children living with HIV in Luwero District to determine the factors that are associated with VLS among these children showed that most of the CLHIV considered in the study (72.9%) have suppressed viral load results compared to 27.1% who were not virally suppressed at the time of the study. Results below highlighted per specific objective provide insights into what could be the possible reasons to the low VLS rate.

Child-Related Factors Associated with VLS for CLHIV

The first specific objective of the study was to determine the child-related factors that are associated with VLS CLHIV whose caretakers are virally suppressed. The relevant child-related characteristics that were considered in this study were: age, gender, religion, education level, parental status, age at diagnosis of HIV, duration on ART, ART treatment regimen, other medical conditions, other treatment regimens, belonging to PSS club for teenagers living with HIV, and belonging to a differentiated service delivery model.

As seen in Table 12.1, an equal number of children (50%) were 5 years and above and less than 5 years. Most, 69.4% were female. More than three-quarters (76.1%), belonged to a Christian religion (76.1%) and had primary level of education or less (75.4%). The majority of them had at least one of the biological parents alive (69.7%), and 69.7% of them had an HIV diagnosis done when they were younger than 2 years. The majority of them had been on ART for more than 2 years (82.6%). More than half (56.4%) were on ABC/3TC/DTG, and 57.5% had been on their current ART regimen for less than 24 months. The majority of them (81.6%) did not have any other medical condition at the time of the study. Most of them (68.2%) did not belong to a teen PSS club for teenagers living with HIV, but 87.8% of them belonged to a differentiated service delivery model, and 47.5% had been on multi-month dispensing for a period of 2 years and above, with slightly less than half (49.0%) being on FBG. To ascertain the child-related factors that are associated with VLS CLHIV whose caretakers are virally suppressed, the different child-related factors were first subjected to Pearson's Chi-Square test to determine those that had a significant association with VLS, with the level of significance being $p \leq 0.05$. Table 12.2 shows the summary of the results.

The child-related factors that were found to be significantly associated with VLS CLHIV were: the child's gender, religion, educational level, parental status, age at diagnosis of HIV, ART regimen, duration of on the current regimen, other medical conditions the child has had in the past 6 months, belonging to teen PSS club for teenagers living with HIV, belonging to a differentiated service delivery model and being on multi-month dispensing. The other child-related factors (age, and duration on ART) were not found to be significantly associated with VLS CLHIV as they had a P-value greater than 0.05. The different child-related factors that were found to be significantly associated with VLS CLHIV were subjected to logistic regression analysis to obtain the crude odds ratios (COR) and the corresponding 95% confidence interval (CI), and then those that were significant were

Table 12.1 Characteristics of children considered for this study

Characteristics	Frequency ($N = 402$)	Percent (%)
Child's age		
5 years and above	201	50
<5 years	201	50
Child's gender		
Female	279	69.4
Male	123	30.6
Child's religion		
Christian religion	306	76.1
Other religions	96	23.9
Child's educational level		
Primary or less	303	75.4
Higher than primary	99	24.6
Child's parental status		
At least one of the biological parents alive	280	69.7
None of the biological parents are alive	122	30.3
Child's age at diagnosis of HIV		
Within 2 years	280	69.7
After 2 years of age	122	30.3
Child's duration on ART		
Less than 2 years	70	17.4
More than 2 years	332	82.6
ART regimen the child is on at the moment		
TLD	135	33.6
TDF/3TC/DTG	40	10
ABC/3TC/DTG	227	56.4
Duration of on the current regimen of the child		
Less than 24 months	231	57.5
24 months or more	171	42.5
Other medical conditions the child has had in the past 6 months		
None	328	81.6
TB, malaria, etc.	74	18.4
Other treatment regimens the child has been on in the past six months		
None	348	86.6
Antimalarial, anti-TB drugs, etc.	54	13.4
Child belongs to teen PSS club for teenagers living with HIV		
Yes	128	31.8
No	274	68.2
Child belongs or attends any differentiated service delivery model		
Yes	353	87.8
No	49	12.2
Child on multi-month dispensing		
Less than 2 years	155	38.6
2 years and above	191	47.5
Not applicable	56	13.9
Type of DSDM		
FBIM	149	37.1
FBG	197	49
Not applicable	56	13.9

further subjected to multivariate analysis to obtain the adjusted odds ratio (AOR) and the corresponding 95% CI. The level of significance for all statistical analyses was set at $P \leq 0.05$. As can be observed in the study results in Table 12.3, on bivariate analysis, all associated child-related factors (gender, religion, educational level, parental status, age at diagnosis of HIV, ART regimen the child is on, duration of on current regimen, other medical conditions, belonging to teen PSS club, belonging to a differentiated service delivery model, and being on multi-month dispensing) were still found to be significantly associated with VLS CLHIV as they had a P-value less than 0.05. On multivariate analysis, however, the child's educational level, ART regimen, other medical conditions, and belonging to differentiated service delivery models were not found to be significantly associated with VLS CLHIV as $P \geq 0.05$.

Based on multivariate analysis, the female CLHIV had lower odds [AOR: 0.19 (0.07–0.55)] of being virally suppressed. In essence, the male CLHIV were more than five times more likely to attain viral suppression than their female counterparts. Similarly, those who belonged to Christian religions had lower odds [AOR: 0.05 (0.02–0.21)] of being virally suppressed, implying that those who belonged to other religions were more than 17 times more likely to attain viral suppression than their counterparts who belonged to Christian religions. Those who had at least one of the biological parents still alive were more than 10 times [AOR: 10.90 (4.14–28.64)] more likely to be virally sup-

Table 12.2 Child-related factors associated with VLS CLHIV

	Virally suppressed					
Factor	Yes		No		*χ2*	*P-value*
	Freq.	Perc (%)	Freq. (%)	Perc (%)		
Child's gender						
Female	184	62.8	95	87.2	22.196	<0.001
Male	109	37.2	14	12.8		
Child's religion						
Christian religion	211	72.0	95	87.2	10.021	0.001
Other religions	82	28.0	14	12.8		
Child's educational level						
Primary or less	234	79.9	69	63.3	11.738	<0.001
Higher than primary	59	20.1	40	36.7		
Child's parental status						
At least one of the biological parents alive	232	79.2	48	44.0	46.42	<0.001
None of the biological parents alive	61	20.8	61	56.0		
Child's age at diagnosis of HIV						
Within 2 years	219	74.7	61	56.0	13.256	<0.001
After 2 years of age	74	25.3	48	44.0		
ART regimen the child is on at the moment						
ABC/3TC/DTG	159	54.3	48	44.0	14.509	0.002
Other regimen (AZT/3TC/DTG, TLD, etc.)	134	45.7	61	56.0		
Duration of on the current regimen of the child						
Less than 24 months	184	62.8	47	43.1	12.587	<0.001
24 months or more	109	37.2	62	56.9		
Other medical conditions the child has had in the past 6 months						
None	226	77.1	102	93.6	14.305	<0.001
TB, malaria, etc.	67	22.9	7	6.4		
Child belongs to teen PSS club for teenagers living with HIV						
Yes	108	36.9	20	18.3	12.544	<0.001
No	185	63.1	89	81.7		
Child belongs or attend any differentiated service delivery model						
Yes	251	85.7	102	93.6	4.647	0.02
No	42	14.3	7	6.4		
Being on multi-month dispensing						
Yes	244	83.3	102	93.6	7.346	0.023
No	49	16.7	7	6.4		

pressed than those who had none of the biological parents still alive (Table 12.3). The children who had their HIV diagnosis done when they were under 2 years of age were more than three times [AOR: 3.63 (1.43–9.19)] more likely to be virally suppressed than those who had their HIV diagnosis done after 2 years of age. Those who had been on their ART regimen for a period of less than 24 months were 65% less likely to achieve viral [AOR: 0.35 (1.12–7.75)] than those who had been on this regimen for 24 months and above, in essence implying that those who had been on the current ART regimen for above 24 months were three times more likely to be virally suppressed than those who had been on it for less than 24 months. The children who belonged to PSS club for teenagers living with HIV were more than 21 times [AOR: 21.03 (6.00–73.69] more likely to be virally suppressed than those who didn't belong to such support groups. Those who were on multi-month dispensing were 53% [AOR: 1.53 (0.45–1.64)] more likely to be virally suppressed than those who were not on multi-month dispensing.

Table 12.3 Results of bivariate and multivariate analysis of child-related factors associated with VLS CLHIV

	Virally suppressed						
Factor	Yes	No	*Df*	COR (CI: 95%)	*P-Value*	AOR (CI: 95%)	*P-value*
	Freq (%)	Freq (%)					
Child's gender							
Female	184 (62.8)	95 (87.2)	1	0.249 (0.135–0.457)	<0.001	0.198 (0.070–0.559)	0.002
Male	109 (37.2)	14 (12.8)					
Child's religion							
Christian religion	211 (72.0)	95 (87.2)	1	0.379 (0.205–0.702)	0.002	0.058 (0.016–0.210)	<0.001
Other religions	82 (28.0)	14 (12.8)					
Child's educational level							
Primary or less	234 (79.9)	69 ((63.3)	1	2.299 (1.419–3.727)	0.001	1.426 (0.423–4.807)	0.567
Higher than primary	59 (20.1)	40 (36.7)					
Child's parental status							
At least one of the biological parents alive	232 (79.2)	48 (44.0)	1	4.833 (3.016–7.747)	<0.001	10.900 (4.148–28.644)	<0.001
None of the biological parents alive	61 (20.8)	61 (56.0)					
Child's age at diagnosis of HIV							
Within 2 years	219 (74.7)	61 (56.0)	1	2.329 (1.469–3.692)	<0.001	3.630 (1.434–9.192)	0.007
After 2 years of age	74 (25.3)	48 (44.0)					
ART regimen the child is on at the moment							
ABC/3TC/DTG	159 (54.3)	48 (44.0)	1	0.675 (0.539–0.846)	0.001	0.370 (0.141–0.976)	0.055
Other regimen (ART/3TC/DTG, TLD, etc.)	134 (45.7)	61 (56.0)					
Duration of on the current regimen of the child							
Less than 24 months	184 (62.8)	47 (43.1)	1	0.227 (1.424–3.482)	<0.001	0.351 (1.124–7.750)	0.028
24 months or more	109 (37.2)	62 (56.9)					
Other medical conditions the child has had in the past 6 months							
None	226 (77.1)	102 (93.6)	1	0.231 (0.103–0.522)	<0.001	4.224 (0.975–18.302)	0.054
TB, Malaria, etc.	67 (22.9)	7 (6.4)					
Child belongs to teen PSS club for teenagers living with HIV							
Yes	108 (36.9)	20 (18.3)	1	2.598 (1.514–4.458)	0.001	21.030 (6.001–73.695)	0.012
No	185 (63.1)	89 (81.7)					

Table 12.3 (continued)

Factor	Virally suppressed Yes Freq (%)	No Freq (%)	*Df*	COR (CI: 95%)	*P-Value*	AOR (CI: 95%)	*P-value*
Child belongs or attends any differentiated service delivery model							
Yes	251 (85.7)	102 (93.6)	1	0.410 (0.178–0.943)	0.036	0.726 (0.554–1.934)	0.425
No	42 (14.3)	7 (6.4)					
Being on multi-month dispensing							
Yes	244 (83.3)	102 (93.6)	1	0.672 (0.480–0.941)*	0.021	1.529 (0.450–1.640)	0.003
No	49 (16.7)	7 (6.4)					

Caretaker-Related Factors Associated with VLS CLHIV

The second specific objective of the study was to establish the caretaker-related factors that are associated with VLS among CLHIV whose caretakers are virally suppressed. The relevant caretaker-related characteristics that were considered in this study were: age, gender, religion, education level, relation with the CLHIV, marital status, occupation, average monthly income, disclosure of positive HIV sero status to the child, duration on ART, ART treatment regimen, other medical conditions, other treatment regimens, belonging to PSS club for teenagers living with HIV, belonging to differentiated service delivery model, alcohol and drug use. Table 12.4 shows the summary of results.

The study's demographic findings revealed that 73.4% of caretakers were aged 25+ years, predominantly female (80.8%), and mostly biological parents (62.4%). Most had living biological parents (61.9%), with 26.1% having lost their mother. Caretakers' characteristics included: 71.6% with primary/no education, 69.2% married, 67.7% peasants, and 84.6% earning <120,000/= monthly. Notably, 72.9% had lived with HIV for ≥5 years, 57.2% disclosed their status to children, and 77.9% to significant others. Additionally, 64.7% were on ART for ≥5 years, 47.3% on TDF/3TC/DTG regimen, and 77.9% on their current regimen for >24 months. Caretaker characteristics included; 52.5% caretaking for ≤2 years, 76.1% without other medical conditions, 79.6% in differentiated service delivery models (45.8% in FBIM, 59.2% on multi-month dispensing). Additionally, 48.3% had previous non-suppressed viral loads, 35.3% consumed alcohol, and 8.5% used substances (khat, bhangi, mira, etc.). To identify caretaker-related factors associated with viral load suppression (VLS) in CLHIV, Pearson's Chi-Square test ($p \leq 0.05$) was conducted, revealing significant associations between various caretaker factors and VLS, as summarized in Table 12.5.

The child-related factors that were found to be significantly associated with VLS CLHIV were: the caretaker's age, gender, biological parents being alive, formal educational level, occupation, duration with HIV, disclosure of HIV sero status to significant others, duration on ART, duration on the current ART regimen, duration of taking care of the CLHIV, other medical conditions and alcohol consumption. The other caretaker-related factors (relationship with the CLHIV, marital status, income, and disclosure of HIV sero status, etc.) were not found to be significantly associated with VLS CLHIV as they had a P-value greater than 0.05. The different caretaker-related factors that were found to be significantly associated with VLS CLHIV were subjected to logistic regression analysis to obtain the COR and the corresponding 95% CI and then those that were significant were further subjected to multivariate analysis to obtain the AOR and the corresponding 95% CI. The level of significance for all statistical analyses was set at $P \leq 0.05$. Table 12.6 shows the summary of the results. In Table 12.6, on

Table 12.4 Caretaker characteristics of study respondents

Characteristics	Frequency (*N* = 402)	Percent (%)
Caretakers' age		
Less than 25 years	107	26.6
25 years and above	295	73.4
Caretakers' gender		
Male	77	19.2
Female	325	80.8
Caretakers' relation with the CLHIV		
Biological parent	251	62.4
Other relatives	151	37.6
Biological parents alive		
Yes	249	61.9
No	153	38.1
Dead parent		
Mother	105	26.1
Father	48	11.9
Not applicable	249	61.9
Caretakers' formal education level		
Primary or no education	288	71.6
Secondary and above	114	28.4
Caretakers' marital status		
Married	278	69.2
Not married	124	30.8
Caretaker's occupation		
Peasantry	272	67.7
Commercial business or employment	130	32.3
Caretakers' average monthly income (Uganda shillings)		
Less than 120,000	340	84.6
120,000 and above	62	15.4
Caretakers' duration with HIV		
Less than 5 years	109	27.1
5 years or more	293	72.9
Caretaker disclosed HIV sero status to the child		
Yes	230	57.2
No	172	42.8
Caretaker disclosed HIV sero status to significant others		
Yes	313	77.9
No	89	22.1
Caretaker's duration on ART		
Less than 5 years	142	35.3
5 years or more	260	64.7
Caretakers' ART treatment		
TLD	190	47.3
TDF/3TC/DTG	143	35.6
AZT/3TC/DTG	69	17.2
Duration on current ART regimen		
24 months or less	89	22.1
More than 24 months	313	77.9
Duration of taking care of the CLHIV		
2 years or less	211	52.5
More than 2 years	191	47.5
Other medical conditions caretaker has		
No medical condition	306	76.1
Malaria, cough, back pain, other conditions	96	23.9
Caretaker belongs to differentiated service delivery model		
Yes	320	79.6
No	82	20.4
DSD model caretaker belongs to		
FBIM	184	45.8
FBG	49	12.2
FTDR	61	15.2
Not applicable	82	20.4
CDDP	26	6.5
On multi-month dispensing		
Yes	238	59.2
No	164	40.8
Has non-suppressed viral load result		
Yes	194	48.3
No	208	51.7
Take alcohol		
Yes	142	35.3
No	260	64.7
Smokes or uses any drugs (khat, bhangi, mira, etc.)		
Yes	34	8.5
No	368	91.5

bivariate analysis, the different caretaker-related factors that were still found to be significantly associated with VLS CLHIV as they had a P-value less than 0.05 are shown. On subjecting to multivariate analysis, however, the caretaker's occupation, caretaker's duration with HIV, caretaker's disclosure of HIV sero status to significant others, duration on current ART regimen, other medical conditions, and alcohol consumption were not found to be significantly associated with VLS CLHIV as they had a P-value greater than 0.05. According to the results of multivariate analysis, CLHIV whose caretakers were younger than 25 years of age had lower odds [AOR: 0.56

Table 12.5 Caretaker-related factors associated with VLS CLHIV

	Virally suppressed					
Factor	Yes		No		$\chi 2$	*P-Value*
	Freq.	Perc (%)	Freq. (%)	Perc (%)		
Caretakers' age						
Less than 25 years	87	29.7	20	18.3	5.234	0.014
25 years and above	206	70.3	89	81.7		
Caretakers' gender						
Male	49	16.7	28	25.7	4.123	0.031
Female	244	83.3	81	74.3		
Biological parents alive						
Yes	168	57.3	81	74.3	9.71	0.002
No	125	42.7	28	25.7		
Caretakers' formal education level						
Primary or no education	192	65.5	96	88.1	19.875	<0.001
Secondary and above	101	34.5	13	11.9		
Caretaker's occupation						
Peasantry	218	74.4	54	49.5	22.442	<0.001
Commercial business or employment	75	25.6	55	50.5		
Caretakers' duration with HIV						
Less than 5 years	95	32.4	14	12.8	15.41	<0.001
5 years or more	198	67.6	95	87.2		
Caretaker disclosed HIV sero status to significant others						
Yes	238	81.2	75	68.8	7.111	0.007
No	55	18.8	34	31.2		
Caretaker's duration on ART						
Less than 5 years	87	29.7	55	50.5	14.995	<0.001
5 years or more	206	70.3	54	49.5		
Duration on current ART regimen						
24 months or less	75	25.6	14	12.8	7.496	0.004
More than 24 months	218	74.4	95	87.2		
Duration of taking care of the CLHIV						
2 years or less	169	57.7	42	38.5	11.679	<0.001
More than 2 year	124	42.3	67	61.5		
Other medical conditions caretaker has						
No medical condition	231	78.8	75	68.8	4.399	0.048
Malaria, cough, back pain, other conditions	62	21.2	34	31.2		
Take alcohol						
Yes	88	30.0	54	49.5	13.233	<0.001
No	205	70.0	55	50.5		

(1.78–7.97)] of being virally suppressed. In essence, CLHIV whose caretakers were 25 years of age and above were close to two times more likely to attain viral suppression than their counterparts whose caretakers were younger than 25 years of age. Further, CLHIV whose caretakers were male were 63% less likely [AOR: 0.38 (0.18–0.81)] to be virally suppressed, implying that those whose caretakers were female were more than two times more likely to attain viral suppression than their counterparts whose caretakers were male. CLHIV whose caretakers had primary education level were 88% less likely [AOR: 0.12 (0.05–0.26)] of being virally suppressed, implying that CLHIV whose caretakers had secondary educational level and above were more than eight times more likely to attain viral suppression than those whose caretakers had primary level education and below (Table 12.6).

Table 12.6 Results of the bivariate and multivariate analysis of the caretaker-related factors associated with VLS CLHIV

	Virally suppressed						
Factor	Yes	No	*Df*	COR (CI: 95%)	*P-Value*	AOR (CI: 95%)	*P-Value*
	Freq.	Freq. (%)					
Caretakers' age							
Less than 25 years	87 (29.7)	20 (18.3)	1	0.879 (1.089–3.244)*	0.024	0.564 (1.779–7.967)*	0.001
25 years and above	206 (70.3)	89 (81.7)					
Caretakers' gender							
Male	49 (16.7)	28 (25.7)	1	0.581 (0.343–0.985)*	0.044	0.378 (0.176–0.814)*	0.013
Female	244 (83.3)	81 (74.3)					
Biological parents alive							
Yes	168 (57.3)	81 (74.3)	1	0.465 (0.285–0.757)*	0.002	0.434 (0.538–3.820)	0.471
No	125 (42.7)	28 (25.7)					
Caretakers' formal education level							
Primary or no education	192 (65.5)	96 (88.1)	1	0.257 (0.137–0.482)*	<0.001	0.119 (0.054–0.260)*	<0.001
Secondary and above	101 (34.5)	13 (11.9)					
Caretaker's occupation							
Peasantry	218 (74.4)	54 (49.5)	1	2.960 (1.873–4.681)*	<0.001	2.696 (1.444–5.034)	0.062
Commercial business or employment	75 (25.6)	55 (50.5)					
Caretakers' duration with HIV							
Less than 5 years	95 (32.4)	14 (12.8)	1	3.256 (1.765–6.005)*	<0.001	0.837 (1.379–7.847)	0.659
5 years or more	198 (67.6)	95 (87.2)					
Caretaker disclosed HIV sero status to significant others							
Yes	238 (81.2)	75 (68.8)	1	1.962 (1.190–3.235)*	0.008	0.625 (0.347–1.125)	0.117
No	55 (18.8)	34 (31.2)					
Caretaker's duration on ART							
Less than 5 years	87 (29.7)	55 (50.5)	1	0.415 (0.264–0.651)*	<0.001	0.263 (1.434–7.425)*	0.005
5 years or more	206 (70.3)	54 (49.5)					
Duration on current ART regimen							
24 months or less	75 (25.6)	14 (12.8)	1	2.335 (1.257–4.337)*	0.007	23.923 (8.013–71.422)	0.053
More than 24 months	218 (74.4)	95 (87.2)					
Duration of taking care of the CLHIV							
2 years or less	169 (57.7)	42 (38.5)	1	0.174 (1.386–3.410)*	0.001	0.221 (1.464–6.230)*	0.003
More than 2 year	124 (42.3)	67 (61.5)					
Other medical conditions caretaker has							
No medical condition	231 (78.8)	75 (68.8)	1	1.689 (1.032–2.765)*	0.037	1.746 (0.826–3.694)	0.145
Malaria, cough, back pain, other conditions	62 (21.2)	34 (31.2)					
Take alcohol							
Yes	88 (30.0)	54 (49.5)	1	0.437 (0.279–0.686)*	<0.001	0.183 (0.033–1.029)	0.054
No	205 (70.0)	55 (50.5)					

Still, according to multivariate analysis, CLHIV whose caretakers had been on ART for a duration of less than 5 years were 74% less likely to be virally suppressed, implying that CLHIV whose caretakers had been on ART for a duration of 5 or more years were three times more likely to attain viral suppression [AOR: 0.26 (1.43–7.43)]. Further, CLHIV whose caretakers had been with them for a period of less than 2 years were 78% less likely [AOR: 0.22 (1.46–6.23)] to achieve viral suppression than their counterparts whose caretakers had been with them for less than 2 years.

Household-Related Factors Associated with VLS CLHIV

The relevant household-related characteristics that were considered in this study were: place of residence, distance to ART facility, number of people in the household, number of CLHIV in the household, access to food, family support, stigma and discrimination, violence, belonging to support organizations as illustrated in Table 12.7.

The study found that most respondents (69.7%) resided in rural areas, with 88.1% living within 10 km of an ART facility. Household characteristics included: 74.9% having 5+ members, 65.9% with <5 children under 18, 79.6% with <3 PLHIV, and 78.6% with <3 CLHIV. Additionally, 74.9% had 3+ meals/day, 82.8% didn't go without food, and 86.8% easily accessed healthcare. However, 80.8% received no support from others, 26.4% experienced HIV-related stigma (mainly from family, 16.2%), and 34.3% suffered violence (24.1% non-physical). Notably, 68.2% did not belong to support organizations (OVC, CSO, etc.). To ascertain the household-related factors that are associated with VLS CLHIV whose caretakers are virally suppressed, the different household characteristics were first subjected to Pearson's Chi-Square test to determine those that had a significant association with VLS, with the level of significance being $p \leq 0.05$. Table 12.8 shows the summary of the results.

The child-related factors that were found to be significantly associated with VLS CLHIV were: the number of people in the household, number of

Table 12.7 Household characteristics of study respondents

Characteristics	Frequency ($N = 402$)	Percent (%)
Urban/peri-urban	122	30.3
Rural	280	69.7
Distance from home to ART facility		
10 kilometres or more	48	11.9
Less than 10 kilometres	354	88.1
No. of people in the household		
Less than 5	101	25.1
5 or more	301	74.9
No. of children under the age of 18 years in the household		
Less than 5	265	65.9
5 or more	137	34.1
No. of people in the household who have HIV		
Less than 3	320	79.6
3 or more	82	20.4
No. of CLHIV in household		
Less than 3	316	78.6
3 or more	86	21.4
Meals taken by household members		
3 or more	301	74.9
Less than 3	101	25.1
Household goes without food sometimes		
No	333	82.8
Yes	69	17.2
Finds it easy to access healthcare services for household members		
Yes	349	86.8
No	53	13.2
Caretaker or CLHIV missed ART treatment appointment in the last 6 months		
Yes	82	20.4
No	320	79.6
Receive support from other family members, friends, other people or organizations		
Yes	77	19.2
No	325	80.8
Kind of support		
Financial	42	10.4
Other support	35	8.7
Not applicable	325	80.8
Caretaker or CLHIV experienced HIV-related stigma and discrimination		
Yes	106	26.4
No	296	73.6
Source of stigma		
Family member	65	16.2
Others	41	10.2
Not applicable	296	73.6

(continued)

Table 12.7 (continued)

Characteristics	Frequency ($N = 402$)	Percent (%)
Have suffered violence in the past 6 month		
Yes	138	34.3
No	264	65.7
Form of violence		
Physical	41	10.2
Other forms of violence	97	24.1
Not applicable	264	65.7
Household belongs to support organization (OVC, CSO, etc.)		
Yes	128	31.8
No	274	68.2

children under the age of 18 years in the household, number of people in the household who have HIV, number of CLHIV in household, households going without food sometimes, caretaker or CLHIV missing ART treatment appointment in the last 6 months and receiving support from other family members, friends, other people or organizations. The other household-related factors (residence, distance, meals, ease of access to healthcare services, etc.) were not found to be significantly associated with VLS CLHIV as they had a *P*-value greater than 0.05. The different household-related factors that were found to be significantly associ-

Table 12.8 Household-related factors associated with VLS CLHIV

	Virally suppressed					
Factor	Yes		No		$\chi 2$	*P-Value*
	Freq.	Perc (%)	Freq. (%)	Perc (%)		
No. of people in the household						
Less than 5	87	29.7	14	12.8	11.989	0.001
5 or more	206	70.3	95	87.2		
No. of children under the age of 18 years in the household						
Less than 5	203	69.3	62	56.9	5.44	0.014
5 or more	90	30.7	47	43.1		
No. of people in the household who have HIV						
Less than 3	244	83.3	76	69.7	8.986	0.004
3 or more	49	16.7	33	30.3		
No. of CLHIV in household						
Less than 3	220	75.1	96	88.1	7.969	0.004
3 or more	73	24.9	13	11.9		
Household goes without food sometimes						
No	231	78.8	102	93.6	12.137	<0.001
Yes	62	21.2	7	6.4		
Caretaker or CLHIV missed ART treatment appointment in the last 6 months						
Yes	68	23.2	14	12.8	5.256	0.013
No	225	76.8	95	87.2		
Receive support from other family members, friends, other people or organizations						
Yes	70	23.9	7	6.4	15.656	<0.001
No	223	76.1	102	93.6		

ated with VLS for CLHIV were subjected to logistic regression analysis to obtain the COR and the corresponding 95% CI and then those that were significant were further subjected to multivariate analysis to obtain the AOR and the corresponding 95% CI. The level of significance for all statistical analyses was set at $P \leq 0.05$. Table 12.9 shows the summary of the results.

According to multivariate analysis still, CLHIV who hailed from households that had less than 3 CLHIV had lower odds [AOR: 0.26 (0.09–0.69)] of achieving viral suppression than their counterparts who hailed from households with 3 or more CLHIV. Those who hailed from households that do not go without food sometimes were two times [AOR: 2.03 (0.006–0.11)] more likely to achieve viral suppression than their counterparts who hailed from households that sometimes went without food. The CLHIV who hailed from households that received support (financial or otherwise) from other family members, friends, other people or organizations were 11 times [AOR: 11.07 (3.83–25.09)] more likely to achieve viral suppression than those who hailed from households that did not receive such support.

Table 12.9 Multivariate analysis, household-related factors associated with VLS CLHIV

	Virally suppressed						
Factor	Yes	No	*Df*	COR (CI: 95%)	*P-value*	AOR (CI: 95%)	*P-value*
	Freq. (%)	Freq. (%)					
No. of people in the household							
Less than 5	87 (29.7)	14 (12.8)	1	1.693 (1.245–2.302)*	0.001	1.364 (0.630–2.954)	0.430
5 or more	206 (70.3)	95 (87.2)					
No. of children under the age of 18 years in the household							
Less than 5	203 (69.3)	62 (56.9)	1	1.710 (1.087–2.690)*	0.02	2.526 (1.065–5.991)	0.135
5 or more	90 (30.7)	47 (43.1)					
No. of people in the household who have HIV							
Less than 3	244 (83.3)	76 (69.7)	1	2.162 (1.297–3.604)*	0.003	0.619 (0.305–1.256)	0.184
3 or more	49 (16.7)	33 (30.3)					
No. of CLHIV in household							
Less than 3	220 (75.1)	96 (88.1)	1	0.408 (0.216–0.772)*	0.006	0.259 (0.096–0.697)*	0.007
3 or more	73 (24.9)	13 (11.9)					
Household goes without food sometimes							
No	231 (78.8)	102 (93.6)	1	2.256 (0.113–0.578)*	0.001	2.027 (0.006–0.113)*	<0.001
Yes	62 (2.2)	7 (6.4)					
Caretaker or CLHIV missed ART treatment appointment in the last 6 months							
Yes	68 (23.2)	14 (12.8)	1	2.051 (1.100–3.825)*	0.024	0.760 (0.195–2.963)	0.693
No	225 (76.8)	95 (87.2)					
Receive support from other family members, friends, other people or organizations							
Yes	70 (23.9)	7 (6.4)	1	4.574 (2.032–10.297)*	0.001	11.069 (3.829–25.097)*	0.001
No	223 (76.1)	102 (93.6)					

Discussion

The results of our study among 402 virally suppressed caretakers of CLHIV in Luwero District provide valuable insights into the factors hindering optimal viral load suppression among these vulnerable children. Despite caretakers' viral suppression, only 72.9% of CLHIV achieved VLS, falling short of the 95% target outlined in the third 95-95-95 PEPFAR framework. This study identified key child, caretaker, and household factors contributing to this disparity, underscoring the need for targeted interventions to bridge this gap and improve treatment outcomes for CLHIV. These factors are discussed below.

Child-Related Factors Associated with Viral Load Suppression of CLHIV whose Caretakers Are Virally Suppressed

The study found out child-related factors that were significantly associated with VLS among CLHIV whose caretakers were virally suppressed, and this included gender, religion, live biological parents, HIV diagnosis, duration on ART regimen, belonging to a PSS club for teenagers living with HIV, and being on multi-month dispensing. Children living with HIV (CLHIV) showed gender disparities in viral load suppression (VLS). Males were over five times more likely to achieve VLS than females, suggesting caretakers may prioritise boys' care. This finding aligns with Teasdale et al. (2018) but contrasts Onyango et al. (2023). The discrepancy may stem from varying study settings. Nonetheless, these results emphasize the need for caretakers and healthcare workers to equally support both male and female CLHIV in achieving VLS, regardless of gender. Children living with HIV from Christian backgrounds had significantly lower odds of achieving viral load suppression (VLS), with non-Christian CLHIV being over 17 times more likely to attain VLS. This disparity may stem from certain Christian sects discouraging medication adherence, including antiretroviral therapy (ART). Contrary to Humphrey et al.'s (2019) findings, which reported higher VLS rates among Christian CLHIV, this study suggests religious affiliation impacts VLS outcomes. The conflicting results likely reflect differences in study settings. Those who had at least one of the biological parents still alive were more than 10 times more likely to be virally suppressed than those who had none of their biological parents still alive. Probably the parents took good care of their CLHIV that could have enabled these children to adhere to their ART regimen, thereby increasing their odds of attaining VLS. This finding is supported by Sitienei and Pillay (2019) whose study also reported higher odds of VLS among children who had one of their biological parents still alive. It is, however, contrasted by Teasdale et al. (2018) whose study found no significant relationship between parental status and VLS among CHLIV, and the difference with the current study finding can still be attributed to differences in study settings.

Children who had their HIV diagnosis done when they were under 2 years of age were three times more likely to be virally suppressed than those who had their HIV diagnosis done after 2 years of age. This probably results from the fact that early diagnosis of HIV in children is essential for attaining VLS as was also reported in the study by Mulinge et al. (2024) in Kenya where viral load non-suppression among CLHIV was commonest among those who were diagnosed with HIV when they were older than 2 years. The finding is contrasted by Khetan et al. (2022) who highlighted that there was no significant relationship between age at diagnosis of HIV and VLS among CLHIV. CLHIV on antiretroviral therapy for 24 months or more had significantly higher odds of achieving viral load suppression (VLS) compared to those on ART for less than 24 months. Specifically, long-term ART adherents were over twice as likely to attain VLS. This underscores the importance of sustained ART adherence, as prolonged treatment facilitates viral suppression. Research by Jiamsakul et al. (2017) supports this finding, demonstrating that VLS can be achieved within 2 years of starting and consistently adhering to ART. Extended ART duration allows for better viral management, reduced transmission risk, restored immune function, and improved overall health outcomes. Long-term ART adherence also enhances the effectiveness of treatment, reducing the development of drug-resistant viral strains and minimizing treatment failure.

The children who belonged to a PSS club for teenagers living with HIV were 21 times more likely to be virally suppressed than those who didn't belong to such support groups. Probably, belonging to such groups facilitated peer support from other CLHIV, which probably promoted adherence to ART, hence increasing the odds of such children attaining VLS. This finding implies that it is essential for children (and others) living with HIV to belong to social support groups that promote social support and enhance adherence to ART, thereby promoting VLS as was also reported by Khetan et al. (2022). Children living with HIV receiving multi-month dispensing (MMD) of ART were nearly twice as likely to achieve viral load suppression compared to those without MMD. This significant association is attributed to MMD's patient-centred benefits, including reduced clinic visits, lower travel costs, and minimized time off work or school. By alleviating logistical barriers, MMD enhances treatment adherence, ultimately increasing the likelihood of VLS. This finding aligns with Onyango et al.'s (2023) research, highlighting MMD's positive impact on ART adherence and viral suppression. By streamlining treatment access, MMD supports VLS, which improves health outcomes, and supports long-term viral management, making it a valuable strategy for optimizing CLHIV care.

Caretaker-Related Factors Associated with Viral Load Suppression of CLHIV Whose Caretakers Are Virally Suppressed

The study revealed significant associations between several caretaker factors and viral load suppression among CLHIV whose caretakers are virally suppressed, specifically: caretaker age, gender, occupation, ART duration, and duration of caregiving. CLHIV with caretakers aged 25 or older were nearly twice as likely to achieve viral load suppression compared to those with caretakers under 25. This suggests older caretakers may be more responsible and provide better care, enhancing VLS odds. This finding aligns with Van Wyk and Davids' (2019) study, which reported higher VLS rates among CLHIV with older caretakers. However, it contrasts Tsikhutsu et al.'s (2022) findings, which showed no significant link between caregiver age and VLS. The discrepancy may stem from differences in study settings, highlighting the importance of context in understanding caregiver age's impact on CLHIV outcomes. Children living with HIV with male caretakers had significantly lower viral load suppression rates. In contrast, CLHIV with female caretakers were over twice as likely to achieve VLS. This disparity suggests female caretakers may provide more effective care, contributing to better health outcomes. Research by Williams et al. (2018) supports this finding, highlighting female caregivers' positive impact on CLHIV's viral suppression. This gender dynamic may stem from societal caregiving norms, where women often assume primary caregiving roles, fostering stronger caregiver-child relationships and adherence to treatment.

CLHIV with caretakers having secondary education or higher were over eight times more likely to achieve VLS compared to those with caretakers having primary education or less. This significant association suggests higher education enhances caretakers' capacity to provide effective care, treatment literacy, and adherence to healthcare instructions. Better-educated caretakers can manage complex treatment regimens, recognize symptoms, and navigate healthcare systems. Van Wyk and Davids (2019) support this finding, reporting higher VLS rates among CLHIV with better-educated caregivers. However, Onyango et al.'s (2023) study contrasts this, highlighting potential variations in study settings or populations.

The CLHIV whose caretakers had been on ART for 5+ years were nearly four times more likely to achieve viral load suppression (VLS) compared to those with caretakers on ART for less than 5 years. Prolonged ART exposure likely enhances caretakers' confidence in its effectiveness, promoting adherence and proper ART administration among CLHIV. Probably being on ART for this long (5 years or more) gave them the confidence that ART is effective in controlling the challenges related to HIV, which could have caused such caregivers to ensure proper ART uptake among CLHIV, thereby increasing the odds of VLS among these kids. Ashburn et al.'s (2021) study supports this finding, highlighting caretaker ART duration as a key factor in CLHIV's VLS. Children living with HIV with consistent caretakers for over 2 years were four times more likely to achieve VLS compared to those with caretakers for less than 2 years. This prolonged caretaking fosters familiarity with treatment schedules, dosages, and frequencies, deeper understanding of child behaviour, needs, and health patterns, effective care coordination, adherence support, and medication management. Moreover, it cultivates stronger emotional bonds through daily routines, enhancing trust and communication. Exavery et al.'s (2020) study supports this finding, highlighting long-term caretaking's positive impact on CLHIV's VLS. Stable caretaker relationships ensure continuity of care, minimize treatment disruptions, promote caregiver-child trust, facilitate open communication, enable caregivers to recognize and address treatment challenges, and mitigate caregiver turnover, reducing stress and treatment instability.

Household-Related Factors Associated with Viral Load Suppression of CLHIV Whose Caretakers Are Virally Suppressed

Viral load suppression of CLHIV whose caretakers are virally suppressed was significantly associated with three household factors: number of CLHIV in household, households going without food sometimes, and receiving support from other family members, friends, other people or organizations. Household factors play a crucial role in VLS outcomes, highlighting the importance of addressing household-level challenges to optimize treatment success. The study revealed that CLHIV from households with three or more CLHIV were nearly four times more likely to achieve VLS compared to those from households with fewer than three CLHIV. This association suggests that peer support within households with multiple CLHIV facilitates adherence to ART, ultimately enhancing VLS outcomes. This finding aligns with Maena et al.'s (2021) research, which highlighted the importance of social and moral support among CLHIV living together, reducing stigma, promoting treatment reminders, and increasing caregiver responsibility. The presence of multiple CLHIV in a household appears to foster a supportive environment, encouraging adherence and contributing to improved VLS rates.

The study revealed that CLHIV from households that don't go without food sometimes were two times more likely to achieve VLS compared to their counterparts from households that sometimes went without food. This association suggests that food security promotes adherence to ART, leading to improved VLS outcomes. The consistent availability of food supplies likely enhances overall health and well-being, supporting the effectiveness of ART to fight the virus. This finding is consistent with Van Wyk and Davids' (2019) research, which emphasized the critical role of food security and access to food in achieving VLS among CLHIV. CLHIV from households that received support (financial or otherwise) from other family members, friends, other people, or organizations were 11 times more likely to achieve VLS than those from households without such support. This external support likely enhanced household ability to ensure continuous access to and uptake of ART, fostering adherence and ultimately leading to viral suppression. This finding aligns with Exavery et al.'s (2020) research, highlighting external support's role in promoting ART uptake among children and increasing the likelihood of CLHIV achieving viral suppression.

Conclusion

This study investigated factors associated with viral load suppression (VLS) among children living with HIV (CLHIV) in Luwero District, Uganda, whose caretakers had achieved viral suppression. The findings revealed a suboptimal VLS rate of 72.9% among CLHIV, falling short of the 95% target outlined in the third 95-95-95 PEPFAR framework. Child-related factors, including gender, religion, presence of live biological parents, early HIV diagnosis, duration on ART, participation in support groups, and multi-month dispensing, significantly influenced VLS outcomes. Additionally, caretaker factors such as age, gender, education level, ART duration, and consistent caregiving played crucial roles in VLS achievement. Household factors, including number of CLHIV, food security, and external support, also contributed to VLS outcomes.

The study's findings underscore the need for targeted interventions addressing these factors to improve VLS rates among CLHIV. Specifically, promoting equal support for male and female CLHIV, addressing religious affiliations that may hinder ART adherence, and ensuring consistent caregiving and access to support groups are essential. Moreover, enhancing food security, providing external support, and streamlining ART access through multi-month dispensing can significantly improve VLS outcomes. By addressing these factors, healthcare providers and policymakers can optimize treatment outcomes, reduce HIV-related morbidity, and promote resilient health for CLHIV. The study's results inform evidence-based strategies to bridge the gap between current VLS rates (72.9%) and the 95% target, ultimately contributing to improved health and well-being for CLHIV in Luwero and other similar settings in Uganda.

Recommendation

To address the suboptimal viral load suppression rate of 72.9% among children living with HIV in Luwero District, Uganda, targeted interventions are crucial. Caretakers play a vital role in ensuring consistent caregiving, equal support, and facilitating access to support groups and multi-month ART dispensing. Health workers should provide regular adherence counselling, monitor viral load outcomes, and offer support for children with unsuppressed viral loads.

A collaborative approach is necessary to improve viral load suppression outcomes. This includes strengthening community-based support services, enhancing food security initiatives, and increasing access to multi-month dispensing and ART services. Policymakers should develop policies promoting food security, economic empowerment, and HIV/AIDS education, while reviewing policies addressing religious affiliations hindering ART adherence. Effective collaboration between healthcare providers, community health workers, and caretakers is crucial, along with continuous training and guideline updates for healthcare workers. Community leaders should be engaged in HIV/AIDS sensitization efforts, and policymakers should allocate sufficient funding for community-based support services and ART access.

Acknowledgement The authors would like to thank all the participants of this study, Luwero District Health Team and the Luwero ACORD Uganda team for their support, which led to the successful completion of this study. Special thanks to Mr. Lwenge Mathias, Luba Henry and Nurse Nalutaali Rachel Lydia for their support and guidance.

References

Ashburn, K., Chouraya, C., Khumalo, P., Mpango, L., Mthethwa, N., Machekano, R., Guay, L., & Mofenson, L. M. (2021). A randomized controlled trial evaluating the effects of a family-centered HIV care model on viral suppression and retention in care of HIV-positive children in Eswatini. *PLoS One, 16*(8), e0256256.

Bulage, L., Ssewanyana, I., Nankabirwa, V., Nsubuga, F., Kihembo, C., Pande, G., Ario, A. R., Matovu, J. K., Wanyenze, R. K., & Kiyaga, C. (2017). Factors associated with virological non-suppression among HIV-positive patients on antiretroviral therapy in Uganda, August 2014–July 2015. *BMC Infectious Diseases, 17*(1), 326.

Exavery, A., Charles, J., Barankena, A., Kuhlik, E., Mubyazi, G. M., Tani, K., Ally, A., Minja, E., Koler, A., Kikoyo, L., & Jere, E. (2020). ART use and associated factors among HIV positive caregivers of orphans and vulnerable children in Tanzania. *BMC Public Health, 20*(1), 1251.

Ferrand, R. A., Briggs, D., Ferguson, J., Penazzato, M., Armstrong, A., MacPherson, P., Ross, D. A., & Kranzer, K. (2016). Viral suppression in adolescents on antiretroviral treatment: Review of the literature and critical appraisal of methodological challenges. *Tropical Medicine & International Health, 21*(3), 325–333.

Gordon, T. P., Talbert, M., Mugisha, M. K., & Herbert, A. E. (2022). Factors associated with HIV viral suppression among adolescents in Kabale district, South Western Uganda. *PLoS One, 17*(8), e0270855.

Hader, S. (2019). *Impact of HIV on children and youth [Internet].* https://www.unaids.org/sites/default/files/media_asset/12122019_UNAIDS_PCB45_PPT_DXD-PRG.pdf

Humphrey, J. M., Genberg, B. L., Keter, A., Musick, B., Apondi, E., Gardner, A., Hogan, J. W., & Wools-Kaloustian, K. (2019). Viral suppression among children and their caregivers living with HIV in western Kenya. *Journal of the International AIDS Society, 22*, 4.

Jiamsakul, A., Kariminia, A., Althoff, K. N., Cesar, C., Cortes, C. P., Davies, M.-A., Do, V. C., Eley, B., Gill, J., Kumarasamy, N., Machado, D. M., Moore, R., Prozesky, H., Zaniewski, E., & Law, M. (2017). HIV viral load suppression in adults and children receiving antiretroviral therapy—Results from the IeDEA collaboration. *JAIDS Journal of Acquired Immune Deficiency Syndromes, 76*(3), 319–329.

Khetan, P., Liu, Y., Dhummakupt, A., & Persaud, D. (2022). Advances in pediatric HIV-1 cure therapies and reservoir assays. *Viruses, 14*(12), 2608.

Kish, L. (1965). *Survey sampling*. John Wiley & Sons. ISBN 0-471-10949-5.

Maena, J., Banke-Thomas, A., Mukiza, N., Kuteesa, C. N., Kakumba, R. M., Kataike, H., Kizito, S., Babirye, J. A., & Nakalega, R. (2021). Determinants of viral load non-suppression among adolescents in Mbale District, Eastern Rural Uganda. *AIDS Research and Therapy, 18*(1), 91.

Ministry of Health. (2022). *Ministry of Health releases the preliminary results of Uphia 2020.* https://www.health.go.ug/2022/02/10/ministry-of-health-releases-the-preliminary-results-of-uphia-2020/

MOH. (2022). *Consolidated guidelines for the prevention and treatment Of HIV and AIDS in Uganda.*

Mulinge, M. M., Kibui, N. K., Kimani, H., Wainaina, J., Bwana, P., Omondi, M., Wafula, K., Wamalwa, D. C., Omondi, E. O., Nduati, R. W., & Mwau, M. (2024). Factors associated with viral load non-suppression among treatment-experienced pre-teenage children living with HIV in Kenya: A nationwide population-based cohort study, 2015–2021. *EClinicalMedicine, 68*, 102454.

Nabukeera, S., Kagaayi, J., Makumbi, F. E., Mugerwa, H., & Matovu, J. K. B. (2021). Factors associated with virological non-suppression among HIV-positive children receiving antiretroviral therapy at the joint clinical research Centre in Lubowa, Kampala Uganda. *PLoS One, 16*(1), e0246140.

Nasuuna, E., Kigozi, J., Muwanguzi, P. A., Babirye, J., Kiwala, L., Muganzi, A., Sewankambo, N., & Nakanjako, D. (2019). Challenges faced by caregivers of virally non-suppressed children on the intensive adherence counselling program in Uganda: A qualitative study. *BMC Health Services Research, 19*(1), 150.

Onyango, B., Mokaya, R., Wasianga, J., Wao, H., Achwoka, D., Onyango, N., & Kadengye, D. T. (2023). Factors associated with viral load suppression among orphans and vulnerable children and adolescents living with HIV in Kenya. *PLOS Global Public Health, 3*(3), e0000794.

Oryokot, B., Kazibwe, A., Kagimu, D., Oluka, A. I., Kato, D., Miya, Y., Etukoit, M. B., & Namusoke-Magongo, E. (2023). Improving retention and HIV viral load suppression among adolescents living with HIV in TASO Soroti and TASO Mbale centers of excellence using operation triple zero model: A before and after study protocol. *Implementation Science Communications, 4*(1), 65.

Sitienei, E. C., & Pillay, J. (2019). Psycho-educational and social interventions provided for orphans and vulnerable children at a community-based organisation in Soweto, South Africa. *African Journal of AIDS Research, 18*(1), 1–8.

Ssewamala, F. M., Dvalishvili, D., Mellins, C. A., Geng, E. H., Makumbi, F., Neilands, T. B., McKay, M., Damulira, C., Nabunya, P., Sensoy Bahar, O., Nakigozi, G., Kigozi, G., Byansi, W., Mukasa, M., & Namuwonge, F. (2020). The long-term effects of a family based economic empowerment intervention (Suubi+Adherence) on suppression of HIV viral loads among adolescents living with HIV in southern Uganda: Findings from 5-year cluster randomized trial. *PLoS One, 15*(2), e0228370.

Teasdale, C. A., Sogaula, N., Yuengling, K. A., Wang, C., Mutiti, A., Arpadi, S., Nxele, M., Pepeta, L., Mogashoa, M., Rivadeneira, E. D., & Abrams, E. J. (2018). HIV viral suppression and longevity among a cohort of children initiating antiretroviral therapy in Eastern Cape, South Africa. *Journal of the International AIDS Society, 21*, 8.

Tsikhutsu, I., Bii, M., Dear, N., Ganesan, K., Kasembeli, A., Sing'oei, V., Rombosia, K., Ochieng, C., Desai, P., Wolfman, V., Coakley, P., Lee, E. H., Hickey, P. W., Livezey, J., & Agaba, P. (2022). Prevalence and correlates of viral load suppression and human immunodeficiency virus (HIV) drug resistance among children and adolescents in South Rift Valley and Kisumu, Kenya. *Clinical Infectious Diseases, 75*(6), 936–944.

UBOS, & ICF. (2016). *Uganda demographic and health survey.*

Uganda Bureau of Statistics [UBOS] and ICF. (2017). *2016 Uganda demographic and health survey key findings*. UBOS and ICF.

Uganda HIV and AIDS Facts Sheet. (2021). *Facts on HIV and AIDS in Uganda 2021 (Based on Data ending 31st December 2020).*

Uganda HIV and AIDS Facts Sheet. (2022). *Facts on HIV and AIDS in Uganda 2022 (Based on Data ending 31st December 2021).*

UNAIDS. (2022). *Report on the Global AIDS Epidemic Available from:* http://www.unaids.org/documents/20101123_globalreport_em.pdf

UNAIDS. (2024). *UNAIDS 2024 epidemiological estimates. FACT SHEET-Global HIV statistics.*

UNICEF. (2021). *UNICEF Data: Monitoring the situation of children and women.* Global and regional trends [Internet]. https://data.unicef.org/topic/hivaids/global-regional-trends/

UPHIA. (2017). *Uganda population-based HIV impact assessment 2016–2017. Summary sheet: Preliminary findings.* https://www.afro.who.int/sites/default/files/2017-08/UPHIA%20Uganda%20factsheet.pdf

Van Wyk, B. E., & Davids, L.-A. C. (2019). Challenges to HIV treatment adherence amongst adolescents in a low socio-economic setting in Cape Town. *Southern African Journal of HIV Medicine, 20*, 1.

Wang, X., & Cheng, Z. (2020). Cross-sectional studies. *Chest, 158*(1), S65–S71.

Williams, M., Van Rooyen, D. R. M., & Ricks, E. J. (2018). Provision of antiretroviral therapy for children in Nelson Mandela Bay: Health care professionals' challenges. *African Journal of Primary Health Care & Family Medicine, 10*(1).

13 Awareness, Knowledge, and Utilisation of Pre-Exposure and Post-Exposure Prophylaxis Against HIV Among Young Women in Mbarara City

Samuel Ssanyu Balamaga and Kizito Omona

Abstract

With an HIV prevalence of 8.1%, Mbarara is one of the areas struggling with high HIV infection rates in Uganda. The study was aimed at determining the level of awareness, knowledge, and utilisation of pre-exposure and post-exposure prophylaxis against HIV among adolescent girls and young women in Mbarara City. It was a community-based survey conducted among 150 randomly selected adolescent girls and young women aged 15–25 years in Mbarara City. The majority of the study respondents, 103 (68.7%), had their sex debut before the age of 18 years, and most, 59 (39.3%), have had more than 4 sexual partners in life. Slightly more than half, 78 (52.0%), had 1–2 sexual partners at the time of the study. Only 49 (32.7%) and only 64 (42.7%) had heard about pre-exposure prophylaxis (PrEP) and post-exposure prophylaxis (PEP), respectively. Most adolescent girls and young women in Mbarara City are sexually active but with limited knowledge about PrEP and PEP, which predisposes them to a high risk of HIV.

S. S. Balamaga (✉) · K. Omona
Faculty of Health Sciences, Uganda Martyrs University, Kampala, Uganda
e-mail: balamaga.samuel@stud.umu.ac.ug

Keywords

HIV infection · Adolescent girls · Abstinence · Slums · Epidemic · Public health

Introduction

As of 2023, 42.3 million people had died of human immunodeficiency virus (HIV) with the greatest burden being experienced in low- and middle-income countries (UNAIDS/WHO, 2024). Africa accounts for more than two-thirds of the persons living with HIV (PLHIV) globally (UNAIDS, 2024). In Uganda, AIDS cases were first recognised in the early 1980s, and several interventions have been put in place to prevent the escalation of this epidemic (Slutkin et al., 2006; Wang et al., 2016). Despite the interventions, Uganda still faces a high burden of HIV cases (Ijaiya et al., 2023). Unfortunately, adolescent girls and young women (AGYW) are at a higher risk of HIV infection (30% of all new HIV infections occur in this category of people) (Bajunirwe et al., 2020). There is therefore a need to prioritize efforts for combating HIV infection among AGYW. Pre-exposure prophylaxis (PrEP) and post-exposure prophylaxis (PEP) are among the key measures for preventing sexual transmission of HIV among the high HIV risk populations and communities (Tumwebaze et al., 2023). This particularly because other HIV-prevention measures

P. Nyadoi et al. (eds.), *Sustainable Development Goals in Africa*, Sustainable Development Goals Series, https://doi.org/10.1007/978-3-032-20742-5_13

such as abstinence from sexual intercourse, being faithful to one sexual partner, and condom use have become ineffective in preventing HIV transmission especially among the young people. For example, Reis et al. (2019) highlighted that inconsistent condom use is one of the predisposing factors to new HIV infections among young people. On its part, PrEP reduces the risk of HIV transmission through sexual intercourse by about 99% (Desai et al., 2017), while an equal proportion (99%) has been reported as the benefit of reducing the risk of HIV seroconversion after events with high risk of exposure to HIV (such as unprotected anal or vaginal sex, needlestick injuries, or sharing needles).

The Ugandan government through the Ministry of Health (MoH) promotes the use of PrEP (Witte et al., 2022) and PEP (Alitubeera et al., 2021) among the measures for promoting the prevention of HIV among persons at substantial risk of HIV infection (like AGYW), with guidelines in place about the same. However, these (PrEP and PEP) are relatively new measures, and some studies elsewhere have pointed to poor knowledge and awareness about these measures. For example, a recent study by Lunkuse et al. (2022) among AGYW in Kampala City, Uganda, indicated that only 23.2% of the study respondents were aware of oral PrEP and only 3.9% of them had ever heard about injectable PrEP. It is also worth noting that information is scanty regarding awareness and knowledge about non-occupational PEP, with much of available information on PEP relating to healthcare practitioners and students of healthcare training institutions. This study examined the level of awareness, knowledge, and utilisation of pre-exposure and post-exposure prophylaxis against human immunodeficiency virus (HIV) among adolescent girls and young women in Mbarara City.

Methodology

The study was carried out in Mbarara City, Southwestern Uganda, which is one of the recently created cities in Uganda. Mbarara is one of the fastest-growing cities, and is located about 230 kilometres southwest of Kampala. The study targeted AGYW aged 15–25 years residing in the different slum areas who willingly and freely consented to participate in the study. The sample size was determined using Kish formula (1965) for a cross-sectional study. A consideration of 8.5% was made to cater for nonresponse, giving a sample size of 150.

Study participants were selected from six slum areas of Katete, Nyamitanga, Kakoba, Kamukuzi, Kijungu, and Kakiika using simple random sampling. Local Council (LC) 1 records of each slum area were used to generate a list of AGYW which was then used as the sampling frame. The names of the AGYW on the sampling frame were written on small pieces of paper of the same shape, colour, size, and texture. The pieces of paper were placed in a small box and shaken. The LC Chairperson was requested to pick a piece of paper from the box without returning. The box was shaken again, and a second piece of paper picked. This process was repeated until the required sample size was obtained.

Data was collected using self-administered and researcher-administered questionnaires. Self-administered questionnaires were completed by respondents who were able to read and understand questions and write answers in English. The researcher-administered questionnaires involved respondents who had limited reading or writing abilities in English. Quality control measures entailed validity, reliability, and training of research assistants. The data were entered into SPSS, version 26 for Windows, and descriptive statistics were generated.

Results

The demographic characteristics of respondents involved in this study are presented in Table 13.1.

Out of 150 respondents, 138 (92.0%) were aged 18–24 years, 51 (34.0%) were of primary level of education, 51 (34.0%) were sex workers, 31 (20.7%) were students and 50 (33.3%) were earning less than UGX 100,000 on average per month. In essence, the majority of the AGYW in Mbarara belong to a low socioeconomic class, which predisposes them to high risk of infection with HIV.

As can be observed in Table 13.2, 60 (40.0%) respondents started engaging in sexual intercourse at the age of 15–17 years, with the majority of them, 103 (68.7%), having their sex debut before 18 years of age, and a sizeable portion of them, 5 (3.3%), had their sex debut at less than 10 years of age. Most of them, 59 (39.3%), have had more than 4 sexual partners from the time they started engaging in sexual relations, and more than half of them, 78 (52.0%), had 1–2 sexual partners at the moment, while a good portion of them, 28 (18.7%), had more than four sexual partners. Most of them, 35 (23.3%), reported that they were having sexual intercourse several times daily or at least once weekly, and the latest sexual encounter was the previous day, 41 (27.3%). Unfortunately, less than a half, 73 (48.7%), used protection (condom) during their latest sexual encounter. These findings denote that majority of the AGYW in Mbarara are highly sexually active and are at increased risk of infection with HIV, yet a big proportion of them indulge in unsafe sexual intercourse.

Table 13.1 Demographic characteristics of respondents

Characteristic	Frequency (n = 150)	Percent (%)
Age (complete years)		
Less than 18	12	8.0
18–24	138	92.0
Highest education level		
None	18	12.0
Primary	51	34.0
O level	44	29.3
A level	14	9.3
Certificate	11	7.3
Diploma	3	2.0
Bachelor's degree and above	9	6.0
Source of income		
None (student)	31	20.7
None (others)	16	10.7
Professional	22	14.7
Peasant	21	14.0
Sex worker	51	34.0
Other	9	6.0
Average monthly income (UGX)		
None (student)	31	20.7
None (others)	18	12.0
Less than 100,000	50	33.3
100,000–200,000	32	21.3
More than 200,000	19	12.7

Results in Table 13.3 show that 150 (100.0%) respondents had heard about HIV and were aware of the means of protecting themselves from being infected. Using condoms consistently and correctly was the most mentioned means of protecting self from infection with HIV, 93 (62.0%). Using PEP was mentioned by only 6 (4.0%) of the respondents, while none of them made mention of PrEP, an indicator of their poor knowledge about PrEP and PEP. Further, only 49 (32.7%) of them had ever heard about or could correctly define/describe PrEP, and only 64 (42.7%) of them had ever heard about PEP, but only 30 (20.0%) were in position to give the correct definition/description of PEP. Only a small proportion of them were aware of where to find PrEP or PEP services. These findings further point to limited knowledge about PrEP and PEP among the AGYW in Mbarara.

The results in Table 13.4 point to poor utilisation of PrEP or PEP by AGYW to prevent infection with HIV, and this can be related to poor knowledge about these methods of prevention of HIV infection. As can be observed, only 1 (0.7%) reported having used PrEP and only 4 (2.7%) reported having used PEP (Table 13.4).

Table 13.2 Sexual behaviour of respondents

Characteristic	Frequency (*n* = 150)	Percent (%)
Age at started engaging in sexual intercourse (sex debut)		
Less than 10 years	5	3.3
10–14 years	38	25.3
15–17 years	60	40.0
18 years and above	36	24.0
Not applicable (I have never had sex)	11	7.3
Number of sexual partners since the start of engaging in sex		
1–2	47	31.3
3–4	33	22.0
More than 4	59	39.3
None (have never played sex)	11	7.3
Number of sexual partners at the moment		
None	17	11.3
1–2	78	52.0
3–4	27	18.0
More than 4	28	18.7
Frequency of engagement in sex		
Once a day	11	7.3
Several times daily	35	23.3
Once weekly	35	23.3
Several times weekly	19	12.7
Once monthly	23	15.3
Several times monthly	10	6.7
Not applicable	17	11.3
Latest sexual encounter		
Today	24	16.0
Yesterday	41	27.3
A few days back	27	18.0
A week ago	19	12.7
More than a week ago	8	5.3
More than a month ago	9	6.0
More than several months ago	5	3.3
Not applicable (not having sex)	17	11.3
Used protection (condom) during latest sexual encounter		
Yes	73	48.7
No	66	44.0
Not applicable (have never had sex)	11	7.3

Table 13.3 Respondents' awareness, knowledge, and attitude towards HIV/PrEP/PEP

Characteristic	Frequency (*n* = 150)	Percent (%)
Ever heard about HIV		
Yes	150	100.0
Gives correct definition/explanation of HIV		
Yes	139	92.7
No	11	7.3
Considers self at risk of HIV infection		
Yes	89	59.3
No	61	40.7
Have heard about PrEP		
Yes	49	32.7
No	101	67.3
Correctly defines/describes PrEP		
Yes	49	32.7
No	0	–
Not applicable (have never heard about PEP)	101	67.3
Places where PrEP services can be obtained from		
Government health centres	38	25.3
Private health facilities	6	4.1
Pharmacies	5	3.3
Not applicable/not aware	101	67.3
Have ever heard about PEP		
Yes	64	42.7
No	86	57.3
Correctly defines/describes PEP		
Yes	30	20.0
No	34	22.7
Not applicable (has never heard about PEP)	86	57.3
Know where PEP can be got		
Government health centre	43	28.6
Private health facility	13	8.7
Pharmacies/drug shops	4	2.7
Other places	4	2.7
Not applicable/not aware	86	57.3
Aware of means of protecting self from HIV infection		
Yes	150	100.0
No	0	–
Known ways of protecting self from HIV infection		
Abstinence from sexual intercourse	29	19.3
Being faithful to one partner	22	14.7
Using condoms consistently and correctly	93	62.0
Using post-exposure prophylaxis	6	4.0

Table 13.4 Respondents' practices about HIV prevention and PrEP/PEP

Characteristic	Frequency ($n = 150$)	Percent (%)
Methods currently being used to protect self from HIV infection		
Abstinence	19	12.7
Being faithful to one regular sexual partner	17	11.3
Using condoms consistently and regularly	97	64.7
Using pre-exposure prophylaxis	1	0.7
Using post-exposure prophylaxis	4	2.7
None	12	8.0
Would use PrEP for protection against HIV infections		
Yes	85	56.7
No	31	20.7
Not sure	34	22.7
Would use PEP for the protection against HIV if given chance		
Yes	118	78.7
No	26	17.3
Not sure	6	4.0

Discussion

Further, only 49 (32.7%) of them had ever heard about or could correctly define/describe PrEP, and only 64 (42.7%) of them had ever heard about PEP, but only 30 (20.0%) were in position to give the correct definition/description of PEP. These findings point to a low level of awareness and knowledge about both PEP PrEP, which can probably be attributed to these two HIV prevention measures being relatively new (not among the original ABC strategy—Abstinence, Being faithful and Condom use), which was used for decades to control the HIV epidemic in Uganda. This is probably why only 6 (4.0%) of the respondents mentioned while none of them made mention of PrEPas some of the measures for preventing HIV infection (Table 13.3), a further indicator of their poor knowledge about PrEP and PEP. These findings are consistent with other research findings which have pointed to poor knowledge of community members about PrEP (Stankevitz et al., 2019; Lunkuse et al., 2022) and PEP (Ayieko et al., 2021). These findings imply that the majority of the young people (AGYW) cannot make use of these services (PrEP and PEP) yet they are very effective at preventing the transmission of HIV. With AGYW being among those who are disproportionately affected by HIV, it is essential that health officials, leaders, and those involved sensitize AGYW about the availability of PrEP and PEP to increase their utilisation of measures for combating HIV among these people.

The above findings, however, contradict Shamu et al. (2021) whose study among young people in two South African districts indicated a relatively high level of awareness (44.6%) about PrEP. Furthermore, the findings contradict Makhado et al. (2022) whose study among healthcare practitioners and students revealed a high level of awareness and knowledge about PEP. The study found that only 1 (0.7%) reported using PrEP and only 4 (2.7%) reported using PEP, yet a majority of the study respondents were highly sexually active. These findings are in line with previous research studies which have also highlighted poor utilisation of PrEP (Lunkuse et al., 2022) and PEP (Ayieko et al., 2021). These findings, in essence, imply the need for healthcare professionals and those involved in the prevention of HIV to strategize and advance measures to support sexually-active AGYW to benefit from PrEP and PEP since these have been proved as highly effective measures for preventing the transmission of HIV.

Conclusion

The findings show limited utilisation and knowledge of PrEP and PEP among adolescent girls and young women in Mbarara City. There is need for government and partners in the area to improve awareness and utilisation of pre- and post-exposure prophylaxis by adolescent girls and young women. The Mbarara City leadership, especially the health department and the HIV Focal Person, could use low-cost means such as social media and community gatherings to reach out to AGYW in the community.

References

Alitubeera, P. H., Mutanda, J. N., Aggrey, M., Kobusingye, O. C., Biribawa, C., Tusubira, A., Eyu, P., & Kiwanuka, N. (2021). Prevalence, correlates of occupational percutaneous injuries and use of post exposure prophylaxis against HIV, hepatitis B among health workers in Kampala, Uganda-may 2016. *Journal of Interventional Epidemiology and Public Health, 4*, 13.

Ayieko, J., Petersen, M. L., Kabami, J., Mwangwa, F., Opel, F., Nyabuti, M., Charlebois, E. D., Peng, J., Koss, C. A., Balzer, L. B., Chamie, G., Bukusi, E. A., Kamya, M. R., & Havlir, D. V. (2021). Uptake and outcomes of a novel community-based HIV post-exposure prophylaxis (PEP) programme in rural Kenya and Uganda. *Journal of the International AIDS Society, 24*(6), e25670. https://doi.org/10.1002/jia2.25670

Bajunirwe, F., Semakula, D., & Izudi, J. (2020). Risk of HIV infection among adolescent girls and young women in age-disparate relationships in sub-Saharan Africa. *AIDS, 34*(10), 1539–1548. https://doi.org/10.1097/QAD.0000000000002582

Desai, M., Field, N., Grant, R., & McCormack, S. (2017). State of the art review: Recent advances in PrEP for HIV. *BMJ (Clinical Research Ed.)., 359*, j5011. https://doi.org/10.1136/bmj.j5011

Ijaiya, M. A., Olowu, A., Oguntade, H. A., Anjorin, S., & Uthman, O. A. (2023). HIV research output in African countries between 1986–2020. *PLOS Global Public Health, 3*(6), e0000544. https://doi.org/10.1371/journal.pgph.0000544

Lunkuse, J. F., Kamacooko, O., Muturi-Kioi, V., Chinyenze, K., Kuteesa, M. O., Price, M. A., & Mayanja, Y. (2022). Low awareness of oral and injectable PrEP among high-risk adolescent girls and young women in Kampala, Uganda. *BMC Infectious Diseases, 22*, 467. https://doi.org/10.1186/s12879-022-07398-z

Makhado, L., Musekwa, O. P., Makhado, T. G., & Otsheleng, R. (2022). Healthcare practitioners and students' PEP knowledge, attitude and adherence in Southern Africa. *Health SA, 27*, 2036. https://doi.org/10.4102/hsag.v27i0.2036

Reis, R. K., Melo, E. S., Fernandes, N. M., Antonini, M., Neves, L. A. D. S., & Gir, E. (2019). Inconsistent condom use between serodifferent sexual partnerships to the human immunodeficiency virus. *Revista latino-americana de enfermagem, 27*, 1–13.

Shamu, S., Shamu, P., Khupakonke, S., Farirai, T., Chidarikire, T., Guloba, G., & Nkhwashu, N. (2021). Pre-exposure prophylaxis (PrEP) awareness, attitudes and uptake willingness among young people: Gender differences and associated factors in two South African districts. *Global Health Action, 14*(1), 1886455. https://doi.org/10.1080/16549716.2021.1886455

Slutkin, G., Okware, S., Naamara, W., Sutherland, D., Flanagan, D., Carael, M., Blas, E., Delay, P., & Tarantola, D. (2006). How Uganda reversed its HIV epidemic. *AIDS and Behavior, 10*(4), 351–360. https://doi.org/10.1007/s10461-006-9118-2. PMID: 16858635; PMCID: PMC1544374.

Stankevitz, K., Schwartz, K., Hoke, T., Li, Y., Lanham, M., & Mahaka, I. (2019). Reaching at-risk women for PrEP delivery: What can we learn from clinical trials in sub-Saharan Africa? *PLoS One, 14*(6), e0218556.

Tumwebaze, M., Rubaihayo, J., & Harold, M. (2023). Appraisal of existing HIV/AIDS prevention and control measures and presentation of innovative strategies to end HIV/AIDs epidemic by 2030. *Open Journal of Epidemiology, 13*(3), 178–194.

UNAIDS. (2024). FACT SHEET—Global HIV statistics.

UNAIDS/WHO. (2024). *HIV statistics, globally and by WHO region, 2024.*

Wang, H., Wolock, T. M., Carter, A., Nguyen, G., Kyu, H. H., Gakidou, E., Hay, S. I., Mills, E. J., Trickey, A., Msemburi, W., & Coates, M. M. (2016). Estimates of global, regional, and national incidence, prevalence, and mortality of HIV, 1980–2015: The global burden of disease study. *The Lancet HIV, 3*(8), 361–387.

Witte, S. S., Filippone, P., Ssewamala, F. M., Nabunya, P., Bahar, O. S., Mayo-Wilson, L. J., Namuwonge, F., Damulira, C., Tozan, Y., Kiyingi, J., & Nabayinda, J. (2022). PrEP acceptability and initiation among women engaged in sex work in Uganda: Implications for HIV prevention. *EClinicalMedicine*, 44.

14 An Investigation of Maternal Depression and Child Health Among South Sudan Refugees in Palabek Settlement, Uganda

Christopher Kugonza and Peter Celestine Safari

Abstract

Government and non-state actors have increasingly put more focus on maternal and child health interventions. However, there has been little attention on maternal mental health, yet this directly impacts child health. This study was carried out in Palabek refugee settlement in Lamwo district in northern Uganda focusing on women of reproductive age (15–49 years) with children below the age of 5 years. Results showed higher prevalence of stunting among children of mothers with depressive symptoms. Interventions to address maternal depression and its impacts on children and the wider community are needed.

Keywords

Stunting · Nutrition · Domestic violence · Mental disorders

Introduction

Since the coming into force of the Sustainable Development Goals (SDGs), there has been concerted effort globally toward addressing malnutrition. Goal 2 of the SDGs is primarily focused on ending hunger, achieving food security, and improving nutrition. Nevertheless, stunting among children under five remains persistent (WHO, 2024). Studies have found a strong relationship between stunting in children and maternal depression (Elgar et al., 2004; Surkan et al., 2011). In this study, we investigated the relationship between maternal depression and child health among refugees in Palabek settlement.

Methodology

The study was carried out in Palabek refugee settlement in Lamwo district in northern Uganda; the settlement hosts more than 32,000 South Sudanese refugees. Infrastructure is still being developed because the settlement is relatively new. Refugees seem to be integrating well with the host community, as many of them are from the same ethnic group. The local communities hosting refugees are themselves in a precarious socioeconomic situation, impacted by food insecurity and malnutrition, suffering from limited access to basic social services and economic infrastructure, as well as scarce livelihood opportunities.

In this study, we used both qualitative and quantitative approaches which helped reduce the limitations associated with using only one approach. The study focused on women of

C. Kugonza · P. C. Safari (✉)
Uganda Martyrs University, Kampala, Uganda
e-mail: psafari@umu.ac.ug

P. Nyadoi et al. (eds.), *Sustainable Development Goals in Africa*, Sustainable Development Goals Series, https://doi.org/10.1007/978-3-032-20742-5_14

reproductive age (15–49 years) and children below 5 years. We determined the sample size using the Krejcie and Morgan (1970) table to be 379 mother and child pairs. Additionally, the study involved four key informants (one District Health Officer, a Village Health Team Member, and two representatives of the refugee welfare committee). Cluster sampling was employed in the study; the settlement was divided into clusters represented by zones. The first level of sampling involved obtaining a simple random sample of clusters. The names of the zones were written on papers and put together in a bowl while folded, then half of the zones were randomly picked from the bowl and these zones/clusters were considered for sampling. Having drawn the sample of clusters, individual respondents (or households) were selected from each zone/cluster by the same simple random method. More specifically, all individuals (or households) on each of the selected zones/clusters were listed and, with the help of a table of random numbers, a sample of respondents from each zone/cluster was chosen. The sample distribution among selected zones was proportionate to their total populations, meaning a zone with the highest population got the biggest proportionate part of the sample and vice versa.

We collected data through individual household interviews with mothers using a structured questionnaire; this tool was designed to screen mothers for depressive symptoms, infant and young child feeding practices, such as exclusive breastfeeding, early initiation/continued breastfeeding, and complementary feeding. The caregiver tool also integrated immunization, diarrhea, and diet indicators for children. Another tool focusing on anthropometry (measuring height, weight, and age of children) was also used to screen children less than 5 years for stunting and underweight. This tool was administered according to the Ministry of Health guidelines on the Integrated Management of Acute Malnutrition.

Data Analysis

Quantitative data from depression screening and caregiver interviews was analyzed in form of descriptive statistics (standard deviation, mean, median) using MS excel and anthropometric data for children under 5 years was analyzed using WHO Anthro—software developed for the application of the WHO Child Growth Standards in monitoring growth and motor development in individuals and populations. Child nutrition was quantified in terms of weight-for-age or height-for-age; underweight and stunting are defined using WHO criteria: more than two standard deviations (SD) below the mean weight-for-age and mean height-for-age (<−2SD). The association between variables was tested using Pearson correlation coefficient; regression analysis (bivariate analysis) was used to determine how change in the independent variable predicts change in the dependent variable. Qualitative data was analyzed through thematic analysis; with the use of affinity mapping where data was organized to find ideas and concepts, build overarching themes in the data, find plausible explanations for findings, and draw conclusions. Affinity mapping helped to discover embedded patterns of thinking by sorting and clustering language-based information into relationships (Huberman & Miles, 2013).

Results and Discussion

Demographics

The study included 381 women with children under 5 years of age, from 4 and 7 randomly selected zones and blocks (Table 14.1) and, education and marital status of the mothers as shown in Table 14.2.

Majority (54.3%) of the respondents were Catholic; 32.5% Anglican, 2.4% SDA, 0.5% Muslim, and 10.2% had other religious affilia-

Table 14.1 Number of respondents per block

	Block 1	Block 2	Block 3	Block 4	Block 5	Block 6	Block 7	Total
Zone1		27	21	27				75
Zone2	28	1	9					38
Zone3		21		44	23	40		128
Zone5	42			40			58	140
Total	70	49	30	111	23	40	58	381

Table 14.2 Education and marital status of the respondents

Row labels	No education	Primary	Secondary	Tertiary	Grand total
Divorced	1.0%	3.4%	0.8%	0.3%	5.5%
Married	24.4%	52.0%	8.1%	0.0%	84.5%
Single	2.4%	2.4%	0.3%	0.0%	5.0%
Widowed	2.9%	2.1%	0.0%	0.0%	5.0%
Grand total	30.7%	59.8%	9.2%	0.3%	100.0%

tions. In terms of their education levels, 59.8% had stopped at primary education, 9.2% attended secondary education, only 0.3% had reached tertiary level, while 30.7% did not attend school at all (Table 14.2). About 85% of the respondents were married, 5.5% divorced, 5% single, and 5% widowed.

Association Between Maternal Depression and the Health of Children Under 5

Nutritional Status

As recommended by WHO, assessment of nutritional status was based on comparison of indices (WfH, HfA, WfA) for the children in the survey against their mothers' mental health. Children who fall below minus two standard deviations (<−2SD) from the median are regarded as moderately malnourished, while those who fall below minus three standard deviations (<−3SD) are considered severely malnourished. A total of 381 children (206 girls and 174 boys) aged 6–59 months old were included in the study.

Stunting

The stunting (HfA) indicator measures the number of children whose linear growth has been impaired by chronic malnutrition over a prolonged period of time (during pregnancy or their first years of life). It assesses to what degree (Z–score) a child's height for age (HfA) deviates from the height of a child of the same age and sex in the 2006 WHO growth standards. The prevalence for stunting (<−2 SD) is interpreted as <20% low prevalence, 20–29% medium prevalence, 30–39% high prevalence, and ≥40% very high prevalence.

The overall prevalence of stunting among children below 5 years was 28.6% (Table 14.3); it was higher in boys (31.9%) than in girls (25.9%). Stunting was lowest among children aged 6–11 months old (18.5%), followed by children aged 12–23 months (32%), and those aged 24–35 months (33.3%).

In terms of the comparison between stunting in children and mothers' mental health, depression was categorized in various levels of severity; those without symptoms at all, those with minimal depression (1–4), mild depression (5–9), moderate depression (10–14), moderately severe (15–19), and severe depression (20–27). Each of the respondents was scored on a scale of 0–27, and their individual classification has been compared against the nutrition indicators of their respective children. Overall, stunting among children of mothers who presented depressive symptoms was 29.4% compared to 16.7% among

Table 14.3 Nutrition indicators disaggregated by age and sex of the children in Palabek settlement

Set 1: Sexes combined		Underweight				Stunting				Wasting			
Age groups	*N*	Weight-for-age %				Length/height-for-age %				Weight-for-length/height %			
		%<−3SD	%<−2SD	Mean	SD	%<−3SD	%<−2SD	Mean	SD	%<−3SD	%<−2SD	%<+1SD	%<+2SD
Total (0–60)	380	7.4	16.1	−0.67	1.29	13.7	28.6	−1.17	1.65	6.4	13.3	25.1	11.9
(0–5)	15	0	13.3	−0.48	1.24	23.1	30.8	−1.31	1.68	8.3	16.7	50	33.3
(6–11)	33	15.2	27.3	−0.53	1.33	3.7	18.5	−0.28	1.44	22.2	33.3	18.5	3.7
(12–23)	51	11.8	19.6	−0.47	1.32	20	32	−1.46	1.66	8	16	20	10
(24–35)	79	10.1	21.5	−0.84	1.3	20	33.3	−1.55	1.66	10.5	15.8	26.3	10.5
(36–47)	92	5.4	10.9	−0.66	1.38	13.8	35.6	−1.26	1.74	3.4	10.1	29.2	14.6
(48–60)	110	3.6	11.8	−0.73	1.19	7.6	20	−0.89	1.52	0.9	7.4	22.2	11.1
Set 2: Males													
Age groups	*N*	Weight-for-age %				Length/height-for-age %				Weight-for-length/height %			
		%<−3SD	%<−2SD	Mean	SD	%<−3SD	%<−2SD	Mean	SD	%<−3SD	%<−2SD	%<+1SD	%<+2SD
Total (0–60)	174	6.9	14.4	−0.72	1.28	18.1	31.9	−1.4	1.68	6.2	13.6	27.2	13.6
(0–5)	8	0	12.5	−0.42	1.39	28.6	42.9	−1.71	1.94	16.7	16.7	50	50
(6–11)	19	10.5	21.1	−0.52	1.43	7.1	7.1	0.05	1.68	14.3	28.6	14.3	7.1
(12–23)	24	4.2	8.3	−0.4	1.23	25	37.5	−1.79	1.44	4.3	8.7	17.4	8.7
(24–35)	40	12.5	25	−0.9	1.4	28.9	42.1	−1.88	1.7	10.3	17.9	35.9	15.4
(36–47)	33	6.1	12.1	−0.99	1.13	20	40	−1.8	1.37	3.2	9.7	35.5	12.9
(48–60)	50	4	8	−0.66	1.21	6.4	21.3	−0.96	1.64	2	10.2	20.4	12.2
Set 3: Females													
Age groups	*N*	Weight-for-age %				Length/height-for-age %				Weight-for-length/height %			
		%<−3SD	%<−2SD	Mean	SD	%<−3SD	%<−2SD	Mean	SD	%<−3SD	%<−2SD	%<+1SD	%<+2SD
Total (0–60)	206	7.8	17.5	−0.64	1.3	10.2	25.9	−0.98	1.61	6.5	13	23.5	10.5
(0–5)	7	0	14.3	−0.54	1.16	16.7	16.7	−0.83	1.34	0	16.7	50	16.7
(6–11)	14	21.4	35.7	−0.55	1.2	0	30.8	−0.63	1.08	30.8	38.5	23.1	0

(12–23)	27	18.5	29.6	−0.53	1.43	15.4	26.9	−1.16	1.81	11.1	22.2	22.2	11.1
(24–35)	39	7.7	17.9	−0.78	1.19	10.8	24.3	−1.21	1.57	10.8	13.5	16.2	5.4
(36–47)	59	5.1	10.2	−0.47	1.48	10.5	33.3	−0.98	1.86	3.4	10.3	25.9	15.5
(48–60)	60	3.3	15	−0.78	1.19	8.6	19	−0.83	1.44	0	5.1	23.7	10.2
Set 4: Overall, sexes combined by cluster													
Cluster	***N***	**Weight-for-age %**				**Length/height-for-age %**				**Weight-for-length/height %**			
		%<−3SD	**%<−2SD**	**Mean**	**SD**	**%<−3SD**	**%<−2SD**	**Mean**	**SD**	**%<−3SD**	**%<−2SD**	**%<+1SD**	**%<+2SD**
Total	380	7.4	16.1	−0.67	1.29	13.7	28.6	−1.17	1.65	6.4	13.3	25.1	11.9
Not at all	25	4	12	−0.75	1.06	12.5	16.7	−0.94	1.36	8.3	16.7	4.2	4.2
Minimal depression	73	12.3	13.7	−0.59	1.38	8.8	17.6	−0.79	1.58	11.8	14.7	25	10.3
Mild depression	123	8.1	12.2	−0.72	1.22	14.8	29.6	−1.25	1.62	6	13.7	24.8	11.1
Moderate depression	99	4	22.2	−0.57	1.41	19.1	29.8	−1.18	1.9	4.2	13.5	28.1	16.7
Moderately severe	47	6.4	17	−0.7	1.22	11.6	44.2	−1.53	1.41	4.4	8.9	33.3	13.3
Severe depression	13	7.7	23.1	−1.28	1.13	0	38.5	−1.54	1.44	0	8.3	16.7	0

*Values presented in Table 14.3 are based on WHO standards

children of mothers who did not present any symptoms (Table 14.3).

Wasting

Wasting is used to measure the prevalence of severe acute undernutrition by assessing to what degree (Z score) a child's weight for height (WfH) deviates from the weight of a child of same height and sex in the 2006 WHO growth standards. Its prevalence is interpreted as: <5% acceptable, 5–9% poor, 10–14% serious, ≥15 critical. It is measured by dividing the number of children (6–59 months) whose WfH was < −2 SD by total number of children assessed.

Overall, 13.3% of the children were found to be with moderate acute malnutrition (MAM); it was highest (33.3%) among children aged 6–11 months, 16% among those aged 12–23 months, 15.8% among those aged 24–35 months, 10% among 36–47 months old, and 7.4% among 48–60 months old (Table 14.3). Severe acute malnutrition (SAM) was at 6.4% overall although it was still highest (22.2%) among children aged 6–11 months and 24–35 months just like MAM. The prevalence of wasting was about the same for boys and girls at 13.6% and 13%, respectively (Table 14.3). Wasting (<–2SD) was recorded highest (16.7%) in children of mothers who did not present any depressive symptoms, 14.7% among children of mothers who had minimal depression, 13.7% among those with mild symptoms, 13.5% among those with moderate symptoms, and 8.9% and 8.3% among children of mothers with moderately severe and severe symptoms, respectively.

Underweight

The underweight indicator assesses to what degree (Z–score) a child's weight for age (WfA) deviates from a weight of the child of the same age and sex in the 2006 WHO growth standards. The prevalence of underweight is interpreted as follows: <10% low prevalence, 10–19% medium prevalence, 20–29% high prevalence, and ≥30% very high prevalence. This was measured by diving the number of children whose WfA was <−2 SD by the total number of children (6–59 months) who participated in the survey.

Underweight among the sampled children was found to be at 16.1% overall (Table 14.3). Underweight was highest (27.3%) among children aged 6–11 months, 21.5% among 24–35 months old, 19.6% among 12–23 months, and lower in children above 3 years; and 11.8% and 10.9% among those aged 48–60 months and 36–47 months, respectively.

Children of mothers who did not present any depressive symptoms had the lowest prevalence of underweight cases at 12%, 12.2% among mothers with mild symptoms, 13.7% among mothers with minimal symptoms, 17% among mothers with moderately severe symptoms, 22.2% among children of mothers with moderate symptoms, and 23.1% among the severe cases.

Child Health

Diarrhea

At least 32% of the children of surveyed mothers had suffered from diarrhea in the past 2 weeks before the survey (Table 14.4). About 90% of these diarrhea cases were among children of mothers who also presented depressive symptoms.

About 87% of the caregivers also reported that their children had passed loose stool 3 or more times in a day, 23.8% of the mothers whose children had diarrhea did not wash their hands with water and soap or ash at least 3 out of 5 critical times in the past 2 days prior to the survey. The critical times include after going to the toilet, before preparing food, after attending to a child who has defecated, before feeding a child, and before eating. The respondents were asked after

Table 14.4 Prevalence of diarrhea

Depression cases	Child had diarrhea		
	No	Yes	Grand Total
Mild	22.7%	11.9%	34.7%
Minimal	15.9%	5.1%	21.0%
Moderate	18.5%	9.7%	28.1%
Moderately severe	9.4%	3.7%	13.1%
Severe	1.1%	2.0%	3.1%
Grand total	67.6%	32.4%	100.0%

how long they had sought for treatment when the diarrhea started; overall, they took an average of 2 days before seeking for first treatment. There was a slight difference in the duration taken when compared with the mothers' mental health; those who did not present any depressive symptoms reported taking an average of 1.7 days before seeking treatment compared to 2 days among those who presented symptoms.

Meal Frequency

In relation to meal frequency, we assessed the number of meals that were given to children aged 6–23 months the previous day; there was no difference between mothers with depressive symptoms and those without, as both groups reported giving their children an average of 3 meals per day.

Diet Diversity

We also assessed the diet diversity of children aged 6–23 months who had consumed at least four out of seven pre-defined food groups the previous day or night. It is an indicator of a diet's micronutrient adequacy, an important dimension of its quality. Only 15% of them were found to have a sufficient diet diversity (had eaten food items from at least 4 or more food groups the previous day). The rate was 14% among children of mothers with depressive symptoms and 30.8% among children of mothers without any symptoms.

Mosquito Net Utilization

Respondents were asked if all their children under 5 years had slept under a mosquito net the previous night; 91% of them reported that their children had slept under a net. 53.3% of them also reported that their child had experienced a fever in the past 2 weeks prior to the survey and 94.6% of them said they had sought for treatment for the child's fever from health personnel. Majority of them (68.2%) reported that they had sought for treatment within 24 hours, 30.7% sought treatment after more than 24 hours while just 1% reported not seeking treatment from health personnel at all. On the prevalence of fever among children in the past 2 weeks, mothers without depressive symptoms reported the lowest rate (40%) compared to 61% among those who presented depressive symptoms.

Conclusions

This study has unearthed the occurrence of maternal depression, child health, and nutrition problems in Palabek refugee settlement. The results give evidence of stunting being higher among children of mothers with depressive symptoms. There is thus need for interventions to address maternal depression and studies to determine the prevalence, causes, and impacts of this problem on children and the wider community.

References

Elgar, F. J., McGrath, P. J., Waschbusch, D. A., Stewart, S. H., & Curtis, L. J. (2004). Mutual influences on maternal depression and child adjustment problems. *Clinical Psychology Review, 24*(4), 441–459.

Huberman, A. M., & Miles, M. B. (2013). *Innovation up close: How school improvement works.* Springer Science & Business MediaBerlin: Springer.

Krejcie, R.V., & Morgan, D.W., (1970). Determining Sample Size for Research Activities. Educational and Psychological Measurement.

Surkan, P. J., Kennedy, C. E., Hurley, K. M., & Black, M. M. (2011). Maternal depression and early childhood growth in developing countries: Systematic review and meta-analysis. *Bull World Health Organisation, 89*(8), 608–615.

WHO. (2024). *Malnutrition in children.* World Health Organisation. https://www.who.int/data/nutrition/nlis/info/malnutrition-in-children

Status of the National School Health Policy Framework in Uganda

15

Paulinus Byabasaija and Denis Musinguzi

Abstract

A national school health policy is crucial to achieving Sustainable Development Goals (SDGs) 3 (health and wellbeing) and 4 (quality education) in Uganda. The World Health Organisation (WHO) put in place the Health Promoting Schools (HPS) model to guide the formulation and implementation of the school health policy among member countries. This chapter examined the status of Uganda's national school health policy framework and its implementation, based on review of literature including reports and data from public health databases on national school health policy, implementation of the WHO's HPS model and SDGs 3 and 4, for studies conducted between 2019 and 2024. The findings reveal that Uganda has not yet finalised its school health policy nor fully aligned to the WHO's HPS model. The country has two unfinalised national school health policies (2008, 2018) and this has among others led to inconsistent and fragmented school health services delivery, presenting difficulty in their documentation, reporting and monitoring of progress, and impeding the adaptation and implementation of the WHO HPS model and, thus, curtailing the country's progress towards the achievement of SDGs 3 and 4. There is need for a detailed study to help the country identify strategies to help it fast track and complete its school health policy.

Keywords

Institutional repository · Wellbeing · Learners · Healthy attitudes · Adolescent lifestyles

Introduction

Sustainable Development Goals (SDGs) 3 (health and wellbeing) and 4 (quality education) are closely linked together in that learners who are healthy are more likely to attend school regularly and perform better and in turn, education contributes to better health outcomes. This paper examines the implementation practices and gaps of school health services in view of Sustainable Development Goals (SDGs) 3 and 4 and the World Health Organisation's (WHO's) Health Promoting Schools (HPS) model in Uganda. It outlines suggestions to policy makers and implementers and proposes an agenda for future research to inform policy and practice for a holistic approach to tap into the interdependence of good health and quality education.

School health is the comprehensive effort of developing, implementing and evaluating

P. Byabasaija (✉) · D. Musinguzi
Uganda Martyrs University, Kampala, Uganda
e-mail: paulinus.byabasaija@stud.umu.ac.ug

P. Nyadoi et al. (eds.), *Sustainable Development Goals in Africa*, Sustainable Development Goals Series, https://doi.org/10.1007/978-3-032-20742-5_15

health services within the school and school surrounding communities (Kolbe, 2019). School health services (SHS) are services provided by a health worker to students enrolled in primary or secondary education, either within school premises or in a health care centre situated outside the school (WHO, 2021a). School health services also include health education and promotion services provided by a school health teacher and/or senior woman to learners within the school environment. Proper development and growth of children and the quality of their learning depend on a number of factors including their health status (Chaput et al., 2020). A learner who is not healthy or the one who suffers from an undetected health problem such as hearing or vision problems; a learner who is hungry, or who is impaired by drugs/alcohol, will not very well benefit from the process of education (DiPietro et al., 2020; Greenhalgh & Gibson, 2024). Similarly, a student who has not been assisted in the shaping of healthy attitudes, beliefs and habits when still young is more likely to suffer the consequences of unhealthy lifestyles, and hence reduced productivity when they grow up (Arafa et al., 2024; Debbye Omlie, 2024; Roşioară et al., 2024). Health promoting behaviours taught at school not only contribute to psychosocial and physical development and to the health of students, but also spills over to improve awareness about health issues among the parents and local community where the children come from (Turunen et al., 2017; Lesińska-Sawicka et al., 2021; Debbye Omlie, 2024). Promoting healthy behaviours during adolescence is the foundation for adult health and wellbeing (Lesińska-Sawicka et al., 2021). Moving from childhood to adolescence, is a period of dynamic biological development and social changes. This is critical stage in life where risk-taking behaviours are started, health behaviours are formed and gender norms are nurtured (Mastorci et al., 2024). Any health behaviours which arise during adolescence have an impact throughout the life of an individual (Santos et al., 2023). World Health Organisation (WHO) estimates that around 35% of the disease burden globally has origins from adolescent health lifestyles (Vadeboncoeur et al., 2022; Sun et al., 2023; Mastorci et al., 2024). There is, therefore, a strong need for learners' health promotion, prevention, and health care services to ensure individual, family and community healthy living.

Schools play an important part in promoting the health and well-being of the learners. There is no contestation on the symbiotic benefits from the health and wellbeing of the learners and their education outcomes (Lee, 2024). To tap into this mutual benefit, in 1995, the World Health Organisation (WHO) started the global school health initiative. This was to support countries to advance the health of learners through inclusive school health programmes. In the year 2020, WHO again launched the Health Promoting Schools (HPS) model to build on the 1995 global school health initiatives to ensure a whole school approach to promoting health and wellbeing of the learners and their education outcomes (WHO, 2021a, b). The HPS model emphasises a more integrated and holistic approach to health and education (UNESCO and WHO, 2021).

According to the WHO, the HPS model is well aligned with the United Nations 2030 agenda for Sustainable Development, specifically focusing on SDG 3 and 4. While SDG 3 is about good health and wellbeing, SDG 4 is about quality education (Macnab & Mukisa, 2017). The HPS combines these two SDGs and provides 8 standards for countries to adapt/adopt in order to realise an all-inclusive school health programme (WHO and UNESCO, 2021). This model has since been a foundation of global effort to promote health and wellbeing and education outcomes of the learners for long-term economic development. The HPS standard 8 is exclusively about school health services. This outlines a range of health services to be adapted and delivered to the learners (WHO and UNESCO, 2021).

In Uganda, several school health initiatives have been implemented to promote health and well-being of the learners (Kemigisha et al., 2019; Babatunde & Akintola, 2023). These school health interventions are clearly articulated in the Uganda National Minimum Health Care Package (UNMHCP) (MoH, 2021). These include: provision of health education, screening for and treating common ailments, improvement of the school environment, water and sanitation, personal hygiene, promotion of appropriate nutrition practices and adolescent reproductive health. Included in the adolescent reproductive health are, sexual and reproductive health and rights of adolescent boys and girls, as well as sex education in schools, life skills against sexually transmitted infections, unwanted pregnancies and unhealthy/pervert sex lifestyles (MoH, 2021). With support from development partners, Uganda has also positioned itself in investing in the achievement of all SDGs including SDG 3 and 4 (United Nations, 2024; Wanyana et al., 2024). Nevertheless, no study had been conducted to determine the status of the country's national school health policy framework and the extent to which it aligns with the WHO's HPS model recommended for the achievement of SDGs 3 and 4. Therefore in this chapter, we explore the status of Uganda's school health policy framework, its implementation and extent of alignment to the WHO HPS model. We followed Joanna Briggs Institute (2015) methodology, (Peters et al., 2020) in the study. We searched Global Health, PubMed, Google Scholar, WHO's Institutional Repository for Information Sharing (IRIS) and Journal of Public Health in Africa (JPHiA) databases for information. The terms sought for in the publications included, implementation of school health services, delivery of school health services, school health programmes, health promoting schools, national school health policy implementation guidelines, SDG 3 and 4, WHO HPS model and school health promotion. A total of 89 articles were identified, and of these 50 that were published between 2019 and 2024 and had relevant information sought for were selected and analysed.

Results

The Status of the National School Health Policy Framework in Uganda

Mostly through Implementing Partners (IPs), Uganda continues to offer school health programme despite the lack of a finalised national school health policy and implementation guidelines (World Bank Group, 2014; RTI and USAID, 2017). The country has not finalised its two national school health policies and the implementation guidelines (MoH, 2008, 2018; Musasizi, 2021). The lack of national guidelines means that there is no uniform process that schools and partners follow for the SHP implementation (Hongoro et al., 2018). The failure to finalise the national school health policy and guidelines is attributed to lack of or limited political will, weak coordination between health and education sectors, inadequate funding and competing national priorities. However, the Ministry of Health released the school health service package entailing health education, health and support services, social support, and physical environment, in an effort to guide implementers on a range of health services provision to learners. This service package is hitherto the official guide for service providers and development partners.

In spite of a lack of a policy framework, Uganda continues to offer school health programme (World Bank Group, 2014; RTI and USAID, 2017). Findings reveal that lack of finalised national school health policy framework leads to inconsistency in the implementation of school health programme across the country (World Bank Group, 2014). Individual implementers and schools adopt/adapt different approaches of programme delivery, with fragmentation in funding and uneven service package, and this leads to variations in the quality and comprehensiveness of school health programme package provided. This leads to a piecemeal approach and lack of integration of school health programme into the school curriculum, as well as lack of supportive environments for both students and staff (Turunen et al., 2017; Pulimeno et al.,

2020; Olson et al., 2021). Most health interventions in schools are often times based on the initiatives of individual schools and/or implementing partner, notably non-governmental organisations (NGOs), as opposed to a unified national school strategy (Robinson et al., 2024). This approach limits both the service provider to their areas of coverage and priorities, as well as some learners' access to an essential school health package, especially in geographically uncovered or hard-to-reach areas. There is no evidence of a strong institutionalised national framework for coordination between the health and education sectors, and development partners. This has often led to poor government funding for school health programme, especially, on account of other well-structured competing national priorities. To fill in these gaps, health and education development partners support implementation of school health programmes. This has led to a temporary, donor-dependent/driven school health programme, usually delivered in unsustainable disjointed manner (Zia et al., 2020). This has also led to some schools providing a more comprehensive school health package while in others, this is very minimal or non-existent (Levinson et al., 2019; Babatunde & Akintola, 2023). There is also a lack of effective integration of health and education programmes, and this affects the health and wellbeing of the learners and their education outcomes.

Lastly, lack of a national policy framework makes it hard to document, report and establish a standardised monitoring and evaluation (M&E) system to monitor progress in the delivery of school health programme and the progress to achieving SDGs 3 and 4. For example, in Uganda's health and education annual sector performance reports for the years 2018/2019 to 2022/23, there is just a mention of school health in only two reports (MoH, 2023). For the Ministry of Education and Sports, there are some separate reports on school sanitation and hygiene education, nutrition and school feeding, menstrual health and hygiene, sexual and reproductive health/HIV and mental health and psychosocial support; however, these are fragmented with no systematic reporting format to comprehensibly cover health in schools' programme package (MoES, 2024).

Alignment of Uganda's School Health Programme with the WHO'S HPS Model

WHO's eight HPS standards are: national school health policy, individual school health policies and governance, school community partnerships, school health curriculum, school socio-emotional and physical environments, and school health services (Nkamba et al., 2008; WHO, 2023). Our results show that the Uganda SHP is slightly aligned with six out of the eight standards of WHO's HPS model, but with a lot of gaps in breadth and depth. Two of the eight standards of the WHO's HPS model are not in the Uganda SHP. These include school governance and leadership and school and community partnerships. Additionally, there are also variations in individual school and in geographical coverage of these services, as well as in the rural-urban divide. Table 15.1 compares WHO's HPS standards with the components of the Uganda SHP.

Good health and wellbeing (SDG 3) and quality education (SDG 4) were designed to specifically address key global challenges related to health and education, both of which are essential for sustainable human development. One of the intersections between health and education is the health of the learners. A school setting is an essential platform for the delivery of quality health programme to the greatest number of learners. A school health policy is very essential in streamlining the delivery of the SHP. For instance, in the WHO's HPS model, all the other seven standards are hinged on one standard: the national school health policy (WHO and UNESCO, 2021). Uganda has no finalised national school health policy. This absence of a national school health policy framework impedes the implementation, making implementation weak and fragmented. Absence of school policy also affects coordination between ministry of health, education and partners, and this grossly affects documentation, reporting, monitoring

Table 15.1 Comparison of the WHO's HPS standards and the components of Uganda SHP

	WHO's 8 HPS standards	Components of Uganda SHP
1	**Government policies and resources** The whole of government is committed to and invests in making every school a health promoting school	**Two (2008 and 2018) unfinalised of national school health policy exist** School health standards and budgets are available at both ministries of health and education Very weak coordination between ministry of health and education exist
2	**School policies and resources** The school is committed to and invests in a whole school approach to being a health promoting school	**Nutrition services** In some few schools feeding programmes and/or nutrition education to promote healthy eating habits do exist But there are no literature/documents accessed regarding individual school health policies, budgets, guidelines or bi-laws on
3	**School governance and leadership** A whole-school model of school governance and leadership supports a health-promoting school	No literature/documents accessed regarding individual school health policies, budgets, guidelines or bi-laws on
4	**School and community partnerships** The school is engaged and collaborates with the local community for health-promoting school	No literature/documents accessed regarding individual school health policies, budgets, guidelines or bi-laws on
5	**School curriculum** The school curriculum supports physical, social–emotional and psychological aspects of student health	**Health education** In some schools, structured school activities covering personal hygiene, nutrition, physical activity, and disease prevention
6	**School social–emotional environment** The school has a safe, supportive social–emotional environment	**Physical education** Physical activities to promote fitness, motor skills development, teamwork, and social network.
7	**School physical environment** The school has a healthy, safe, secure, inclusive physical environment	**Healthy school environment** In some schools: Provision of clean water, sanitation facilities and school infrastructure. In some schools, learners study under tree sheds
8	**School health services** All students have access to comprehensive school based or school-linked health services that meet their physical, emotional, psychosocial and educational healthcare needs	**Health services** Offered in some districts and in a few schools: Irregular medical examinations, vaccination verification, school nursing care and giving first aid, and in fewer schools, some health screenings to detect potential health issues

and evaluation, and all this hinders progress toward achieving SDGs 3 and 4 (WHO and UNESCO, 2021). As indicated in the results section above, Uganda's SHP package has five components which are, somehow, in line with the WHO's HPS eight standards. Three of the WHO's HPS standards are completely absent. The five aligned ones lack the breadth and depth of what the WHO HPS standards recommend and are haphazardly implemented. There are also gaps in schools and geographical coverage of the HSP. These results give evidence that Uganda, in pursuit of achieving SDGs 3 and 4, faces a number of challenges, including lack of school health policy and implementation guidelines, weak school health programme integration and linkage, low and uneven coverage, unavailability of school health data and reporting, weak planning at all levels, and weak or no monitoring and evaluation, plus lack of or low coordination between line ministries. For Uganda to achieve the aspirations of SDGs 3 and 4, it needs not only to hasten and address these challenges, but also to adapt to the WHO's HPS model, which was purposely coined as a vehicle to support countries achieve these two goals.

This chapter has documented how the absence of a finalised national school health policy and implementation guidelines has created fragmentation in the delivery of school health services in Uganda. It is hard to standardise the documentation, reporting, monitoring and evaluating these services. The absence of the policy and guidelines creates lethargy in adapting and implementing the WHO's HPS model. All these negatively impact progress towards achieving the aspirations of SDGs 3 and 4. By documenting this knowledge gap and proposing the urgent need for the finalisation of the school health policy, this chapter provides critical insights into how Uganda and other countries can improve the school health programme in order to smoothly achieve SDGs 3 and 4. Evidently, the Ministries of Health and Education need to fast-track the finalisation of the School Health Policy, put in place the implementation guidelines, and establish a coordination mechanism to support implementation of the policy. The Ministries need to work with development partners to ensure a unified delivery of the SHP across the country and their documentation, reporting, monitoring and evaluation mechanisms. The country also needs to fast-track comprehensive implementation of the WHO's HPS model so as to make progress towards the achievement of SDGs 3 and 4. There is also need for the academic and research institutions to conduct research to determine the extent to which schools in Uganda are aligned to the eight standards of the WHO's HPS model. Such studies would reveal specific school health programme implementation successes and gaps, and give suggestions to policy makers and implementers on how to bridge the gaps. Such studies should also assess the effectiveness of the current school health programme and quantify the educational benefits arising from investing in school health services and provide policy recommendations on how to strengthen the policy and implementation.

There is also a need for specialised research to identify the political, institutional and financial barriers that might have delayed the finalisation of a national school health policy and implementation guidelines. The results of this study should directly provide insights into the national level structural changes needed to finalise the school health policy and implementation guidelines. There is also a need to investigate and document the best way to document, monitor and evaluate school health services. This will generate information to guide development of robust tools to document and measure the overall school health and education outcomes in Uganda.

Discussion

Status of School Health Policy Framework for Delivery of School Health Services in Uganda

This scoping review has revealed that Uganda has no finalized national school health policy framework and implementation guidelines. Uganda has had two unfinalised national school health policies. The first was in 2008 and the second was in 2018 (MoH, 2008, 2018; Musasizi, 2021). This also means the country has no uniform national guideline that schools and partners can follow for delivering health services (Hongoro et al., 2018). Some of the factors responsible for not finalising these national frameworks point to lack of or limited political will, weak coordination between health and education sectors, inadequate funding and competing national priorities. However, the Ministry of Health has, in 2023, articulated the school health service package (still in draft form) in an effort to guide implementers on a range of services to be provided to the learners across all schools in the country. This service package is now being rolled out by the partners in the selected districts and schools they support.

In spite of this lack of policy framework, Uganda continues to offer school health services (World Bank Group, 2014; RTI and USAID, 2017). Findings reveal that lack of finalized national school health policy framework leads to inconsistency in the implementation of school health services across the country (World Bank Group, 2014). Individual implementers and schools adopt different approaches of service delivery, with fragmented funding and uneven service package, and this may lead to variations in the quality and comprehensiveness of school health services provided. This may also lead to piecemeal and lack of the integration of school health services and health promotion into the school curriculum, and lack of supportive environments for both students and staff (Turunen et al., 2017; Pulimeno et al., 2020; Olson et al., 2021). Most health interventions in schools often times are based on the initiatives of individual schools and/or implementing non-governmental organisations (NGOs) as opposed to a unified national school strategy (Robinson et al., 2024). This approach may limit some learners', especially in geographically uncovered areas, access to essential health services. There is also no evidence of a strong institutionalised national framework for coordination between the health, education sectors and development partners. This may lead to poor government funding for school health programmes, especially on account, partly, of other well-structured competing national priorities. To fill in these gaps, health and education development partners support implementation of school health services. This has led to a temporary, donor-dependent/driven school health programmes, usually delivered in an unsustainable disjointed manner (Zia et al., 2020). This has led to some schools providing more comprehensive school health services, and other schools providing very minimal or non-existent services (Levinson et al., 2019; Babatunde & Akintola, 2023). There is also a lack of effective integration of health and education programmes, and this affects the health and wellbeing of the learners and their education outcomes.

Lack of a national policy framework makes it harder to report and to establish a standardised monitoring and evaluation (M&E) systems to monitor progress in the delivery of school health services and the progress to achieve SDGs 3 and 4. For example, this paper has revealed that, in the annual health and education sector performance reports for the years from 2018/2019 to 2022/23, there is just a mention of school health in only two reports and in others there is total absence of school health services (MoH, 2023). For the Ministry of Education and Sports, there are some separate reports on: school sanitation and hygiene education, nutrition and school feeding, menstrual health and hygiene, sexual and reproductive health/HIV and mental health and psychosocial support (MoES, 2024).

Uganda's School Health Services Alignment to WHO's Model for Achieving SDGs 3 and 4

For school health services, the findings show that school health services in Uganda are aligned with most of the components of the eighth standard of WHO's HPS model, although a few gaps exist (MoH, 2010). The HPS model has 8 standards which are: national school health policy, individual school health policies and governance, school community partnerships, school health curriculum, school socio-emotional and physical environments, and school health services (Nkamba et al., 2008; WHO, 2023). Seven standards are all hinged on one standard: the national school health policy (WHO and UNESCO, 2021). Absence of a national school health policy framework impedes the implementation of all other standards—weak and fragmented. Absence of policy also affects coordination between ministry of health, education and partners, affects reporting, monitoring and evaluation; and all these hinder progress towards achieving SDGs 3 and 4 (Samdal & Rowling, 2012; WHO and UNESCO, 2021).

Conclusion

In summary, good health and wellbeing (SDG 3) and quality education (SDG 4) are specifically to address key global challenges related to health and education, both of which are essential for sustainable human development. One of the intersections between health and education is the health of the learners. A school setting is an essential platform for the delivery of quality health services to the greatest number of learners. This study has revealed that Uganda, in pursuit to achieve SDGs 3 and 4, faces a number of challenges including lack of school health policy and implementation guidelines, weak school health services integration and linkage, low and uneven coverage, unavailability of school health data, weak planning at all levels and weak or no monitoring and evaluation, plus lack of or low coordination between line ministries. For Uganda to achieve the aspirations of SDGs 3 and 4, it needs not only to hasten to address these challenges but also adapt to the WHO's HPS model, which was purposely coined to support countries to achieve these two goals.

The evidence we provide in this chapter pinpoints the need for the establishment of a health and education sectors coordinating mechanism, as not only being crucial in finalising the school health policy and implementation guidelines in Uganda but also for ensuring unified delivery of school health services across the country. It also recommends the fast-tracking of the comprehensive implementation of the WHO's HPS model as a key vehicle in achieving SDGs 3 and 4. We recommend studies be done to determine the following:

- Extent schools in Uganda are aligned to the 8 standards of the WHO's HPS model. Such a study would reveal implementation gaps and give suggestions to policymakers and implementers on how to bridge the gaps.
- Effectiveness of the current school health programme and quantify the educational benefits arising from investing in school health services.
- The political, institutional and financial barriers that might have delayed the finalisation of a national school health policy and implementation guidelines. The results of this type of study will provide insights into the national level structural changes needed to finalise the school health policy and implementation guidelines.
- At national level, the best ways to monitor and evaluate school health services so that they are effective and efficient. This could also support in developing robust tools to measure the overall health and education outcomes.

References

Arafa, A., et al. (2024). Lifestyle behaviors of childhood and adolescence: Contributing factors, health consequences, and potential interventions. *American Journal of Lifestyle Medicine*. [Preprint]. Available at:. https://doi.org/10.1177/15598276241245941

Babatunde, G. B., & Akintola, O. (2023). Beyond access: Can a school health initiative facilitate healthcare services utilisation for school-going children? *International Journal of Environmental Research and Public Health, 20*(15), 6448. Available at:. https://doi.org/10.3390/ijerph20156448

Chaput, J.-P., et al. (2020). 2020 WHO guidelines on physical activity and sedentary behaviour for children and adolescents aged 5–17 years: Summary of the evidence. *International Journal of Behavioral Nutrition and Physical Activity, 17*(1), 141. Available at:. https://doi.org/10.1186/s12966-020-01037-z

Debbye Omlie, B. A. (2024). *Thriving moving forward: A guide to healthy aging*. New York: Gatekeeper Press 7853 Gunn Hwy, Suite 209 Tampa, Fl 33626.

DiPietro, L., et al. (2020). Advancing the global physical activity agenda: Recommendations for future research by the 2020 WHO physical activity and sedentary behavior guidelines development group. *International Journal of Behavioral Nutrition and Physical Activity, 17*(1), 143. Available at:. https://doi.org/10.1186/s12966-020-01042-2

Greenhalgh, M., & Gibson, P. (2024). Introduction. In *All the things they never told you about teaching* (pp. 1–6.Available at:). Routledge. https://doi.org/10.4324/9781032697932-1

Hongoro, C., et al. (2018). Analysis of selected policies towards universal health coverage in Uganda: The policy implementation barometer protocol. *Archives of Public Health, 76*(1), 12. Available at:. https://doi.org/10.1186/s13690-018-0258-4

Kemigisha, E., et al. (2019). Evaluation of a school based comprehensive sexuality education program among very young adolescents in rural Uganda. *BMC Public Health, 19*(1), 1393. Available at:. https://doi.org/10.1186/s12889-019-7805-y

Kolbe, L. J. (2019). School health as a strategy to improve both public health and education. *Annual Review of Public Health, 40*(1), 443–463. Available at:. https://doi.org/10.1146/annurev-publhealth-040218-043727

Lee, I. (2024). Co-benefits from health and health systems to education. *Health Policy, 142*, 105016. Available at:. https://doi.org/10.1016/j.healthpol.2024.105016

Lesińska-Sawicka, M., Pisarek, E., & Nagórska, M. (2021). The health Behaviours of students from selected countries—A comparative study. *Nursing Reports, 11*(2), 404–417. Available at:. https://doi.org/10.3390/nursrep11020039

Levinson, J., et al. (2019). Investigating the effectiveness of school health services delivered by a health provider: A systematic review of systematic reviews.

PLoS One, 14(6), e0212603. Available at:. https://doi.org/10.1371/journal.pone.0212603

Macnab, A. J., & Mukisa, R. (2017). The UN sustainable development goals; using WHO "health promoting schools" to create change. *GHMJ (Global Health Management Journal), 1*(1), 23. Available at:. https://doi.org/10.35898/ghmj-1190

Mastorci, F., et al. (2024). The transition from childhood to adolescence: Between health and vulnerability. *Children, 11*(8), 989. Available at:. https://doi.org/10.3390/children11080989

MoES, K. (2024). *Ministry of Education and Sports: School Health*, https://schoolhealth.education.go.ug/.

MoH. (2008). *Uganda school health policy*. MoH.

MoH. (2010). *The Second National Health Policy promoting people's health to enhance socio-economic development*. Ministry of Health.

MoH. (2018). *Uganda School Health Policy*. Kampala.

MoH. (2021). *Ministry of Health Strategic Plan 2020/21–2024/25*. Kampala.

MoH, K. (2023). *Annual health sector performance reports. Finacial year 2022/2023*. MoH.

Musaasizi B. (2021). Govt to table national school health policy before Cabinet—PM Nabbanja. *The Red Pepper*. 28 October. Retrieved October 2, 2024, from https://redpepper.co.ug/govt-to-table-national-school-health-policy-before-cabinet-pm-nabbanja/111943/

Nkamba, E. M., Tilford, S., & Williams, S. A. (2008). Components of health promoting schools in Ugandan primary schools: A pilot study. *International Journal of Health Promotion and Education, 46*(3), 84–93. Available at:. https://doi.org/10.1080/14635240.2008.10708134

Olson, T., Hegbloom, C., & Egan, C. A. (2021). Whole school approach: Connecting schools to community resources to enhance school health. *Journal of Physical Education, Recreation & Dance, 92*(3), 5–12. Available at:. https://doi.org/10.1080/07303084.2020.1866721

Peters, M. D. J., et al. (2020). Updated methodological guidance for the conduct of scoping reviews. *JBI Evidence Synthesis, 18*(10), 2119–2126. Available at:. https://doi.org/10.11124/JBIES-20-00167

Pulimeno, M., et al. (2020). School as ideal setting to promote health and wellbeing among young people. *Health Promotion Perspectives, 10*(4), 316–324. Available at:. https://doi.org/10.34172/hpp.2020.50

Robinson, R. S., et al. (2024). Reported effects of non-governmental organisations in health and education service provision: The role of government relations and other factors. *Development Policy Review, 42*, 1). Available at:. https://doi.org/10.1111/dpr.12737

Roşioară, A.-I., et al. (2024). Status of healthy choices, attitudes and health education of children and young people in Romania—A literature review. *Medicina, 60*(5), 725. Available at:. https://doi.org/10.3390/medicina60050725

RTI, & USAID. (2017). *USAID/Uganda School Health and Reading Program. Quarterly Report January 1-March 31: 2017*. Kampala.

Samdal, O., & Rowling, L. (eds). (2012). *The implementation of health promoting schools*. Routledge. Available at: https://doi.org/10.4324/9780203119792.

Santos, F., et al. (2023). School-based family-oriented health interventions to promote physical activity in children and adolescents: A systematic review. *American Journal of Health Promotion, 37*(2), 243–262. Available at:. https://doi.org/10.1177/08901171221113836

Sun, J., et al. (2023). Global, regional, and national burden of cardiovascular diseases in youths and young adults aged 15–39 years in 204 countries/territories, 1990–2019: A systematic analysis of global burden of disease study 2019. *BMC Medicine, 21*(1), 222. Available at:. https://doi.org/10.1186/s12916-023-02925-4

Turunen, H., et al. (2017). Health promoting schools—A complex approach and a major means to health improvement. *Health Promotion International, 32*(2), 177–184. Available at:. https://doi.org/10.1093/heapro/dax001

UNESCO, & WHO. (2021). *UNESCO and WHO urge countries to make every school a health-promoting school*, https://www.unesco.org/en/articles/unesco-and-who-urge-countries-make-every-school-health-promoting-school.

United Nations. (2024). *The sustainable development goals in Uganda*. Kampala.

Vadeboncoeur, J. A., et al. (2022). Changing perspectives on adolescence(s). In *Oxford research encyclopedia of education*. Oxford University Press. Available at:. https://doi.org/10.1093/acrefore/9780190264093.013.107

Wanyana, I., et al. (2024). Finding linkages between the sustainable development goals in Uganda: A key to attain health and Well-being for all. *BMJ Public Health, 2*(2), e000529. Available at:. https://doi.org/10.1136/bmjph-2023-000529

WHO. (2021a). *WHO guideline on school health services*. Geneva.

WHO. (2021b). *WHO urge countries to make every school a health—promoting school*. Geneva.

WHO. (2023). *Health promoting schools*.

WHO, & UNESCO. (2021). *Making every school a health-promoting school. Implementation guidance*. World Health Organisation and the United Nations Educational, Scientific and Cultural Organisation. Licence: CC BY-NC-SA 3.0 IGO.

World Bank Group. (2014). *Uganda School Health: SABER Country Report 2014. Systems approach for better education results*. Washington, DC.

Zia, N., et al. (2020). Adaptation and validation of UNICEF/Washington group child functioning module at the Iganga-Mayuge health and demographic surveillance site in Uganda. *BMC Public Health, 20*(1), 1334. Available at:. https://doi.org/10.1186/s12889-020-09455-1

Part V

The Market, Purchasing Power and Impoverishment in Africa

16 Experiences and Lessons from Sustained Poverty Alleviation Interventions Among Rural Communities in Ghana, West Africa

Seidu Abubakari Jeremiah
and Seniwoliba Attia Joseph

Abstract

The chapter presents lessons and experiences obtained from the NGO Jaksally Development Organisation project that employed a multi-pronged approach towards poverty alleviation among 1409 women and 210 men in the Rural-West Gonja District in Ghana. We conducted Organisation Capacity Self-Assessment (OCA) to identify strengths and areas of need of the community. Based on the findings, we designed and implemented empowerment interventions that included village saving and loan associations (VSLAs). We explored the performance of these interventions in terms of poverty reduction and the findings are presented in this chapter. Among others, the findings indicate that the interventions stabilised household incomes, increased financial assets and decreased vulnerability to financial and other shocks. The findings also reveal that money obtained from the VSLAs was invested in farming and petty trade in shea value chain. Income accruing from these investments was used to acquire productive and non-productive assets. The findings also pinpoint the need for bigger loans to enhance achievement of poverty alleviation among these and other vulnerable communities.

Keywords

Impoverished women · Shea · Income stability · Agricultural production

Introduction

The majority of families in Africa lack collateral, the assets promised by a borrower to a lender as security for a loan giving the lender claim to it in the event of non-payment (Booth & Lena, 2006; Ngegba et al., 2016). For this reason, many fail to secure credit from financial institutions (Beyene & Dinbabo, 2019). The problem is pronounced among women in rural settings (Harris, 2009). According to Kwarteng Amaning and Sarfo-Mensah (2019), one of the main sources of funding for marginalised groups, including women, is village savings and loan associations (VSLAs). This concept is widely used across Latin America, Asia and Africa and has been promoted as a vehicle for poverty alleviation in households for example by CARE International (Liou and Chang, 2008; Tzanakis, 2013; Abubakari et al., 2014; Ksoll et al., 2015; Batinge, 2018). The Jaksally Organisation adopted VSLAs in its

S. A. Jeremiah (✉) · S. A. Joseph
Jaksally Development Organisation, Bole, Ghana

P. Nyadoi et al. (eds.), *Sustainable Development Goals in Africa*, Sustainable Development Goals Series, https://doi.org/10.1007/978-3-032-20742-5_16

approach to alleviating poverty among women, youth and men in the rural poor region of Northern Ghana, with financial support from SNV Ghana—the Ghanaian branch of the SNV Netherlands Development Organisation. Nearly 1410 women (i.e., 1409) and 210 men in the Rural-West Gonja District in Ghana were given loans and, trained and guided to invest their loans and savings in income-generating activities (IGAs) such as agribusiness especially the production of shea kernel, butter and shea-based products. In this chapter we present lessons and experiences emerging from the project. In particular, we explored perceptions of the beneficiaries regarding the project's impact on their well-being.

Methodology

Jaksally and the District Cooperative officers undertook a visit to the West Gonja District and selected 12 communities with and without existing VSLA groups (VSLAs, 2017) to pilot the project. The identification of these potential communities was done through stakeholder engagement and community meetings. A needs assessment was then conducted to determine their financial needs and willingness to form a VSLA. This action research arose out of the need to examine the shea development activities that could be used as a tool to eradicate poverty. This provided the opportunity for Jaksally to form VSLAs and train members to expand into cooperative societies. In communities in which VSLAs existed, project officers took the advantage to seek their opinions about the concept. Where the VSLAs did not exist, the project officers introduced the concept and sensitised the members on the formation processes and criteria with the general rules and regulations guiding VSLAs. As a result, semi-structured and unstructured questionnaires known as Organisation Capacity Self-Assessment (OCA) were used to collect data from paticipants following approaches such as focus group discussions and key informants (Creswell, 2014). Meetings were organised with community members, and opinion leaders to introduce VSLA and the cooperative concept and models.

The Village Savings and Loan Groups were then formed, consisting of both men and women, that were in total 1620 (1409 women and 210 men). The OCA questionnaire was administered to identify the strengths and functional areas of the various groups and the training designed to meet their individual needs as exposed by the instrument. The VSLAs were implemented and throughout the period there were awareness creation meetings on why and who can be a member, benefits, prospects and linkages of cooperatives to take advantage of the shea value chain business, in the District. The beneficiaries were from Larabanga, Damongo, Soalepe, Bonyangto, Achubunyo, Jonopkongto, Mempeasem, Kawankura, Tailope, Busunu, Bidimah, Langantere and Sumpini. Data obtained from the beneficiaries on impacts of VSLAs were analysed and emerging themes identified.

Results

Access to Credit and Sustainable Financial Sources

A total of 1619 beneficiaries (1409 women and 210 men) accessed credit from 57 VSLAs, 54 cooperatives and 1 emerging cooperative saving society. This has led to a lot of development activities like the construction of new houses, new small businesses, expansion of table tops, in the area. GH₵560,000.00 is available to these groups as funds to loan from and funds mobilised by these groups from the VSLA. Each member in the VSLAs is getting an average of GH₵500.00 on a year repayment period at agreed interest rates of 6–10% for 6 months.

The VSLAs provide access to basic financial services to many rural poor people, mostly women and it offers members a secure place to save and access loans easily. A woman in Sumpini who used to borrow money from Dagomba women in Tamale to bulk nuts said she now bulks her nuts and owns a motorcycle for conveying nuts. In Busunu, a woman took a loan and invested in shea nuts and processed them into shea butter. Another woman took a loan of GH₵100.00 invested in local soap production from shea and made over GH₵500.00 in 3 months. In Damongo a woman said now she determines prices for her produce and sells them when she likes and not to the dictates of the businesswomen from Tamale. She used to buy five (5) bowls of nuts. With the VSLAs' intervention she now buys 1 bag of nuts. In Tailope, women groups involved in shea-nut bulking and butter trading have increased from 5 to 25. They have bulked over 160 bags (13.6 tons) of nuts that they have sold to buyers in the open market.

Provision of Employment

About 200 small-scale jobs in rural settings have been created by the VSLAs. The jobs created include drivers of vehicles bought by the VSLAs, operators of grinding mills, shopkeepers, farm labourers, produce collectors and truck loaders.

Increased Agricultural Production

As a result of the increased access to credit from the VSLAs, investment in agriculture and micro enterprises has increased. A couple in Busunu has increased their agricultural product bulking from 40 to 100 bags of shea. Women in Langantere and Tailope have started arranging for group shea nuts bulking for the next season. In Mempeasem, groups have mobilised 80 bags and Kawankura has mobilised 90 bags of ground nuts. The VSLAs in collaboration with cooperatives have in Busunu mobilised over 200 bags (17 tons) of shea nuts, and women of Lanagantere have sold 300 tons of yams, and 1 ton of fish. About 35% reductions in charcoal production in the project have been achieved due to VSLA activities.

As a result of availability of finances from the VSLAs, for investment in production activities, Shea production has increased from at least half a bag of nuts mobilised to three bags on average per woman per year. Nut bulking has increased as in the case of the tonnage of bags of nuts that have been supplied to the buyers (PBC). Maize yields have increased, Yam production has doubled in the Soalepe, Bidima and Lanagantere areas.

Stability of Household Incomes

The project has stabilised incomes, increased household financial assets and decreased vulnerability to financial and other shocks in the West Gonja District. The 54 VSLA and Cooperative groups have collectively mobilised 540,000 Ghana cedis (GH₵ 540,000.00). This is reflected in the savings and loan records of the VSLAs groups.

Provision of Access to Education

With the project Jaksally has entered into a partnership with the non-formal education division of the Ministry of Education at the district level and Literacy for Evangelism, a Canadian-based INGO into literacy, to train VSLA on how to read and write. More children from these group members have been sent back to school. As a result of the sensitisation, 1620 members of the VSLAs have undergone training, 285 executive committee members have exhibited high-quality leadership by managing their groups very well, group members meet regularly, and were able to mobilise GH₵540,000.00 within 1 year through their VSLAs savings activities.

Group Participation

Time consciousness in church attendance and meeting times in social activities has increased in the project areas. Participation by VSLA members in religious and political activities has gone up in the project areas. Most of the VSLAs members/beneficiaries can contribute their ideas in the cooperative and union meetings, district assembly meetings, community meetings and unit committees. They bring in some disciplinary actions of getting community members who do not contribute to projects, cooperatives and unions to be levied and fined. The women feel confident and proud that they are now engaging with the buyers and the middlemen.

Discussion

The Jaksally organisation successfully organised community into members of VSLAs and this has created opportunity for the members to be able to access credit and improve their incomes and livelihoods. The study by Nimoh et al. (2013) revealed that loan acquisition is highly serviceable by people who belong to groups, hence the ability of Jaksally to organise both men and women into VSLAs helped in the case of rural poor community in Ghana studied. According to Zeller (1994), members of responsible formal organisations such as Farmer Based Organisations (FBOs) with formal credit experience are viewed positively by lenders for loan approval/reception and credit needs being met. Similarly, Crispin and Kantchewa (2023) maintain that groups formed to save money and foster and encourage a strong culture of saving in the neighbourhood can always obtain minor loans to establish businesses that generate money.

As one of the major achievements, the Jaksally project provided employment opportunities for both men and women as well as inculcated a saving culture among them. Masina (2013) established that women who are in VSLAs can contribute to their families. In these groups, women are encouraged to borrow money to start enterprises of their choice in the end maximizing their income so that they would be able to save and access loans again in their subsequent cycles thereby providing sustainable employment for them. Beneficiaries of the VSLAs project implemented by the Jaksally Organisation mentioned that most of them can now access loans to expand their farms, invest in the shea value chain and have increased their production as evidenced in Busunu, Mempeasem, Kawankura and Lanagantere. Abdulai et al. (2022) found out that before the women joined the VSLAs they could not borrow money from friends and relatives because they were not considered to be creditworthy and as such they could not invest enough money in farming and other income earning activities. The perception of the members of the VSLAs formed by the Jaksally Organisation thus confirm that the VSLAs improved the members' credit worthiness.

Our lessons also align with previous studies, which found that joining savings associations enabled participants to obtain small loans to carry out social and economic activities to enhance their living standards and sustain household incomes. According to the project beneficiaries, they spent the majority of the funds they saved on household food consumption and educational materials for children while concentrating on their small businesses. This also indicates how the VSLAs intervention improved their household wellbeing and enabled beneficiaries engage in increasing investments in farming and agribusiness.

References

Abdulai, I. A., Adams, A.-M., Abdulai, M., & Bukari, S. (2022). Contributions of village savings and loans associations to rural livelihoods' development in the Upper West Region, Ghana. *SN Social Sciences, 2*.

Abubakari, A., Sadik, B. B., & Keisan, A. (2014). Impact of village savings and loans associations on the nutritional status of under-five children: A case study in the Sissala West district of upper west region. *Pakistan Journal of Nutrition, 13*(7), 390–396. https://doi.org/10.3923/pjn.2014.390.396

Batinge, B. K. (2018). Women access to microfinance loans on children formal education: A case of rural northern migrant in the Brong Ahafo region of Ghana. *Journal of Business & Financial Affairs, 7*(3), 1–9. https://doi.org/10.4172/2167-0234.1000347

Beyene, N. L., & Dinbabo, M. F. (2019). An empirical evaluation of the link between women participation in village savings and loans association (VSLA) and poverty reduction in Ethiopia. *Journal of Reviews on Global Economics, 8*, 566–580. https://doi.org/10.6000/1929-7092.2019.08.49

Booth, J., & Lena, C. B. (2006). Loan collateral decisions and corporate borrowing costs. *Journal of Money, Credit and Banking, 38*(1), 67–90. https://doi.org/10.1353/mcb.2006.0011

Coleman, J. S. (1988). Social capital in the creation of human capital. *The American Journal of Sociology, 94*, S95–S120.

Creswell, J. W. (2014). *Research design*. Sage.

Crispin, K., & Kantchewa, P. (2023). An evaluation of the influence of village banking on livelihoods: A case of Mpingu extension planning area in Lilongwe District, Malawi. *International Journal of Scientific Engineering and Science, 7*(11), 35–40.

Harris, S. (2009). *State of the microcredit summit campaign report 2009*. Microcredit Summit Campaign.

Ksoll, C., Lilleor, H. B., Lonborg, J. H., & Rasmussen, O. D. (2015). Impact of village savings and loans associations evidence from a cluster randomized trial. *Journal of Development Economics, 120*, 70–85.

Kwarteng Amaning, T., & Sarfo-Mensah, P. (2019). The impact of savings groups on female agency: Insights from village savings and loans associations in Northern Ghana. *Asian Journal of Agriculture and Rural Development, 9*(2), 133–146.

Liou, T. Y., & Chang, N. Y. (2008). The applications of social capital theory in education. *Hsiuping Journal of Humanities and Social Sciences, 11*, 99–122.

Masina, L. (2013, May 9). *Village banking empowers poor Malawi women*. Voice of America News. Retrieved September 5, 2024, from https://www.voanews.com/articleprintview/1657768.html

Ngegba, M. P., Kassoh, T. L., & Sesay, M. (2016). Impact of village saving and loan association (VSLA) on farm productivity in Lower Banta Chiefdom, Southern Sierra Leone. *International Research Journal of Social Science and Humanities, 1*(1), 29–32.

Nimoh, F., Tham-Agyekum, E. K., & Awuku, M. S. (2013). Factors influencing access of poultry farmers to credit: The case of the Agricultural Development Bank (ADB) in Ga East Municipality, Ghana. *Management, 3*(1), 54–58. https://doi.org/10.5923/j.mm.20130301.11

Tzanakis, M. (2013). Social capital in Bourdieu's, Coleman's and Putnam's theory: Empirical evidence andemergent measurement issues. *Educate, 13*(2), 2–23.

VSLA Associates. (2017). What is VSLA (village savings and loans)? Reaching the very poor: The need for a new microfinance model. VSLA Associates, viewed n.d., from https://www.vsla.net/

Zeller, M. (1994). Determinants of credit rationing: A study of informal lenders and formal credit groups in Madagascar, World Develop 22, Madagascar.

Technical Vocational Education and Training (TVET) as a Solution to Youth Unemployment in Uganda: A Case Study of Wakiso District

17

Yusuf Mayega

Abstract

Technical vocational education and training (TVET) plays a vital role in addressing knowledge and skills challenges to achieving the Sustainable Development Goals (SDGs). In this study, the effectiveness of TVET training and strategies required for improvement were documented. Data obtained from 384 participants in Wakiso District, Uganda, based on a cross-sectional design were analysed. Study participants reported that high costs, administrative inefficiencies and outdated curricula hamper TVET effectiveness. Collaboration with industry as well as flexible teaching methods will be needed to optimize TVET programs. Additionally, longitudinal studies are needed to track career progression and the application of technology in TVET.

Keywords

Skills development · Employment · Partnership · Social stigma

Introduction

Uganda, like many developing countries, has one of the most rapidly growing youth populations in the world. Despite several TVET initiatives, youth in Uganda face challenges in finding decent jobs. This study explores the effectiveness of TVET training in equipping the youth for the labour market. The following questions were answered: (1) How do alumni perceive the effectiveness of TVET training? (2) What hinders the performance of TVET training? And (3) what interventions are needed to enhance the effectiveness of TVET programs?

The study was conducted in Wakiso District, in Central Uganda. Open-ended individual interviews with 367 alumni of TVET institutions and 17 key informants were conducted. The sample size of 384 respondents was determined following Morgan and Krejcie and drawn from selected alumni groups of TVET institutions, TVET institutional heads and employers, the Ministry of Education and Sports, Uganda Business and Technical Examinations Board (UBTEB) and the Directorate of Industrial (DIT). Data analysis involved use of descriptive statistics using the Statistical Package for Social Scientists (SPSS). Additionally, thematic analysis was carried out.

Y. Mayega (✉)
School of Arts and Social Sciences, Uganda Martyrs University, Kampala, Uganda
e-mail: ymayega@umu.ac.ug

P. Nyadoi et al. (eds.), *Sustainable Development Goals in Africa*, Sustainable Development Goals Series, https://doi.org/10.1007/978-3-032-20742-5_17

Results and Discussion

A total of 388 questionnaires were distributed and 17 key informants were contacted. However, 367 questionnaires representing about 95.5% were retrieved and 17 key informant interviews were conducted accounting for the remaining 4.5%. Based on the arguments raised by Mugenda and Mugenda (2003) that any response rate above 50% is adequate to accomplish the study objectives; the researcher went ahead with the analysis of data. The study included both men (51.6%) and women (48.4%).

Age of the Respondents

The age of respondents ranged from 18 to 39 years and above (Fig. 17.1).

Respondents' Education

The educational level of respondents ranged from primary education to diploma (Fig. 17.2).

The Employment Status of Respondents

Most of the respondents (53.6%) stated that they were employed while only 6.5% said that they were not employed (Table 17.1).

Effectiveness of TVET Training

The majority of the respondents stated that they had completed TVET training, the training provided them with relevant skills for labour market, and the overall quality of the program was excellent (Table 17.2).

The key informants, who comprised TVET institution heads, employers, and officials from education sector intimated:

> *TVET is a good program, I really encourage students, parents and the community to embrace the opportunity provided by government for skills acquisition, hands-on training to graduate employability, and economic development* (Informant 1, Kampala district, 3rd March, 2024).

> *Our school produces on the job training in field such as catering bakery, cosmetology, fashion and design* (Informant 2, Wakiso district, 3rd March, 2024).

> *Every year, the college passed out skilled workers and majority of them are either self-employed or working others* (Informant 4, Entebbe, 8th April 2024).

> *The TVET programs in our district have shown remarkable improvement over the past few years. The curriculum is now more aligned with industry needs, and we see a higher number of students securing jobs post-graduation* (Informant 8, Wakiso Town, 8th April 2024).

Others had this to say:

> *The institute is able to determine the courses to offer by scanning market needs* (Informant 5, Wakiso district, 8th April 2024).

> *TVET programs are highly effective in equipping students with practical skills. Many graduates have been able to start their own businesses or find employment in local industries* (Informant 7, Wakiso district, 8th April 2024).

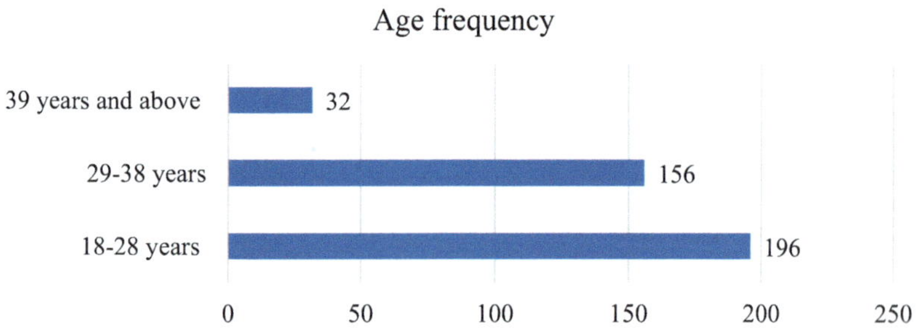

Fig. 17.1 Bar graph illustrating respondent's age group

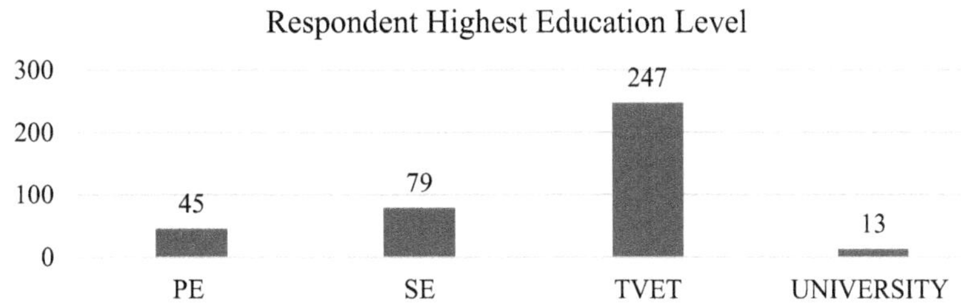

Fig. 17.2 Respondent's highest education

Table 17.1 Employment status of respondents

Category	Frequency	Percentage
Employed	206	53.6
Self-employed	153	39.8
Unemployed	25	6.5
Total	384	100

Table 17.2 Perception of study participants on the effectiveness of TVET training

Category	Frequency	Percentage
Did you complete the TVET training?		
Yes	351	95.6
No	16	4.4
Did the TVET program equip you with relevant skills?		
Yes	339	92.4
No	28	7.6
How would you rate the overall quality of the TVET program?		
Excellent	168	45.8
Fair	26	7.0
Good	60	16.4
Very good	113	30.7
Total	**367**	**100**

Employers are generally satisfied with our graduates. They appreciate the hands-on experience and practical skills our students bring to the workplace (Informant 10, Wakiso district, 8th April 2024).

Another said:

Our graduates are well-prepared for the job market. Many of them secure employment or start their own businesses soon after graduation (Informant 12, Wakiso District, 12th April 2024).

Interestingly, one of the key informants from the Ministry of Education and Sports summarized the role of TVET as an ideal alternative for individuals seeking a practical, hands-on approach to learning and wanting to begin a specific job as soon as possible. TVET can help to close the skills gap in the job market, open up a variety of employment prospects, and promote lifelong learning and career growth. Furthermore, TVET may play an important role in promoting economic growth and development at both the individual and community levels. The officer echoed:

TVET can help close the skills gap by giving students the practical, relevant skills and training they need to succeed in their chosen careers. Many TVET programs are established in partnership with local employers, allowing them to be adjusted to the specific demands of the job market (Informant 9, Kampala, 12th April 2024).

These findings, correspond with the study conducted in central Uganda by Yiga (2022), and Kenya by Mutungi. In Kenya for instance, Mutungi claimed that decision for student to join a public TVET is because of transferable skills that the youths acquire to increase their potential to complete for the available jobs in the market. Yiga (2022) on the contrary with his studies set in central part of Uganda, underscored that unemployment continues because skill levels and learner capabilities are not aligned with work market demands.

What Hinders the Performance of TVET Training?

Inadequate funding

Financial support to student and TVET's institution is one reason I can identify as failing TVET program to bridge unemployment (Informant 13, Kampala, 12th April 2024).

One of the biggest challenges we face is inadequate funding. Many TVET institutions struggle to maintain their facilities and equipment due to lim-

ited financial resources (Informant 17, Kampala, 12th April 2024).

Our facilities are outdated, and we struggle to provide students with the tools they need to learn effectively. Upgrading infrastructure is essential (Informant 14, Kampala, 12th April).

Lack of robust partnerships

There is a lack of strong partnerships between TVET institutions and the industry. This hinders students' exposure to real-world practices and reduces their employability (Informant 9, Kampala, 12th April 2024).

We need more collaboration with industry experts to ensure that the training provided aligns with current market demands. Without this, our programs can become outdated quickly (Informant 13, Kampala, 12th April 2024).

Students do not get enough practical, hands-on experience during their training. This makes it difficult for them to adapt quickly to real-world work environments (Informant 9, Kampala, 15th April 2024).

We need more support from the government and private sector to create internship opportunities and provide students with real-world experience. Collaboration is essential for the growth of TVET (Informant 16, Kampala, 12th April 2024).

Current curriculum is outdated

The challenges range from poor curriculum which does not put the learners or graduate to compete in labor market, to inadequate teaching and facilities from hands-on training and job, and lack of policy will from government (Informant 11, Kampala, 12th April 2024).

The curriculum needs continuous updating to keep pace with technological advancements. Many TVET programs are still teaching outdated skills that are no longer relevant (Informant 16, Kampala, 12th April 2024).

While technical skills are well-taught, graduates often lack soft skills such as communication, teamwork, and problem-solving. These are critical for workplace success (Informant 13, Kampala, 12th April 2024).

Regulatory frameworks are often rigid and slow to adapt to the changing needs of the industry. This makes it difficult for TVET institutions to innovate and update their programs (Informant 15, Kampala, 12th April 2024).

Staff commitment

Basing on my observation so far, staff members have been reporting late to work and little has been done to put them into check. In my view, this is challenging (Informant 10, Kampala, 12th April 2024).

I have observed that the staff of my department seem to be immersed in what they think is best. There is little effort put into following rules. I am actually considering cracking the whip anytime soon to bring sanity into my unit (Informant 14, Kampala, 12th April 2024).

Finding and retaining qualified instructors is a significant challenge. Many experienced professionals prefer working in the industry where they can earn higher (Informant 15, Kampala, 12th April 2024).

Ensuring that instructors are both motivated and capable is essential for maintaining high standards in TVET education.

There is stigma associated with vocational education, with many people viewing it as a less prestigious option compared to university education. This affects enrolment and funding (Informant 13, Kampala, 12th April 2024).

The social stigma attached to vocational education is a significant barrier to the growth and effectiveness of TVET programs. This perception leads to lower enrolment rates and reduced funding, as vocational education is often seen as a less desirable alternative to university education. Addressing this stigma is essential for increasing the appeal of TVET programs and ensuring they receive the necessary support and resources. There is need for efforts towards changing societal attitudes towards vocational education through awareness campaigns and by showcasing the success stories of TVET graduates. Additionally, efforts should be made to align TVET programs with the aspirations of young people by emphasizing the pathways they offer to stable and well-paying careers.

Many students are not adequately prepared for TVET programs due to gaps in their prior education. This makes it challenging to achieve high levels of competence (Informant 17, Kampala, 12th April 2024).

The preparedness of students entering TVET programs is another critical factor affecting their success. The responses indicate that many students come to TVET with significant gaps in their prior education, which hinders their ability to achieve the necessary competence levels. This issue suggests a need for remedial programs or foundational courses that can help bridge these gaps and ensure that all students have the basic skills needed to succeed in TVET. There is need for development of such preparatory programs to enhance student readiness and improve overall outcomes in TVET education.

The results concur with the works of Agole et al. (2022), Mutebi and Ferej (2023), Charles et al. (2023), and Yiga (2022). These researchers stress the importance of TVET in Uganda and the challenges hindering TVET from closing the unemployment gaps existing in the country. Agole et al. (2022) for instance focus on aligning TVET with labour markets since challenges related to curriculum design are cited as problems to TVET. Mutebi and Ferej (2023) agree with the current study by arguing that problems why TVET has failed to close the unemployment gap is limited political will, corruption, and poor conceptualization of TVET training among learners, trainers and the community. Agole et al. (2022) postulated that Uganda still lacks sufficient number of trainers with the required trainers' competences, limited industry participation, and inadequate research support services. Other challenges include: poor geographical distribution Kintu et al. (2019) and location of TVET institutions (Yiga, 2022).

Strategies for Effective Implementation of TVET Programs

During discussions with heads of TVET Institutions and Ministry of Education and Sports officials, the following suggestions were made:

> *Involving experienced business experts offers numerous benefits. They contribute technological expertise and experience. For instance, they can provide industrial attachment opportunities, promote talented individuals for employment, and advocate for sponsorship packages from multilateral companies for financially disadvantaged students* (Informant 9, Kampala, 21/5/2024).

Others said:

> *I believe updating all the curricula from primary level onwards to include more hands-on, practical training would significantly improve employability. Incorporating the latest industry technologies and trends into the coursework can help graduates stay relevant* (Informant 5, Kampala, 21/5/2024).

> *Creating more internship and apprenticeship opportunities would provide students with valuable real-world experience. This could be done by establishing partnerships with local businesses that are willing to mentor and train students* (Informant 12, Kampala, 22/5/2024).

One graduate in formal employment emphasized the relevance of this:

> *This company provided valuable industrial training for me. I gained valuable abilities through collaboration with colleagues and supervisors that I could not have learned at a college training course. After graduating from college, I applied for a position with the manager. He wanted my supervisor's recommendation. It was a wonderful experience that led to my current employment* (Informant 3. Nansana municipality, 23rd/4/2024).

Positively, an official from the Education sector stated:

> *One of the top priorities on the government's and our development pattern's agendas is raising the caliber of TVET graduates. For example, we now have funding for the Higher Education, Student Financing (HESF) initiative at the Ministry of Education and Sports through our arrangement with the African Development Bank (AfDB). In order to meet the demands of the labor market and boost national productivity, the five-year initiative intends to support Uganda's human capital skills development and capacity building, particularly in education, science, and technology. The goal of the HESF project is to increase research and skill development that leads to self-employment and job creation while maintaining equal access to excellent and relevant education* (Informant 9, Kampala, 21/5/2024).

An official at from the Education sector suggested Income Generating units (IGUs) as another way of facilitating TVET graduates transition to the world of work. He said:

> *We at the ministry have always advised TVET institutions to turn their training workshops into vibrant Income Generating Units (IGUs). It offers*

> *the institutions and their students a lot of benefits. I remember when I was a student at former Uganda Polytechnic Kyambogo, now Kyambogo University in the late 90s, the mechanical and production engineering workshops received a lot of work from the industrial areas in and outside Kampala. The technicians in the workshops would engage me and my colleagues to help them complete the work on time. We would get paid for our input but most importantly, we learnt how to negotiate with the customers, improved our technical skills, and a good number of our colleagues were recommended for employment in the customers' organisations. I was told recently that such units were closed. Why close them? He wondered.* (Informant 4, 22/04/2024).

One of the employers in Wakiso, Kampala, and Central region of Uganda suggested internship programs for graduates while waiting for employment. He said:

> *Creating more internship and apprenticeship opportunities would provide students with valuable real-world experience. This could be done by establishing partnerships with local businesses that are willing to mentor and train students* (Informant 6, Wakiso District 25th/04/2024).
>
> Another had this to say:
>
> *I strongly believe that sending children to university may be a waste of time, as the education system struggles to keep up with current job market demands. I've discussed this with my 12-year-old daughter and suggested we pull her out of the traditional system to start working instead. Many parents should consider this, as businesses are growing and scaling but lack qualified workers. University graduates often need retraining, so hiring directly from senior six and training them in our academy might be more effective. This approach ensures they are work-ready, similar to a student we hired straight from senior six who now runs a successful TV show. We spend too many unnecessary years in school learning irrelevant material* (Informant 13, Wakiso District 26th/04/2024).

During the interviews, some graduates who were now in formal employment recalled how they got their jobs. One of them said:

> *The industrial training I underwent when I joined this company was really important for me. I worked so hard with colleagues and supervisors and I acquired a lot of skills* (Informant 7, Wakiso Town 25/04/2024).

Another had this to say: "*The company wrote to our institution requesting for finalists they could employ. I was lucky; my name was among those submitted. After a few weeks, they invited me for training. The training took four months, after which five of us were selected out and invited to negotiate contracts with the company. That is how I got employed here*" (Informant 10, Wakiso District 25/04/2024).

The following insights were provided by the employers and TVET officials on strengthening collaboration between TVET institutions and industries be to better address the workforce requirements: "*Regular meetings and forums between TVET institutions and industry leaders would help ensure that training programs align with current industry needs. This dialogue is essential for keeping the curriculum relevant and up-to-date.*" (Informant 6, Wakiso District 25th/04/2024).

> *We should establish industry advisory boards that include representatives from key sectors. These boards can provide direct input on curriculum development and help identify the skills that are in high demand* (Informant 4, Kampala 22/04/2024).
>
> *Offering incentives such as tax breaks or subsidies to companies that collaborate with TVET institutions could encourage more businesses to get involved. This financial support could offset the costs associated with training and mentoring students* (Informant 13, Wakiso District 26th/04/2024).

The employers, heads of TVET institutions and officials from the Ministry of Education and Sports provided the following major factors that support the effectiveness of TVET programs in preparing graduates for the job market:

> *The effectiveness of TVET programs largely depends on the quality of instructors. Teachers need to have current industry experience and be trained in the latest technologies and methodologies* (Informant 4, Kampala 22/04/2024).
>
> *Access to modern, up-to-date equipment and facilities is crucial. Students must be able to practice their skills using the tools and technology they will encounter in the workplace* (Informant 13, Wakiso District 26th/04/2024).
>
> *Strong industry partnerships are key. When businesses are actively involved in shaping the curriculum and providing training opportunities, it*

> *ensures that graduates are ready to meet the demands of the job market* (Informant 6, Wakiso District 25th/04/2024).

The following policy changes or support mechanisms were suggested by employers, heads of TVET institutions and officials from the Ministry of Education and Sports as means to encourage more employers to hire TVET graduates:

> *Implementing a tax breaks or subsidies for companies that hire TVET graduates would be a strong incentive. This financial benefit could encourage more employers to consider these candidates* (Informant 13, Wakiso District 26th/04/2024).

> *Government grants or subsidies for companies that create apprenticeship programs could help bridge the gap between training and employment. These programs provide graduates with valuable work experience while benefiting employers* (Informant 6, Wakiso District 25th/04/2024).

> *Establishing a certification system that is recognized by the entire employment industry can help standardize the quality of TVET graduates. Employers would have more confidence in the skills of certified graduates, knowing they meet a certain standard of competency* (Informant 4, Kampala 22/04/2024).

During discussions with employers, heads of TVET institutions, TVET graduates, and Ministry of Education and Sports officials, it was emphasized that involving experienced industry experts as guest lecturers could greatly benefit TVET programs. These experts can provide industrial attachment opportunities, promote talented students for employment, and advocate for sponsorships from multinational companies for financially disadvantaged students. Additionally, updating all curricula to incorporate more hands-on training and current industry technologies was suggested as a vital improvement to enhance employability. Creating more internship and apprenticeship opportunities through partnerships with local businesses would also provide students with valuable experience, preparing them better for the job market.

Furthermore, officials highlighted the importance of government initiatives like the Higher Education Student Financing (HESF) project, which can be leverage to support the TVET program. They also suggested turning training workshops into vibrant Income Generating Units (IGUs) to provide practical experience and financial benefits for students and institutions. Collaboration between TVET institutions and industries can be strengthened through regular meetings, establishing industry advisory boards, and offering incentives like tax breaks or subsidies for companies that collaborate with TVET programs. These measures would ensure that training aligns with industry needs and encourages businesses to invest in TVET graduates.

Major factors supporting the effectiveness of TVET programs include the quality of instructors, access to modern equipment, and strong industry partnerships. Ensuring instructors have current industry experience and providing students with up-to-date tools are crucial for preparing graduates for the job market. Policy changes suggested to encourage employers to hire TVET graduates include implementing tax breaks or subsidies for companies that hire TVET graduates, government grants for companies that create apprenticeship programs, and establishing a recognized certification system to standardize quality of TVET graduates. These measures would boost employers' confidence in TVET graduates' skills and competencies.

Conclusions

The study explored the effectiveness of TVET training and has documented strategies required to enhance the performance of TVET programs. By implementing the suggested strategies, stakeholders can improve TVET's quality and effectiveness, leading to better employment outcomes for youth and contributing to economic development. There is need for longitudinal studies to track the career progression of TVET graduates and investigate the role of emerging technologies in TVET training.

Acknowledgements I highly appreciate the team of Research Assistants which assisted me in data collection and analysis. Sincere thanks to the various respondents who provided the insights which are synthesized in this report. Special thanks go to my employers at Office of the President, Cabinet Secretariat, who funded my studies and supported me unwaveringly, most especially the Commissioner Policy Development & Capacity Development, Dr. Moki Muhammad Abubakar, for the extra guidance and support. I thank God for the blessings.

References

Agole, P., Kerre, B., Ochieng, D., & Okaka, W. (2022). Confronting the challenges of university technical vocational education and training in Uganda. *African Journal of Education, Science and Technology, 7*, 2.

Charles, K., Song, Z., & Djaya, H. G. N. (2023). TVET and socio-economic development in Uganda: Lessons from Chinese TVET. https://doi.org/10.5281/zenodo.10214473

Mugenda, O. M., & Mugenda, A. G. (2003). *Research methods, quantitative and qualitative approaches*. ACT.

Mutebi, R., & Ferej, A. (2023). A review of TVET quality assurance practice in Uganda. *East African Journal of Interdisciplinary Studies, 6*(1), 182–196. https://doi.org/10.37284/eajis.6.1.1327

Yiga, S. (2022). Assessment methodologies and determinants of employability and skills level among Technical and Vocational Education and Training (TVET) graduates in Central Uganda. *International Journal of Vocational and Technical Education, 14*(2), 40–47.

The Cooperative Movement in Uganda: An Examination of Performance of Agricultural Cooperatives

18

Nathan Hagobi Hasahya, Marie G. Nakitende, Simeon Wanyama, and Gideon Nkurunziza

Abstract

This chapter explores the evolution of agricultural cooperatives, main determinants influencing their performance, together with approaches for their restoration and strengthening in Uganda. It reviews the historical path of cooperative performance, outlining the drivers of both success and decline, drawing on existing studies and examples from selected cooperative unions, guided by Ronald Coase's (Economica, 4, 386–405, 1937) theory of the firm. Through these case illustrations, the chapter highlights the role of cooperatives and the reasons for their varying outcomes, which are largely shaped by internal organizational practices, such as shared institutional values and transparent leadership, alongside state involvement, the depth of social networks, and community integration. Consequently, for stakeholders seeking to reestablish cooperatives in Uganda, emphasis should be directed toward fostering member-based and locally owned entities that are democratically managed, engaged in high-value agricultural production and marketing, and guided by a collective vision with accountable and responsible leadership.

Keywords

Economic war · Expelled expatriates · Cooperative Bank · Asians · Custodian Boards

Background

This chapter traces cooperative performance trajectory to analyze the key factors influencing the functioning of agricultural cooperatives, with a critical review of strategies for their renewal and strengthening in Uganda. Although a wide range of studies exist, cooperatives are still not well comprehended as institutions (Cuevas & Fischer, 2006). The International Cooperative Alliance (ICA, 2010) defines cooperatives as independent associations of individuals who come together in view of addressing shared socioeconomic and cultural needs as collectively managed initiatives. Nweze (2001) further explains cooper3atives as being enterprises formed and member governed, operating primarily on service rather than profit motives, where ownership, decision-making, and usage are closely aligned. They are recognized as

N. H. Hasahya (✉) · M. G. Nakitende · S. Wanyama
Faculty of Business Administration and Management, Uganda Martyrs University (UMU), Kampala, Uganda
e-mail: hasahya.nathan@stud.umu.ac.ug

G. Nkurunziza
Department of Business and Management Sciences, Makerere University Business School (MUBS), Kampala, Uganda

P. Nyadoi et al. (eds.), *Sustainable Development Goals in Africa*, Sustainable Development Goals Series, https://doi.org/10.1007/978-3-032-20742-5_18

having a significant role in poverty alleviation and in driving both economic growth and development (Develtere et al., 2008; Birchall & Simmons, 2004), explaining why they were introduced in Africa during the colonial era and why they continue to be regarded as vital today, despite limited scholarly investigation into their ongoing relevance (Wanyama et al., 2008; Wanyama et al., 2009 and Wanyama et al., 2011). In Uganda, although some policymakers view cooperatives as outdated, the country recognizes their importance in enhancing agricultural productivity and efforts are being made to revive them (World Bank, 2011; Dennis et al., 2016; ActionAid, 2013).

This chapter examines two dominant perspectives on cooperatives: one that sees them as tools for elite exploitation (Brett, 1970a, b; Bates, 2005) and another that views them as vehicles for Africa's economic liberation (Helmen, 1990; Kwakyewah, 2016). We adopted Ronald Coase's (1937) theory of the firm, which posits that firms are created to minimize transaction costs by internalizing exchanges that would otherwise occur in the market. This theory is relevant to agricultural cooperatives in Uganda, as it explains their capacity to efficiently manage production and transaction costs by pooling resources. Coase's framework clarifies why cooperatives face challenges such as inefficiencies and mismanagement, providing a theoretical basis for evaluating their performance. We employed both descriptive and theoretical approaches, carried out a synthesis, and this chapter provides an in-depth account of the condition of Uganda's cooperative movement and the performance of agricultural cooperatives in particular. The subsequent sections present a cooperative's historical background and Factors of Success and Failure with Selected Cooperative Unions as examples on which to draw conclusions and give recommendations on the way forward.

Historical Perspective of the Cooperative Movement

The cooperative movement's origin is traced to the eighteenth century, coinciding with the Industrial Revolution. As populations shifted from rural farms to expanding urban centers, households increasingly depended on shops for food since they could no longer rely on subsistence production. Ordinary workers had limited influence over food quality and their living conditions, while wealthier classes accumulated more power. Cooperatives emerged as an avenue for safeguarding the interests of marginalized groups—farmers, consumers, laborers, and producers. The first formally recognized cooperative society was established in 1844 in Rochdale, England, by a group of individuals known as the "Rochdale Pioneers," at a time when Britain was undergoing rapid industrial transformation (KNFC website). Shortly afterward, Germany witnessed cooperative formations in 1846 during an agricultural crisis and famine in Europe. Although similar forms of cooperation had existed earlier, the Rochdale model is regarded as the foundation of modern cooperatives, since it was here that the guiding principles of the movement were articulated (Gibson, 2005a, b; Levin, 2005). By around 1850, Germany also pioneered new cooperative approaches, while in England consumer groups began pooling resources to purchase goods directly from wholesalers, as a form of protest against storeowners who adulterated products for profit.

In Africa, the development of cooperatives generally followed two main phases. The first phase stretched from the immediate post-independence period of the 1960s to the 1990s, characterized by strong state involvement through policies, laws, and programs that promoted cooperatives as tools for advancing national development. The second phase emerged during the liberalization period in the mid-1990s onward, which allowed cooperatives more independence to operate as business-oriented entities responsive to market dynamics (Wanyama et al., 2009). Among African countries, the English-speaking nations were the earliest adopters of cooperatives. In Ghana, Father John McNulty introduced cooperatives around 1959 with the aim of assisting rural communities to enhance their livelihoods (Jesee, 2016). Uganda, Nigeria, Kenya, and Tanzania soon followed. Non-English-speaking African nations began to recognize the importance of cooperatives mainly in the 1960s (Jesee, 2016), particularly through farmer-led organizations focusing on export crops, such as cotton, coffee, and pyrethrum.

According to Munyami (2014), Ghana's cooperative experience served as an important reference point for other African countries. Historically, cooperation among people has existed since primitive times when groups worked together, for example in hunting, where collective effort was necessary for survival (Cobia, 1989). After independence, African governments enacted laws and policies giving them significant control over cooperatives. Promotion, regulation, and supervision of these organizations were often handled by specialized government departments or ministries. This granted cooperatives monopolistic roles in sectors like agricultural marketing, where they became the exclusive agents of State Marketing Boards mandated to purchase cash crops of cotton, coffee, and pyrethrum, process them, and prepare them for sale abroad. Although the financial benefits to cooperatives were limited, this monopoly ensured farmers joined them in order to access markets.

In addition, state-led agricultural credit programs were delivered through cooperatives, providing another incentive for farmer participation. However, over time, governments used cooperatives not only for economic objectives but also as tools for political patronage, including offering employment to political allies (Wanyama, 2011). This integration into state politics weakened their voluntary and democratic nature. The situation was compounded by donor agencies' support from the cooperative movements and credit unions of the Nordic, American, and Canadian communities, which often channeled financial support to African cooperatives through state institutions. This reinforced government dominance over the cooperative movement, while at the same time making cooperatives instruments for collecting agricultural produce and distributing farm inputs on behalf of state marketing boards (Braverman, 1991).

In Kenya, cooperatives were initially introduced to augment settlers' socioeconomic well-being. Firstly, Lumbwa Cooperative Society was formed in 1908 by European farmers to collectively support agricultural activities and benefit from economies of scale (KUSCO, 2014). Following independence, the Kenyan government recognized the strategic role of cooperatives in advancing development. In recent decades, cooperatives in the country have become an important socioeconomic driver of the rural economy to alleviate poverty.

During the formative period between 1913 and 1952, cooperatives in Uganda started in 1913, reacting against the exploitative dominance of European and Asian traders in the cotton and coffee trade. At first, the colonial administration opposed these initiatives, describing them as premature and potentially disruptive. An attempt to introduce a cooperative bill between 1935 and 1937 failed after strong resistance from influential private business groups, forcing early cooperatives to function informally until the coming into force of the Cooperative Societies Decree in 1946. The ordinance granted formal legal status to cooperatives, although a number of them hesitated to register, fearing it would subject them to excessive state control (Karanja, 2001). Between 1953 and 1961, Uganda saw significant growth in its cooperative movement. Governor Sir Andrew Cohen sought to shift cooperatives from being protest organizations to collaborative ones. In 1952, he appointed the Dreschfield Commission to investigate why many cooperatives had resisted registering under the existing ordinance. The commission's findings led to amendments in the law, granting cooperatives more autonomy. As a result, by 1959, 273 cooperatives had registered, and by the end of 1961, Uganda had 1662 primary cooperative societies and 21 cooperative unions, including the Uganda Cooperative Alliance, with a membership of over 252,000 then.

This period represented a major shift in the growth of cooperatives, as unions gained increasing influence over key sectors such as coffee and cotton production (Monitor, September 25, 2013; New Vision, October 22, 2015). After Uganda attained independence in 1962, the government used cooperatives as a strategy to uplift rural populations. A significant milestone was the coming into force of the 1963 Cooperative Societies Act, which reinstated registrars' position and combined it with the role of Commissioner for Cooperative Development. The legislation empowered departmental officials to implement policies and enforce regulations, unlike in the past when their recommendations were often disregarded. Overall, the state sought to consolidate cooperative systems, with particular focus on driving agricultural-led economic growth (ICA, 2019).

Recognizing the role of education in advancing government policy, a cooperative Training College was first established at Bukalasa in 1963, from which a well-established institution was constructed at Kigumba. The first cooperative college remained under the main college until it was eventually transferred and taken over by the Ministry of Agriculture, Animal Industry and Fisheries (MAAIF). The Cooperative Development Bank setup in 1964 aimed at mobilizing members' investments in order to provide financing for growth. During this period, cooperatives were granted exclusive rights of cotton ginning. In 1965, the value of agricultural produce marketed by cooperatives, comprising cotton, coffee, livestock products, and other cereal crops, had reached UGX 325,311,500; from this, 423,923 bales were for cotton, of which 267,420 bales were managed (about 61%), coupled with handling 40% of Robusta coffee worth UGX 60,000,000 and 90% of Arabica coffee worth UGX 30,000,000 (Kisaka, 2023; Ochieng, 2022; Tumusiime, 2023; Ssenyonga, 2022; Mutebi, 2023; Nabimanya, 2022). The cooperative movement experienced significant expansion across multiple sectors such as fishing, dairy farming, livestock trade and their products, housing, handicrafts, and financial investment schemes. The rapid development largely depended on strong state backing in the form of direct support and subsidized services (Kyazze, 2010).

One of the most significant changes in Uganda's cooperative framework is reflected in the evolution of agricultural marketing cooperatives. Prior to the liberalization policies of the 1990s, cooperatives in Uganda followed a centralized and tiered model: farmers sold their produce to primary societies, which in turn delivered it to cooperative unions from where they then supplied their products to the marketing boards that were responsible for securing export outlets. However, liberalization greatly weakened cooperative unions, prompting local communities, parish-based farmer associations, and smaller farmer groups to rebuild solidarity and establish new cooperative structures at the grassroots. This led to the creation of Growers' Cooperative Societies (GCSs), which collected produce from individual farmers and aggregated it for collective marketing (Abenakyo, 2023; Kagina, 2022; Nabirye, 2023; Mubiru, 2022; Mugisha, 2023a, b; Ssentamu, 2022).

To enhance their bargaining capacity in setting fair produce prices, several GCSs within a subcounty combined efforts to form Area Cooperative Enterprises (ACEs). These acted as mini-unions for the GCSs, with the responsibility of securing better markets and negotiating improved prices. Unlike the old system, ACEs had the flexibility to sell to a variety of buyers (private traders, larger cooperative unions, or even international organizations), provided they could secure competitive returns for members' produce. As many developing countries struggle with the problem of agricultural stagnation, renewed emphasis has been placed on restoring agricultural cooperatives. Scholars argue that cooperatives play a crucial role in helping farmers coordinate their activities, gain access to credit and other agricultural finance, improve productivity, strengthen collective bargaining and market linkages, and increase household earnings; all of which contribute to poverty reduction (Helmen, 1990; Kwakyewah, 2016).

In Uganda, where 80% of the populace resides in the countryside and nearly 70% of households depend on farming, there has been growing recognition of the need to revive agricultural cooperatives as vehicles for rural transformation and economic progress. National development blueprints articulate the objective.

Factors of Success and Failure of Cooperatives–Cooperative Unions

The establishment of the farmers' association in 1913 in Uganda marked a beginning of the country's cooperative movement, aimed at protecting local farmers from the exploitative practices of foreign firms which had dominated both the local and the foreign markets for agricultural produce, especially coffee and cotton. These key crops became central in the operations of the cooperative system, drawing attention from both colonial and post-independence administrations. Initially, however, the colonial authorities viewed cooperatives as premature and potentially destabilizing, and consequently denied them legal recognition and access to credit or financial services. An attempt to pass a cooperative bill in the legislature during the 1930s faced solid opposition emanating from influential remote stakeholders and was subsequently reversed, compelling early cooperatives to function informally until the 1946 Co-operative Societies Ordinance was enacted. Several authors attest that the growth from 1662 primary cooperative societies in 1961 to over 2500 by 1971, with membership rising from 252,378 to over 750,000, reflects a significant increase in both the scale and reach of the cooperative movement (Opobo, 2015; Kasamba, 2022; Kato, 2021; Kawuma, 2022a, b, Muwanguzi, 2022; Nabwiso, 2022; Namubiru, 2022; Kakumba, 2023; Kiggundu, 2023; Lwanga, 2023; Muwanga, 2023; Sebunya, 2023; Ssekitoleko, 2023). This expansion indicated a growing trust in cooperatives as effective organizational business forms for addressing agricultural and economic challenges.

The substantial rise in the volume of produce managed by cooperatives, from 14,300 tons in 1951 to 89,308 tons in 1960, demonstrates their effectiveness in aggregating and marketing agricultural commodities. By handling 61% of the nation's cotton produce, 40% and 90% of Robusta and Arabica coffee, respectively, cooperatives contributed greatly to Uganda's agricultural sector in terms of productivity. With an annual turnover exceeding £9 million and assets estimated at USh. 500,000,000 million by 1971, cooperatives clearly had a major economic impact, reflecting their capacity to invest, influence markets, and contribute to rural development. This growth also highlights the importance of sound management, appropriate regulation, and continued support to sustain cooperative performance. Beyond their economic functions, cooperatives proved crucial in mobilizing and managing significant financial and material resources. The establishment and ownership of 53 cotton ginneries and 31 coffee factories by cooperative unions signifies a high level of control over key aspects of the agricultural value chain. This ownership likely contributed to better value addition and profit retention within the cooperative sector.

The rise in cooperative membership and the increased handling of agricultural produce suggest that cooperatives were successful in empowering farmers. By providing a collective platform for marketing and processing, cooperatives likely helped smallholder farmers achieve better prices and improved market access. The substantial turnover and asset base of cooperatives indicate that they became integral to Uganda's economy. Their financial and operational scale would have contributed to economic stability and growth, particularly in rural areas. However, with rapid growth, cooperatives likely faced challenges related to governance, management, and sustainability. Ensuring effective management and transparency would have been crucial for maintaining trust and operational efficiency. The success of cooperatives during this period set a strong foundation for their future development. The period also illustrated the potential for cooperatives to serve as key instruments for national economic development, particularly in rural areas where they could continue to foster agricultural expansion and community progress.

Kwapong and Korugyendo note a decline in performance of cooperatives beginning 1971, upon the assumption of power through a coup by Idi Amin, when he ousted the then President Dr. Milton Obote. Amin launched his so-called economic war and expelled foreigners in 1972. Their subsequent leaving combined with the running to exile by the experienced cooperative union leaders and managers, further weakened the operating system. Political and economic self-interest increasingly infiltrated cooperative operations and, coupled with widespread financial mismanagement, led to the collapse of cooperatives. This period coincided with the collapse of the agricultural markets, as controlled crop prices, particularly for cotton and coffee, fell sharply, prompting farmers to switch to maize and other cereals. Cotton produce plunged from 466,775 in 1970 to 32,160 bales in 1980. Smuggling of coffee beans locally referred to as "kasse" and the proliferation of numerous processing plants further reduced cooperatives' market share from nearly 100% to around 37% in 1979, after the removal of Idi Amin.

These findings underscore the severe effects of political instability and poor economic governance on Uganda's cooperative movement. The decline in cooperative efficiency and agricultural output highlights the need for strong institutional governance, ethical leadership, and resilient frameworks to maintain cooperative systems during periods of political and economic uncertainty. The economic liberalization of the early 1990s dealt a further blow to cooperatives. The closure of the Cooperative Ministry, privatization of the Cooperative Bank coupled with liberalization of the agricultural marketing activities stripped cooperatives of their assets, financial muscle including the social capital, ending their historic crop management monopoly of produce marketing and export operations and depriving them of their original operations. This left cooperatives ill-equipped to compete with experienced multinational corporations and agile private traders (Birchall & Simmons, 2004). Their performance declined even further following the 1991 Cooperative Societies Statute, which increased government oversight and intervention in cooperative affairs. This led to significant politicization in the election of governing boards, resulting in the selection of incompetent and self-serving individuals. The maladministration by these boards caused issues such as adulteration of cotton quality and internal conflicts driven by political authority. Consequently, cooperatives lost their market position, cotton prices fell, and the resulting financial losses led to the sale of assets to settle debts.

Amelia Kyambadde (2018), then minister, emphasized that the Ugandan Government had taken a step in prioritizing cooperative revitalization due to their vital contribution to socioeconomic development. Cooperatives support employment creation, increase household incomes, enhance agricultural productivity, promote value addition, and foster social cohesion. By May 30, 2018, the Ministry had registered 17,967 cooperatives across sectors such as agricultural marketing, transport, housing, health, education, fishing, energy, dairy and beef, multipurpose activities, and financial cooperatives/SACCOs. By February 2020, the cooperative movement had grown to 21,346 registered societies, encompassing approximately 5.6 million members; a significant rise from the 10,746 societies registered in January 2011 (Kyambadde, 2018).

Uganda's agricultural cooperative performance has notably improved in recent years, driven by government-led revitalization initiatives. The growth in cooperative numbers and their expansion into multiple sectors underscores their crucial role in fostering socioeconomic development. By learning from past successes and failures, as well as from experiences of other African cooperatives, the government has implemented strategies to strengthen the cooperative sector. Key measures include creating an enabling environment through robust policies and legal frameworks to back up cooperatives as autonomous business units, building self-governing institutions, offering practical education and training, and providing clear operational guidelines (Uganda Cooperative Alliance, 2023). These strategies aim to enhance autonomy, governance, technical capacity, and sustainability, thereby improving the effectiveness of the cooperative movement in agricultural development and broader socioeconomic progress.

A development economist, Stephen Ssembogga, contends that while there are commendable deliberate efforts for cooperative revival, additional interventions are needed to boost competitiveness. He recommends reinstating the Cooperative Bank to provide low-interest credit to cooperatives, which could stimulate affordable financing across agriculture and other non-traditional sectors. Additionally, he emphasizes that cooperatives should uphold upright governance by prioritizing member benefits and integrating farmer education, production, agroprocessing, and marketing to remain competitive in a liberalized market. The liberalization reforms of the 1990s significantly challenged cooperative societies in Uganda. Only a few, such as Bugisu Cooperative Union (BCU), Nyakatonzi Growers Cooperative Union Limited (NGCU), and Banyankole Kweterana Cooperative Union, survived the upheavals, while most others failed to adapt. Successful cooperatives relied heavily on strong, focused leadership. Boards democratically elected by representatives from primary societies instilled accountability, ensuring continuity of service and fostering member loyalty even under liberalization pressures (Okello & Ahikire, 2013).

Reports from the Center for Basic Research & Action Aid (2013a, b, c) indicate that cooperative survival and success depend on robust leadership, member-centered approaches, innovation, technical support, and strategic partnerships. For example, UCSCU's collaboration with WOCCU allows the cooperative to benefit from international expertise in debt management and socioeconomic stability. Conversely, failures are often linked to poor management, weak member engagement, financial mismanagement, lack of innovation, and insufficient institutional support. Comprehensive strategies are therefore necessary to address operational weaknesses, strengthen member participation, and secure both financial and technical resources to ensure sustainability.

Indebtedness and misappropriation have posed major challenges for cooperatives. For instance, South Bukedi Cooperative Union in Tororo faced severe financial difficulties due to poor management and inability to repay loans, resulting in the seizure of assets, such as vehicles, buildings, and the Nyakesi Ginnery. Reports also indicate that funds accruing from the sale of Dabani and Busolwe Ginneries, totaling Ushs 240,000,000 million were misappropriated (Center for Basic Research & Action Aid, 2013a, b, c). Mismanagement has similarly led to significant losses in other unions. At its peak, the Bunyoro Growers' Union held extensive assets, including a cotton ginnery, petrol station, tannery, carpentry workshop, vehicles, and large land and livestock holdings, yet mismanagement undermined its operations and asset value. However, most of these assets were eventually lost due to poor management and the need to sell off property to settle debts. Similarly, the Masaka Growers Cooperative Union (MGCU) faced severe challenges due to mismanagement, political interference, and financial difficulties. Despite efforts to utilize loan facilities to establish a soft drink factory, the venture collapsed by 1996, and attempts to revive it failed due to the exhaustion of funds and the burden of debt. Political interference, including the appointment of political leaders to the Union's board, further worsened the situation. This means that the cases of the Bunyoro Growers' Union and Masaka Growers Cooperative Union illustrate how mismanagement, coupled with financial and political challenges, can severely impact cooperative operations. The Bunyoro Growers' Union, once a thriving entity with diverse assets, saw its fortunes decline due to mismanagement and the necessity to liquidate assets to cover debts. This scenario underscores the critical importance of sound financial management and strategic oversight in sustaining cooperative organizations.

The Masaka Growers Cooperative Union's experience highlights additional layers of complexity, such as the detrimental effects of political interference. The appointment of politically motivated leaders to the board often leads to decisions driven by personal or political agendas rather than the cooperative's best interests. This, combined with financial mismanagement and the inability to capitalize on loan-funded ventures, can result in the collapse of otherwise viable enterprises. These examples emphasize the need for effective management practices, transparent financial oversight, and the avoidance of political interference to ensure the sustainability and success of cooperatives (Center for Basic Research & Action Aid, 2013a, b, c).

Unfeasible Entities: Sebei Elgon Union is a clear example whose establishment was founded by the belief that its farmers contributed 8% of the total business of BCU. The political motivations behind the formation of the Union did not consider whether this level of contribution was sufficient to support a self-sustaining, independent cooperative.

Inefficient and Ineffective Leadership: The challenges faced by cooperatives due to inefficient and ineffective boards and management are highlighted by several key issues. Ineffective boards and management teams often failed to innovate, seize new opportunities in an open economy, form beneficial partnerships, and maintain a member-focused approach. This mismanagement led to a decline in the cooperatives' performance and member satisfaction. A specific example of such ineffective leadership is seen in the Lango Cooperative Union. The leadership's inability to keep members adequately informed about the union's operations led to dissatisfaction and a decline in trust. Some members alleged that ginneries of the union had been sold to a foreigner-owned cotton company. The union's secretary, however, clarified that the ginneries located at Odokomit, Aboke, Jaber, Aloi, Namasale, Aduku, and Ibuje had been vended in 1997 by the government in a bid to settle financial obligations payable to previous union employees (Center for Basic Research & Action Aid, 2013a, b, c).

Losses from Conflicts: The effects of war on cooperative unions are exemplified by the Banyankole Kweterana Cooperative Union, which experienced asset losses during regional conflicts.

Although the government provided some compensation, the losses and disruptions caused by the wars significantly affected the cooperative's operations and long-term survival. This implies that the impact of war on cooperative unions can be profound, leading to significant asset losses and operational disruptions. For instance, the Banyankole Kweterana Cooperative Union, established in 1986, experienced severe asset losses due to the conflict. Despite the government providing some compensation, it was insufficient to fully address the extensive damage and operational challenges faced by the union.

The war-induced losses not only affected the physical assets but also disrupted the union's ability to maintain effective operations and serve its members. Furthermore, internal factors such as management inefficiencies, inability to adapt to the post-war economic environment, and other organizational weaknesses compounded the difficulties. The decline of the Banyankole Kweterana Cooperative Union demonstrates that although government compensation may provide partial support, it frequently falls short of addressing the complex challenges caused by conflict. Effective recovery from such setbacks requires a combination of financial support and strategic rebuilding efforts, including strengthening management practices, improving organizational resilience, and addressing the broader socioeconomic impacts of the conflict. The experience of the Banyankole Kweterana Cooperative Union highlights the need for comprehensive crisis management and recovery strategies in cooperatives affected by war, to mitigate both immediate damage and long-term operational issues.

The government's liberalization policy was introduced suddenly, despite cooperative unions having relied on subsidies for many years. This sudden shift represented a major change for unions that had previously operated with substantial government support. Prior to liberalization, cooperatives received financial support, favorable credit terms, and other assistance, enabling them to expand and stabilize their operations, particularly in agriculture, where they handled substantial portions of crops like cotton and coffee. Liberalization removed these subsidies, forcing cooperatives to compete in an open market without sufficient preparation. The sudden withdrawal of government support highlighted the cooperatives' vulnerabilities, contributing to their collapse or decline. This situation underscores the need for a phased approach to policy changes, as well as capacity-building measures and market access to support organizations transitioning from subsidy dependence. The experience of Uganda's cooperatives shows that abrupt liberalization can destabilize key sectors, emphasizing the importance of gradual implementation and strategic support (Center for Basic Research & Action Aid, 2013a, b, c). Today, many cooperatives in Uganda face challenges and seek government support to revive their operations. The government's revival initiative seeks to offer financial support, policy guidance, and strategic direction to enable cooperatives to rebuild, thereby promoting economic growth and enhancing wealth creation (Kyambadde, 2018; Ssembogga, 2020a, b; Center for Basic Research & Action Aid, 2013a, b, c).

Conclusion

The disintegration of cooperatives in Uganda contributed to rising income disparities and poverty in rural areas, which in turn accelerated rural-to-urban migration and weakened the rural economy (Halerimana et al., 2010). This situation sparked debates on reviving cooperatives as tools for rural development (DFID, 2008) and prompted the government to reconsider its stance on cooperatives (RoU, 2012). Cooperatives have been identified as vital for addressing income generation and human development challenges in developing countries, with the International Labour Organization estimating that they provide employment for 250,000,000million people worldwide. Historically, cooperatives in Uganda played a significant role in agricultural transformation, enabling farmers to access markets, credit, and quality standards (ActionAid, 2013; Kyazze, 2010). However, many collapsed by the late 1970s due to economic liberalization, political instability, and price declines in key crops like cotton and coffee. Farmers turned to smuggling their produce to neighboring countries, further undermining the cooperatives (Brett, 1970a, b). Conflicts and political instability during the 1970s and 1980s worsened these challenges, with key cooperative areas such as the Luwero Triangle experiencing substantial losses (Mutibwa, 1990; Akandwanaho, 2002). Economic liberalization in the 1990s dealt a final blow to cooperatives through privatization and the dissolution of key cooperative institutions like the Ministry of Cooperatives and the Cooperative Bank. While liberalization boosted national economic growth, rural communities where 80% of Ugandans reside suffered as cooperative assets were left to decay (Republic of Uganda, 2015).

Critics of cooperatives, influenced by New Institutional Economics, argue that cooperatives failed due to flawed incentive structures that fostered leadership corruption and inefficiency (Bates, 2005). The Berg report of the World Bank (1981) emphasized government interventions and policy errors that played a role in causing the agricultural crisis of the 1970s (Sender & Smith, 1984). However, some cooperatives, like Bugisu Cooperative Union, survived and even thrived despite these challenges, questioning the notion that all cooperatives were doomed to fail due to leadership exploitation (Bates, 2005). Furthermore, poverty and inequality were most severe during the era of structural adjustment reforms, not state intervention, which often directed surplus from cooperative activities toward industrialization efforts (Bates, 2005). This complicates the argument that cooperatives primarily failed due to elite exploitation and highlights the need for more nuanced policy approaches to rural development initiatives (Mkandawire, 2001).

The contribution of cooperatives to Africa's development is often viewed from contrasting perspectives. Critics (Brett, 1970a, b; Bates, 2005) contend that cooperatives serve as tools for elite exploitation, whereas supporters (ActionAid, 2013; Helmen, 1990; Kwakyewah, 2016) regard them as instruments of opportunity for advancing African capitalism. However, such one-dimensional views fail to capture the nuanced realities of cooperative performance across the continent. The evidence presented here highlights that cooperatives' success or failure largely hinges on organizational dynamics, leadership, government intervention, and external reforms. The varied experiences of African cooperatives, with some thriving while others collapsed, underscore the importance of context-specific factors, particularly the role of government intervention, leadership values, and organizational stability. The decline of many cooperatives, contrasted with the success of a few, challenges the narrative that cooperatives are no longer relevant for African development. Instead, it reinforces the need for a more nuanced understanding of their roles and challenges in different contexts.

Recommendations

Revival efforts for cooperatives in Uganda should prioritize the development of democratic organizations that are deeply rooted in local communities. These organizations must foster shared values, purpose, and accountability among members to ensure long-term sustainability. The success of cooperatives heavily depends on strong and ethical leadership. Leaders must be accountable to members, transparent in decision-making, and committed to the cooperative's mission. Building leadership capacity and governance structures should be a central focus of cooperative development.

Cooperatives thrive when they are built on social cohesion and mutual trust. Encouraging active member participation and building a sense of ownership are crucial for the survival of cooperatives. Social capital should be harnessed to foster unity and collaboration among members. Cooperatives must aim to integrate into higher-value agricultural production and marketing chains. This allows members to increase their incomes and improve the long-term viability of cooperatives. Agricultural Cooperative Unions should be encouraged to facilitate access to markets and value-added products. The growth and success of cooperatives should be driven by the genuine interests of their members rather than external incentives or handouts. Encouraging self-sufficiency and member-driven motivation will create stronger, more resilient cooperatives that can withstand external shocks. By adhering to these principles, practitioners can help ensure that cooperatives in Uganda and across Africa demonstrate a significant role in the countryside economic growth and development coupled with alleviating deficiency. Members' dedication, accountable leadership, and effective participation in value chains are key factors that will ultimately shape their success.

References

Abenakyo, J. (2023). Resilience and regeneration of grassroots cooperative societies in postliberalization Uganda. *Journal of Rural and Community Development, 18*(2), 58–73.

ActionAid. (2013). *The cooperative movement and the challenge of development: A search for alternative wealth creation and citizen vitality approaches in Uganda*. Publication of Centre for Basic Research, ActionAid and the Uhuru Institute for Social Development.

Akandwanaho, R. (2002). *The impact of political turmoil on agricultural cooperatives in Uganda*. Uganda Cooperative Alliance.

Bates, R. H. (2005). *Markets and states in tropical Africa: The political basis of agricultural policies: With a new preface*. University of California Press.

Birchall, J., & Simmons, R. (2004). What motivates members to participate in co-operative and mutual businesses? A theoretical model and some findings. *Annals of Public and Cooperative Economics, 75*(3), 465–495.

Braverman, A. (1991). *The role of cooperatives in agricultural development. World Bank discussion paper, 113*. World Bank.

Brett, E. A. (1970a). Problems of co-operative development in Uganda. Rural co-operatives and planned change in Africa.

Brett, E. A. (1970b). *The cooperative movement in Uganda: A historical perspective*. Makerere University Press.

Center for Basic Research & Action Aid. (2013a). *The role of cooperatives in Uganda's agricultural sector: Challenges and opportunities*. Center for Basic Research.

Center for Basic Research & Action Aid. (2013b). Report on the performance of Cooperative Unions in Uganda.

Center for Basic Research & Action Aid. (2013c). Revival and performance of cooperatives in Uganda: Challenges and solutions. [Analysis of Cooperative Performance and Government Interventions].

Coase, R. H. (1937). The nature of the firm. *Economica, 4*(16), 386–405.

Cobia, D. W. (1989). *Cooperatives in agriculture*. Prentice Hall.

Cuevas, C., & Fischer, E. (2006). Cooperatives as a way to reduce poverty. In F. O. Wanyama, P. Develtere, & I. Pollet (Eds.), *Reinventing the cooperative: A developmental perspective* (pp. 139–155). International Labour Organisation.

Dennis, M., Juma, M., & Murekezi, A. (2016). Agricultural productivity in Uganda: Policy options and challenges. *Uganda Journal of Agricultural Sciences, 17*(1), 1–12.

Develtere, P., Pollet, I., & Wanyama, F. (2008). Cooperating out of poverty: The Renaissance of the African Cooperative Movement.

DFID. (2008). *Understanding the role of cooperatives in rural development*. Department for International Development.

Gibson, R. (2005a). *The role of cooperatives in community economic development*. Canadian Social Economy Hub, University of Victoria.

Gibson, S. (2005b). The Rochdale pioneers: A co-operative legacy. *International Journal of Cooperative Studies, 2*(2), 19–28.

Halerimana, T., Muwanga, F., & Okwakol, P. (2010). Agricultural cooperatives and rural development in Uganda: A case study. *Ugandan Journal of Agricultural Economics, 3*(1), 45–67.

Helmen, A. (1990). African cooperatives: Past, present and future. *Journal of Co-operative Studies, 27*(2), 45–56.

International Cooperative Alliance (ICA). (2010). What is a cooperative?

International Cooperative Alliance (ICA). (2019). Uganda legal framework analysis.

Jesee, A. (2016). The emergence of cooperatives in Africa: A historical overview. *African Journal of Cooperative Studies, 7*(1), 44–55.

Kagina, S. (2022). Community-based organisations and cooperative revival in Uganda: A post-liberalization analysis. *East African Journal of Social and Economic Studies, 16*(1), 90–105.

Kakumba, U. (2023). Revitalization of cooperative societies in Uganda: A policy perspective

Karanja, G. (2001). *The co-operative movement in Kenya: Historical development and impact*. Co-operative University of Kenya Press.

Kasamba, I. (2022). Underground cooperatives: The fight for legal recognition in Uganda during the late colonial period. *Uganda Historical Review, 12*(3), 98–112.

Kato, J. T. (2021). The role of cooperative banks in economic development: The case of Uganda. *Financial Review, 18*(3), 254–270.

Kawuma, D. (2022a). Assessing the impact of cooperatives on farmers' livelihoods in Uganda.

Kawuma, D. (2022b). Colonial policies and cooperative movements in Uganda: A historical overview. *African Development Studies, 23*(3), 120–135.

Kiggundu, D. (2023). The future of agricultural cooperatives in Uganda.

Kisaka, R. N. (2023). Economic contributions of cooperatives in post-independence Uganda: An analysis of 1960s data. *Journal of African Economic History, 21*(2), 135–150.

KUSCO (Kenya Union of Savings and Credit Cooperatives). (2014). *The role of cooperatives in Kenya's economic development*. KUSCO.

Kwakyewah, C. (2016). Rethinking the role of cooperatives in African development. *Inquiries Journal, 8*(6). Retrieved August 9, 2016, from http://www.inquiriesjournal.com/aid=1419

Kyambadde, A. (2018). Government prioritization of cooperative revival: Strategies and impact [Ministerial statement on cooperative development and support].

Kyazze, L. M. (2010). *Cooperatives: The sleeping economic and social giants in Uganda*. International Labor Organisation (ILO).

Levin, J. (2005). The cooperative principles and the Rochdale pioneers: A historical perspective. *International Journal of Cooperative Studies, 2*(2), 10–18.

Lwanga, J. B. (2023). The legislative journey of the 1946 co-operative societies ordinance in Uganda. *Journal of African Legal History, 22*(2), 56–70.

Mkandawire, T. (2001). Thinking about developmental states in Africa. *Cambridge Journal of Economics, 25*(3), 289–314.

Mubiru, R. (2022). From displacement to empowerment: The impact of growers' cooperative societies on local farmers. *Uganda Agricultural Review, 14*(4), 95–110.

Mugisha, A. (2023a). Economic contributions of cooperatives in Uganda: Historical data and analysis. *Uganda Economic Review, 24*(1), 88–106.

Mugisha, F. (2023b). The impact of economic liberalization on cooperative unions in Uganda: Challenges and adaptations. *African Development Review, 21*(2), 140–155.

Munyami, A. (2014). The role of cooperatives in the socio-economic development of Ghana.

Mutebi, P. (2023). Growth and performance of cooperatives in Uganda: A historical perspective. *Journal of African Development Studies, 27*(3), 175–190.

Mutibwa, P. (1990). The impact of political instability on cooperative development in Uganda. *Journal of African History, 31*(2), 179–195.

Muwanga, C. (2023). The struggle for cooperative legislation in Uganda: Political and economic challenges of the 1930s. *Journal of East African Historical Studies, 17*(1), 34–49.

Muwanguzi, F. (2022). Impact of cooperatives on rural development in Uganda.

Nabimanya, T. (2022). The expansion of Uganda's cooperative movement: Insights from 1960s economic data. *Journal of Economic History and Policy, 13*(1), 50–65.

Nabirye, H. (2023). The role of growers' cooperative societies in agricultural marketing in Uganda. *Journal of Cooperative Economics, 11*(3), 115–130.

Nabwiso, H. (2022). Economic contributions of cooperatives in Uganda.

Namubiru, A. (2022). The impact of the 1946 co-operative societies ordinance on cooperative develop-

ment in Uganda. *East African Journal of Policy and Governance, 14*(4), 145–160.
New Vision published a piece on the contributions of cooperatives, including Bugisu Cooperative Union, on October 22, 2015, discussing the economic impact and revitalization of cooperatives in Uganda (New Vision).
Nweze, N. J. (2001). Cooperatives in Nigeria: The role of cooperative movement in economic development.
Ochieng, J. A. (2022). The role of cooperatives in Uganda's agricultural sector: Historical data and analysis. *Agricultural Economics Journal, 14*(4), 210–225.
Okello, L., & Ahikire, J. (2013) The cooperative movement and the challenge of development: A search for alternative wealth creation and citizen vitality approaches in Uganda. TUI, CBR & ActionAid.
Opobo, M. (2015). The history and impact of cooperative societies in Uganda.
Republic of Uganda. (2015). Cooperative Societies Act (Amendment Bill).
RoU. (2012). *Cooperative Societies Act, 1991 (Amended 2012).* Uganda Legal Information Institute.
Sebunya, M. (2023). The role of cooperatives in economic development: A Ugandan perspective.
Sender, J., & Smith, S. (1984). The development of agricultural co-operatives in Africa: A comparative study. *World Development, 12*(4), 541–561.
Ssekitoleko, E. (2023). Historical perspectives on cooperative legislation in Uganda: From resistance to reform. *Uganda Development Studies Journal, 19*(1), 84–99.
Ssembogga, G. (2020a). *Performance of agricultural cooperatives in Uganda: A case study of the Luwero triangle*. Uganda Cooperative Alliance.
Ssembogga, S. (2020b). Reviving cooperatives in Uganda: The role of financial support and policy [Report on challenges and recommendations for cooperative revival].
Ssenkumba, E. (2023). Government support and cooperative development in Uganda: An analysis of policy impact. *African Policy Journal, 11*(3), 133–150.
Ssentamu, E. (2022). Economic liberalization and its effects on cooperative structures in Uganda: A case study. *Journal of Economic Policy and Research, 13*(1), 75–92.
Ssenyonga, M. (2022). The market share of cooperatives in Uganda's coffee sector: A historical review. *Uganda Coffee Review, 8*(2), 70–85.
Tumusiime, D. (2023). Historical data on coffee and cotton production in Uganda: A cooperative perspective. *East African Journal of Agricultural Studies, 19*(1), 95–110.
Uganda Cooperative Alliance. (2023). *Strategic initiatives for strengthening cooperatives in Uganda*. Retrieved from Uganda Cooperative Alliance.
Wanyama, F. O. (2011). *Cooperatives for African development: Lessons from experience*. United Nations.
Wanyama, F. O., Develtere, P., & Pollet, I. (2008). Encountering the evidence: Cooperatives and poverty reduction in Africa. Working papers on social and cooperative entrepreneurship WP-SCE, 08-02.
Wanyama, F. O., Develtere, P., & Pollet, I. (2009). Reinventing the wheel? African cooperatives in a liberalized economic environment. *Annals of Public and Cooperative Economics, 80*(3), 361–392.
World Bank. (2011). *Uganda: Agriculture sector review*. World Bank.

19 Livelihood Dynamics of Urban Refugees in Uganda: Evidence from Congolese and Somalis in Kampala

Morris Egesa and Denis Musinguzi

Abstract

Uganda receives a large number of refugees from neighbouring countries. Some of the refugees end up in urban areas where they begin new life without any help from the government and development agencies. We investigated the livelihood strategies employed by Somali and Congolese refugees in Kampala City, Uganda. Start-up capital, language barrier, and poor conditions of work emerged as the major livelihood challenges faced by the refugees. Most of the refugees mentioned that they resorted to social networks, personal savings, and aid as start-up capital. There is a need for strengthened coordination between the different actors engaged in livelihood programmes for the refugees.

Keywords

Forced migration · Political violence · Environmental degradation · Start-up capital · Resilience

Introduction

Climate change and environmental degradation, armed conflict, and political, economic, and food crises continue to force people to flee, resulting in large numbers of urban refugees (IOM, 2013). Millions of people have fled their homes in the East and Horn of Africa and the Great Lakes region due to conflict, climate-related events, and political violence (Pavanello et al., 2010). Uganda is one of the largest refugee-hosting countries in the world. While most refugees are hosted in refugee settlements, a small number of the refugees find their way into urban areas, especially within Kampala, Uganda's capital city as well as other cities and towns across the country. Urban refugees do not receive support from government and mainstream development agencies although some have benefitted from capacity training by civil society organizations. It is against this backdrop that this study explored the efforts used by urban refugees originating from DRC and Somalia in meeting their livelihood needs. Specifically, the study set out to determine the livelihood needs of these two refugee communities, efforts used to meet their livelihood needs, and overall level of adequacy.

M. Egesa · D. Musinguzi (✉)
Uganda Martyrs University, Kampala, Uganda
e-mail: dmusinguzi@umu.ac.ug

P. Nyadoi et al. (eds.), *Sustainable Development Goals in Africa*, Sustainable Development Goals Series, https://doi.org/10.1007/978-3-032-20742-5_19

Methodology

We purposively sampled 40 respondents who identified themselves as refugees and had lived in Kampala for more than 5 years. Semi-structured interviews were conducted with 16 household heads to capture an in-depth comprehension of livelihood strategies employed to sustain their families. Four focus group discussions of six members each were conducted. Two focus group discussions were conducted for the Somali refugees and specifically one for female Somali refugees comprising six participants. The second discussion was for males, comprising six refugees, half of whom were youthful refugees. The same procedure was conducted for the Congolese refugees. In addition, we held interviews with six key informants who included the Livelihood Officer at the Office of the Prime Minister, Strategy Coordinator at Kampala Capital City Authority (KCCA), Programme Manager at InterAid Uganda, Executive Director at Young African Refugees for Integrated Development (YARID), and the leaders of the refugees. Analysis of the data was done through identification of key themes using the Atlas.ti software.

Findings and Discussions

Capital a Cornerstone for Refugees' Livelihood

Almost 90% of the refugees who were carrying out small trade considered capital as a key component for their survival; capital was perceived in the form of stock and physical equipment like refrigerators, blenders, sewing machines, and dryers. Capital was highlighted a number of times by all categories of refugees—men, women, and the youth inclusive. One of the Somali refugees the researcher investigated had this to say: Capital is needed for one to conduct business in Kampala. For me, I started with 2 million for my business in Kisenyi and I was paying rent of 500,000 monthly (interview with a male Somali respondent in Kisenyi). This was cemented with a similar comment from a female Congolese respondent selling fresh vegetables at Copper Complex in the central business area. Capital is essential just like it is for any other business; money enables me to buy the fresh vegetables as far as Matugga on Bombo Road (interview with female Congolese refugee at Copper Complex). From the study findings, it was quite clear that capital was the lifeblood for the self-reliance and a dignified life of the two refugee communities. Capital was an enabler for them to purchase merchandise which they would in turn sell for profit in designated areas or through hawking, though this practice was punishable by law in accordance with KCCA resolution.

Access to Market Key in Refugees' Livelihood

This chapter reveals that market is an indispensable need for urban refugees in Kampala. The study was in line with the previous studies that had been done by other scholars on refugee economies, which established that access to market was one of the factors that would make small-scale trade for the refugees thrive and flourish (Jacobsen & Fratzke, 2016). About this particular subject of market, a Congolese lady hawking merchandise had this to say: Market for these products is necessary; nowadays, we have fake jewellery that fades, which in a way discourages customers. So I try as much as I can to convince especially the new customers that the ornaments I am selling are authentic enough (interview with the female Congolese in Kampala). Similarly, a female Somali refugee selling samosas and tea was in agreement with the above view, holding that "Market is also key because if you make samosas with tea and you do not have good marketing skills you are bound to flop in the business". Across the categories of the respondents, it largely came out that market was an important factor that was needed for the refugees selling their products ranging from those involved in selling samosas, juice, tea to those selling clothes like bitenge and jewellery. To this end, the interaction with the staff at KCCA serves to illustrate this view: The Sunday market is a street market;

a section of Luwum Street is converted into a market. It is non-motorized the whole of Sunday, so it only happens on Sunday and it is used as a road for the rest of the weekdays. But apart from the Sunday markets, there are also seasonal markets like Fridays in Kamwokya. These are good opportunities for informal traders, of whom refugees make up a large number, but these are also good opportunities where they can showcase some of the petty trade items (interview with the Strategy Coordinator KCCA, at City Hall Kampala).

Language Facilitates Livelihood Inclusion

Language was indicated as one of the bedrocks for successful livelihood for the urban refugees staying in Kisenyi and Katwe though engaged in different occupations in the city. This chapter equally acknowledges the importance of language for the economic inclusion of refugees who have settled in urban areas, as previous studies and scholars have indicated. Chiswick and Miller (2007) in their paper entitled "Modelling Immigrants' language skills of the host country", contributed to higher attainment in the labour market. This was reflected during the in-depth interview with the male Somali riding a *Tuku Tuku* revealed:

You need to know the language of the local people since not all the time will you come across passengers who are Somalis. Sometimes, I use sign language while dealing with my customers. I have also tried to befriend the Ugandan boda-bodas and tell them that we all human beings looking for survival (interview with the male Somali refugee in Kisenyi.) Elsewhere among the Congolese refugees there was a major deviation in opinion about the aspect of language in an attempt to navigate the labour market in the foreign territory. This remark from the male Congolese working as a toilet attendant at Arua Park business attested to the fact, stating, "You need to know the language of the customers who come here for short or long call otherwise it becomes difficult to communicate" (interview with male Congolese refugee at Arua Park). To navigate in Kampala's urban environment, one needs to be acquainted with the local language so as to strike a balance for survival. This was raised by the staff at YARID, who strongly hinted at this issue: We are in the language and language is a key and a very important aspect and when you look at the situation where the refugees come from most of them are from French-speaking countries and they cannot speak the local language or English so we support them in that (interview with the Executive Director at YARID offices in Nsambya).

Social Networks

The finding of this chapter resonates well with the previous studies on social networks. The study by Amisi (2002) revealed the importance of social networks in a study on Congolese refugee communities. This finding clearly resonated well in this study among the Somalis and Congolese in Kampala. One of the Congolese youth carrying luggage around William Street in Kampala confirmed: "Social network is key, and it reminds me how I ended up in Kampala. I used to hear from Ugandans I was working with in people's farms in Kyaka about opportunities in Kampala" (interview with a male Congolese refugee at William Street in Kampala). During the focus group discussion with the Somali men, it was also highlighted that social networks were a fundamental requirement for the Somali refugees to get access to employment, "Before I got a job life was really difficult, but it is because of the network that I got the job of security guard here in Kisenyi" (focus group discussion with male Somali refugees in Kisenyi). The study in this case was in consonance with the sustainable livelihood approach that stresses how people use social capital during times of shock and stress in order to earn a living.

Skills Development

Having hands-on skills was largely indicated as a key issue that would enhance the chances of urban refugees in having sustainable livelihood interventions. Easton-calabria (2016) echoed livelihood training as a means of achieving self-reliance among the refugees in Kampala. This study in Katwe and Kisenyi gathered information on how skills training, especially among the refugees, was viewed as indispensable for their survival. The interviews that the researcher conducted with the Somali nationals illustrated this, "to become a good chef, one must be equipped with skills in organising and preparing good food that is palatable and appeals to the customers" (interview with a male Somali refugee in Kisenyi). "To do this business some skills are also needed because most people are making the same products, so you need to make good quality products to ensure that you have firm control of the market otherwise you stand to lose out to other competitors" (interview with a female Somali refugee in Kisenyi). This was further supported by the key informants who are at the level of policy implementation at the Office of the Prime Minister in charge of Refugee Affairs and Kampala City Council Authority. The Livelihood Officer made the following remarks: "We have got a big number of women in the urban, take the context of Congolese women. We have the youth but how do we support them in their education—through skills training. Quite a number of Congolese refugees, especially women they like hair dressing and tailoring; the youth could be supported in terms of carpentry and other vocational skills. How we find a mechanism of skilling them so that they are able to learn a skill which they will use in their future or they will use wherever they are staying in any country is very important because they will use those to start small businesses where they can support and fend for themselves" (interview with Livelihood Officer OPM at Sky Hotel). As noted by the officer from the Office of the Prime Minister and confirmed through the responses from the refugees interviewed during this study, these skills have made these urban refugees able to fend for themselves and their families as chefs in local restaurants within the city. The skills training ingredient was equally substantiated by the staff at YARID who considered it an invaluable component for the livelihood inclusion of urban refugees and he shed more light in that prism by stating that: "We have other trainings like women's empowerment projects. Why women's empowerment projects? We are in a situation whereby women are playing a very big role. Many households here are headed by women, women are breadwinners, so we came up with this so that we can support them and they support their immediate families. So, we have the tailoring training and handcraft-making; right now we are in the process of opening up a livelihood centre where we shall have a diversity of skills so that everyone can come and pick the ones that they think are comfortable with" (interview with YARID Official at YARID office). These remarks are a clear pointer to how the skills upgrade is a recipe for effective livelihood programming for the self-settled refugees in Kampala without which refugees become more vulnerable. This chapter therefore argues that it is highly plausible for urban refugees to opt for negative coping mechanisms in the event of a deficiency of cognitive and practical skills that would enable them to keep afloat in the city.

Access to Credit Facilities

This study equally revealed that urban refugees from the two communities did not have easier access to credit facilities from banking institutions and other microfinance institutions. Earlier researchers in this domain had echoed the significance of this aspect in reviving and improving the respective enterprises set up by refugees which would later uplift their well-being in the cities. The findings further revealed that even for the few who accessed the credit facilities, the interest rate and period of repayment were so harsh to them, as disclosed during one of the in-depth interviews with a Somali refugee: "I got a loan from International Office of Migration worth 500,000 but this loan subjected me to immense pressure. Paying sixty-five thousand (65,000) weekly with my small business and big family that I am taking care of is really killing me" (interview with female Somali refugee in Kisenyi). Access to credit facilities is vital for meeting refugee livelihoods.

Efforts Used to Meet Their Livelihood Needs

One of the objectives this study sought to explore was the efforts used by the urban refugees to meet their livelihood need; consequently, the subsequent findings paint a picture of how the self-settled refugees, once in the urban area, mitigated the livelihood challenges in order to earn a living.

Casual Employment/Petty Jobs Survival Assets

Given the complexity of getting employed, the study findings reveal how occasionally urban refugees from the two communities had no choice but to immerse themselves in the informal sector in order to meet the basic necessities of life. Casual employment was one of the ways the refugees used to meet their livelihood needs though the conditions of work were not good conducive at all as the findings of this research portray. Among the Somali respondents that the investigator interacted with half them were working for other people either as *Tuku Tuku* riders, chef or waiters in restaurants, and security guards with the exception of few conducting retail trade. The Congolese were carrying luggage for customers in the city, working as toilet attendants and dancers in the music band tailoring and occasionally selling jewellery. One of them noted, for me I applied for a job three times but the fourth time is when I was offered a place as a waiter and I was willing to work at any pay because life was getting worse for me (interview with Somali male refugee in Kisenyi). Another Somali respondent also confirmed that he works as a night guard security person at City Tyres Petrol station at Bombo road so as to supplement on the meagre income he was getting from his small business in Kisenyi. In his own words, the respondent was captured saying, "I work in the night at City Tyres as a guard, aware that money from the small business cannot meet all my demands" (interview with male Somali refugee in Kisenyi). A Congolese who works as toilet attendant at Arua Park stated that: I only get paid one hundred thousand and because of high cost of living, I no longer stay in Katwe, the salary I get is not enough for rent food and transport. For the last 2 months I have been sleeping here in the toilets and no one knows that, that is how harsh life has been. (interview with Congolese refugee at Arua Park). This study therefore exposed that employment chances were very few and if they existed, they were at lower levels of administrative hierarchy mostly those that required less skills.

Business Enterprises

Refugees of Congolese and Somali origin have been actively involved in enterprises that can generate money for them so that they can be in a position to meet the increasing demands that are associated with urban dwellers, such as food, rent, education and medical needs.

The women of the two nationalities have been actively involved in selling tea, samosas, juice, leafy vegetables, jewellery, and tailoring as a form of business to support themselves. Though they are in business, they presented how they cope with the challenges in their respective occupations and the following seems to shed light on this dimension. One of the female refugee women selling jewellery commented: Due to limited capital and lack of a specific area for transacting business, I usually avoid KCCA enforcement by selling in restaurants and salons that are located in major arcades since the enforcement officers do not enter buildings (interview with a female Congolese refugee in Gaza land Kampala). Relatedly, a male Congolese respondent indicated that: Because I do not have enough capital, I normally put the necklaces in the pocket and when I go to an arcade especially where there is a restaurant or bar, I display the product on sale. This is how I have managed to push on with life. I am a trained mechanic but here in Uganda I have tried to seek for opportunities in that dimension but all my efforts have been futile (interview with a male Congolese refugee at Gaza land Plaza). Generally, from the study it could be discerned that most of the refugees found it easier to manage their own private income-generating activities if supported than seeking paid employment. A critical look at the respondents who took part in this study shows that 60% were running their own employment ventures.

Social Network

Jacobsen (2002) found out that social networks are helping many refugees to enhance their livelihood. Through social networks refugees are able to gain help from their friends and relatives in other countries. The present study of urban refugees in Kampala also established that social networks play an important role in enhancing the livelihood of refugees. The study revealed through the in-depth interviews that networks were part and parcel of the survival of the refugees in the city without which it would be hard to make headway. A case in point is the interview the researcher had with one of the respondents: Networking is what has kept me going forward, we have a group of Congolese women that are selling vegetables that inform me of the availability of stock at affordable rates (interview with Congolese lady selling vegetables at Copper Complex in Kampala). A similar sentiment was shared by other refugees who were men as to how they could not underscore the importance of social networks and most of them had smartphones that could help them link each other to any available offer. A youth Somali respondent indicated that: I was a footballer and I was known within the Somali community in Kisenyi, so one day someone approached me to work for him in a gaming and football centre (focus group discussion with male Somali refugees in Kisenyi August 2019). Although social networking was considered as a way of meeting their livelihood needs, it was not a guarantee that it would link the refugees to employment. While having a session with the male refugees from Congo during the focus group discussion, this important remark was captured: "They say that someone's character reveals him or her, so after staying here at William Street and people had started gauging me by giving me simple assignments. So, they established that I was a faithful and hard-working person and so me too I welcome other people who come here to fend for themselves" (focus group discussion with male Congolese refugees at William Street in Kampala). This paper in a way reveals that social networks per se could not lead to livelihood satisfaction. Personality and character could also be an important attribute for refugees seeking wage or salary employment.

Faith and Prayers

The refugees, especially those of Congolese background, alluded to the fact that had it not been for the support of the church at the time they arrived in Kampala, life would have taken a different path altogether. The church, notably Holy Pentecost in Katwe, was really instrumental in cushioning the initial needs of the refugees who visited the church for remedy. The response extracted from one of the interviews the investigator had with the responses illustrates this:

The church in Katwe was key from the onset providing me with accommodation and that is something that forever I will be grateful to the Almighty. For 6 months I was staying at the church, in this town people can give you food and something to drink but not accommodation the way the church did (interview with Congolese male refugee at William Street in Kampala).

Another refugee dealing in hawking of jewellery in central Kampala business area in her submissions equally echoed the importance of the church in shaping her destiny in a country where everyone seemed to be a stranger: "Holy Pentecost church in Africa which is located in Katwe gave me the start-up capital having stayed at the church when I arrived in Kampala. Also every time our stock would be confiscated by KCCA enforcement officers, I would go back to the church and share my ordeal with them then it would give me some cash to revive the business" (interview with female Congolese refugee). Clearly in this chapter we posit that the church's contribution to the livelihood debate of urban refugees cannot be underestimated. The previous studies on urban refugees exposed the role played by the church towards the livelihood of urban refugees.

Aid from NGOs

From this study, it was discovered that some refugees received some assistance that they used to ensure that they remain relevant in their respective localities in the urban context. Quite a number of refugees from Congo and Somalia pointed to the help they got from InterAid Uganda as a stepping stone to their livelihood inclusion in Kampala. Among the respondents whom the researcher probed as to how they started business, one of them who sells necklaces noted:

The start-up capital that I used I got it from InterAid Uganda, they gave me 320,000 that was meant to be for house rent since the roof of the house I had been staying in had been blown by wind. I used to pay 150,000 per month for the house, but when I got the money from InterAid Uganda, I decided to get a cheaper house for 80,000 monthly, and I decided to invest the rest of the money in the necklace business. My wife also sells necklaces, and that is how we have managed to ensure that food is available, that we pay rent and that our children also manage to go to school (interview with a male Congolese refugee at Gaza land Plaza). From the above submission, it is clear that, given the vulnerability of the respondent, he decided to divert part of the money that was meant to be for housing into his start-up capital and that is how he has managed, in a way, to fend for himself and his family in Katwe. Another refugee from Somalia could not hide her joy when she remarked that: "I got some money from InterAid Uganda, approximately 900,000, and that is the money that I used to start my small business. They also facilitated my certificate in Tailoring and Fashion Design at Uganda Home Economics Institute" (interview with a female Somali refugee in Kisenyi September 2018). This study is in agreement with the previous studies about the livelihood of urban refugees, which pointed out that aid was an important input for those refugees with business acumen, giving the refugees aid in terms of capital would not only make them start small initiatives but also make them more responsible for their own affairs.

Savings and Loans

The refugees who were part and parcel of this study are in agreement with the previous studies that loans and savings have a profound impact on their mainstay in Kampala. Access to credit is indispensable for refugees' needs and to enhance their well-being. One Somali lady selling samosas and tea in Kisenyi acknowledged in the following remarks: In July 2018, the International Office of Migration (IOM) through a Somali agent gave me a loan of 500,000 shillings. I was paying weekly 65,000 until I finished all the loan in September 2018. The money I got was used to boost my small business because the cost of buying business items had gone up (interview with female Somali refugee in Kisenyi). Saving was interestingly considered crucial for some of the refugees that the study zeroed down, one male refugee of Somali origin who had stayed in Kampala for 10 years validated this assertion by acknowledging, "I worked for five years in supermarket as attendant the money I could make was saved for opening up the soda business" (interview with a male Somali refugee in Kisenyi). Unlike the previous studies that took place in the continent, this study differs in its findings because the refugees whom the investigator engaged clearly showed how the money was used for business and not diverted for consumption hence its contribution. Nevertheless, in this chapter we argue that inasmuch as loans and credit facilities would have multiplier effects for the overall business undertakings of the refugee communities, it was limiting in terms of access. According to the Programme Manager at InterAid Uganda, the credit facilities would only be availed to refugees organized in registered entities: They are not ready for business. You see government funds; no youth will receive funds from youth livelihood funds if they are not organized. So, it's the same model that we are implementing, why would I give money to a disorganized person? Yes, we signed a memorandum of understanding with UGAFODE Microfinance to only give credit where applicable to those who are ready to receive it, you know, in business not everyone will be willing to do business. Just like the nationals, it's the reason you are seeing some of us are in this NGO sector, there are those who are in the arcades doing business. There is a sieve of those who qualify, and there are those who try and fail in their first attempt, and there are those who try out and thrive. So not every human being is supposed to do business, so for us, we only level the ground (interview with the Programme Manager Inter Aid Uganda). The foregoing discussion is a pointer to the fact that accessing and availing credit facilities was likely to exclude the already desperate refugees whose lives would have been transformed if they too had access to soft loans to plough into their businesses.

Livelihood Satisfaction

This section sought to determine the level of satisfaction or livelihood needs of urban refugees. Based on the challenges that were structural and occupational, the respondents listed mixed reactions as far as their well-being is concerned. This section has been categorized as those in casual employment and business where most refugees have been seen struggling to make ends meet. While relating the general well-being of refugees, established that for refugees to enjoy a good-quality life, they needed access to economic opportunities and social rights.

Casual Employment

Inasmuch as the refugees were employed in a chain of activities that were meant to improve their well-being and respect in the community, most of them pointed to the indicator that their livelihood needs were not adequate. From those employed in restaurants to those working in informal sectors, especially the Congolese carrying luggage at William Street, evidence suggests that it was because of no choice that they had to take up their occupations. Satisfaction for those employed in informal sectors was linked to working hours, wages or salary earned, and job security. A large number of the respondents expressed dislike for the jobs they were doing because of long hours of work that were not commensurate with the payment. This is illustrated by the following comments: "The KCCA officials should handle us like human beings, since it was not our own making to become refugees" (focus group discussion with the Somali refugees). "I work for long hours from 8 a.m. to 8 p.m. in a restaurant with little pay and my boss is not bothered at all" (interview with a male Somali refugee in Kisenyi). Similarly, such a response was generated from the focus group discussion of female Somali refugees, "the pay that I get at school for my services are very low but I am hopeful that when the number of pupils increases, I shall be okay" (focus group discussion with female Somali refugees in Kisenyi). Satisfaction from the foregoing remarks indicated that for those in paid jobs, it was anchored on the working conditions and the amount of salary that was being paid to them.

Business Ownership

Even those refugees in pursuit of their private business equally expressed despair with the livelihood connections that kept them afloat in the city. All the categories of refugees from DRC and Somalia expressed their level of dissatisfaction owing to the gaps in their routine businesses that were meant to transform their lives from worse to better. Among the Congolese refugee selling vegetables, the following information indicate how inadequate her livelihood needs were: I am not satisfied, and I have never been satisfied at all even for a single day. I have suffered so much and it is because of lack of what to do that we find ourselves doing this. KCCA enforcements officials can end up taking up everything that you have sold and you remain with nothing. They even go to the extent of checking our small money purse in case we are arrested to establish how much money is inside so that you can give them and they release you. And if you do not have money, you can be convicted for about 3 months and then you are released (interview with female Congolese refugee at Copper Complex in Kampala). Another respondent noted, I am not satisfied since the capital I have is little to cater for my needs as a single mother with three children. I need more capital and a tailoring machine to supplement on the money I sell from the clothes (interview with a female Somali refugee in Kisenyi). This remark was further supported by the key informant interview with the staff of YARID in Nsambya who equally shed more light as regards to the level of adequacy of livelihood needs of urban refugees: It is not adequate of course. The majority of the families here and the household, the income they get is not enough for them to make choices at the end of the day. In most cases, it is not enough for them to get something to eat. They rent a very cheap room in Katwe here, without access to education and healthcare. So many people dying in the community because they do not receive proper treatment. You know children are not going to school, the research that were done recently in 2016 showed that only 35 of refugees have access to primary education here in Kampala. That is very low compared to the number of the children, only 12% of them have access to secondary education and these are connected to the livelihood of the income. So, it is not adequate. I can say most of them YARID has just supported them to meet the foods needed and accommodation. Even the accommodation is still a challenge because we have so many complaints of someone being chased out of the house because they cannot afford rent (interview with Executive Director at YARID offices in Nsambya). The frustration of the refugees was buttressed by the remarks of the Congolese refugee that the researcher had interphase with at Arua Park in Kampala. I feel the people who are concerned with the refugees are the ones who are torturing us; they were giving out 145,000 to refugees but selectively. They were even giving the nationals the money yet we who are meant to be beneficiaries of the money as refugees were not being considered. OH, it's a long story and I have gone through a lot here in Uganda. And all these were happening at InterAid because it's the one which recommends any refugee to go for any appointment with UNHCR or OPM. Interview with a male Congolese refugee. Another refugee from Congo sowing clothes at copper complex shared her sentiments about the livelihood hood satisfaction. You see, my satisfaction relates to my ability to take my children to school, feed them, and also develop as a human being. With this challenge of market and high rent charges affecting me I cannot say that my livelihood needs are adequate. When children are chased from schools and they sit at home due to low sales at my work place that makes me feel bad. It can be argued from the perspective of those refugees carrying out trade that satisfaction to them was linked to ability to meet their immediate family obligations and access to inputs like market and capital which were presumed important for sustaining their livelihoods. With the majority of the respondents of the study having a consensus on the level of dissatisfaction of their livelihood needs, the ball is set rolling on how to uplift the lives of the refugees by ensuring that the multifaceted issues that hinder their sustainability in labour market spheres are holistically addressed.

Conclusions and Recommendations

Findings reveal the following:

- The host government, particularly the office of the Prime Minister and NGOs should strengthen awareness creation among the refugees in Kampala about the available opportunities for them by broadcasting on the local TV stations and through other community initiatives using their area leaders of what is in stock for them.

- The host government and UNHCR should ensure that there is more collaboration between all the stakeholders that handle the urban refugees' livelihood intervention. Through having quarterly stakeholder meetings, the concerned parties can be in a position to share notes, detect the livelihood gaps, and come up with concrete strategies that can enable the refugees to contribute to the sustainable development goals.
- The KCCA should expedite the formation of the framework that handles the urban refugee question. Inasmuch as there are no differentiated services for the refugees in Kampala, refugees have unique challenges and a refugee desk could go a long way in capturing their concerns in a bid to contribute towards a sustainable city. More of the initiatives like the Sunday market by KCCA should be extended to other divisions where the refugees carry out their trade so that the refugees capitalize on them and achieve economic independence.
- The office of the Prime Minister, specifically the department of refugees and its implementing partners, should devise alternative means of reaching out to refugees who are in need of start-up capital or micro credit facilities for their businesses. Reaching out only to those registered and organized refugees serves to exclude the majority who may not be aware of the existing facilities.
- The development partners hand in hand with KCCA should find a mechanism of integrating the urban refugees into the value chain of Urban Agriculture and precisely backyard farming that will help boost not only food security but also their economic independence. Currently, Urban Agriculture is being enforced by KCCA in liaison with the Operational Wealth Creation project, but refugee communities are not factored in this economic initiative.
- Research should be undertaken to unpack the impact of skills training and start-up capital that have been advanced to refugee communities in order to ascertain if such undertakings have translated into sustainable livelihood outcomes for the urban refugees.

References

Amisi, B. B. (2002). An exploration of the livelihood s trategies of Durban. *UNHCR*, no. 123.

Chiswick, B. R., & Miller, P. W. (2007). Modeling immigrants' language skills. *Research in Labor Economics, 27*(2974), 75–128. https://doi.org/10.1016/S0147-9121(07)00003-9

Easton-calabria, E. (2016). 'Refugees asked to fish for themselves': The role of livelihoods trainings for Kampala's urban refugees, no. 277.

International Organization for Migration (2013). World Migration Report 2013. Switzerland. ISBN 978-92-9068-668-2 ISN 1561-5502.

Jacobsen, K. (2002). Livelihoods in conflict. *International Migration*, 40(5). http://onlinelibrary.wiley.com

Jacobsen, K., & Fratzke, S. (2016). Buidling livelihood opportunities for refugee populations: Lesson from past practice. *Migration Policy Institute, 19*(3), 43.

Pavanello, S., Elhawary, S., Pantuliano, S., Research Team, Abdi, A., Bashir, A., Hakizayezu, E., & Umer, K. M. (2010). Hidden and exposed: Urban refugees in Nairobi, Kenya.

Organisational Culture and Talent Management in Tertiary Institutions in Uganda

20

Josephine Namugumya,
John Chrysostom Kigozi Munene, Samuel Mafabi,
and James Kagaari

Abstract

Despite the growing importance of talent management, not many studies have been done to identify strategies for it, for many of Uganda's tertiary institutions. This study gathered data from 573 respondents across 86 tertiary institutions, including deans, heads of academic departments, and human resource managers, using a cross-sectional survey and proportionate stratified random sampling. Using data on perceptions of managers, this study examined how organisational culture (recognition of talent, open communication and teamwork) is viewed to influence talent management (mentorship, matching talent to key positions, rewarding performance) in tertiary institutions in Uganda. Preliminary analyses based on a total of 573 respondents from 86 tertiary institutions revealed that organisational culture alone, within limits of the aspects studied particularly aspects of mentorship, performance rewards and placement in key positions, may not influence how top talent is managed. More in-depth research is still needed to determine what aspects of organisational culture may influence what aspects of talent management in tertiary institutions in Uganda.

Keywords

Employee satisfaction · Systems adaptability · Associations · Academic talents

Introduction

Talent is widely recognized as the most valuable asset in enhancing an organisation's performance (Putri et al., 2023). Yet, talent management remains one of the key challenges organisations face today (Collings et al., 2022; Vaiman et al., 2021). To stay competitive, organisations often focus on attracting and identifying highly skilled individuals for critical roles (Gallardo-Gallardo et al., 2020), offering meaningful rewards to retain them and navigate the ongoing "war for talent" (Kwon & Jang, 2022). However, effective talent management also requires a deep understanding of the organisational culture and systems in place. Leaders' decisions are often shaped by the specific context in which they operate (Kemethofer et al., 2022). A supportive culture is characterized by open communication, loyalty

J. Namugumya (✉)
Uganda Martyrs University, Kampala, Uganda

Makerere University, Kampala, Uganda
e-mail: jnamugumya@umu.ac.ug

J. C. K. Munene · S. Mafabi
Makerere University, Kampala, Uganda

J. Kagaari
Kyambogo University, Kampala, Uganda

P. Nyadoi et al. (eds.), *Sustainable Development Goals in Africa*, Sustainable Development Goals Series, https://doi.org/10.1007/978-3-032-20742-5_20

and team spirit, along with a system that promotes connections among staff, creates an environment conducive to talent growth and retention (McMillan, 2008). Thus understanding how operational strategies of organisations take care of this is important, and hence the rationale for our study that examined different tertiary institutions in Uganda. Many organisations, including tertiary education institutions, continue to struggle to attract, select and retain top talent for key roles (Mitchell et al., 2001). This requires organisations including tertiary educational institutions, to focus on the best talent management practices (McDonnell et al., 2017; Matongolo et al., 2018; Sackey & Selassie, 2022).

Managing academic talent effectively requires strong institutional support and well-defined talent strategies to build a strong pool of senior skilled academic staff, and hence the rationale for our study.

To better understand talent management within tertiary education institutions, our study draws on the principles of complex adaptive systems (CAS) theory (McMillan, 2008; Patel, 2024), particularly the aspects of organisational culture and talent management—the belief that organisations thrive when they foster a culture of continuous interactions, learning and adaptation. It is through culture that adaptive behaviour emerges and, on the other hand, adaptable management system supports innovative practices (Schophuizen et al., 2022).

Despite its importance in organisational performance evidenced by the increasing global interest (Saleh & Atan, 2021; Krishnakumar & Kavya Vasu, 2021), not many studies on talent management have been done in East Africa (Fahim, 2018; Anlesinya et al., 2019). In particular, examining how talent management intersects with organisational culture of tertiary institutions in Sub-Saharan Africa, knowledge on this aspect for the regions is still limited. For example, Fahim (2018) and Anlesinya et al. (2019) point out that there has been little research done on talent management in East Africa. In our study, we explored how organisational culture influences talent management among managers of tertiary institutions in Uganda.

Methods

The study employed a positivist paradigm because it looks at reality as objective and can be measured. Organisational culture, system adaptability and talent management were quantified using structured questionnaires. The study employed a cross-sectional research design and data was collected at one point in time. We performed Cronbach's Alpha and composite reliability tests to establish suitability of our data collection instruments (Cronbach, 1951; Hair et al., 2020). Furthermore, the study adopted procedural and statistical remedies to control for common method bias, minimise measurement errors and increase the validity of conclusions (Podsakoff et al., 2012). Self-administered questionnaires were distributed to deans, heads of academic departments and human resource managers. Data were collected from 573 respondents sampled from 86 (out of 128) tertiary institutions in Uganda, 39 (out of 43) of them public and 47 (out of 85) private, using proportionate stratified sampling because the sizes of the stratum were not the same. Data was collected from 3 to 5 respondents purposively sampled from each of the selected institutions. Data were aggregated and analysed at the organisational level.

Analyses

The data were entered into Microsoft Excel and analysed using SPSS version 26. We checked for missing values and outliers using frequency counts. We also assessed normality of the data using graphical and statistical methods and conducted Pearson correlation analysis to determine associations among the study variables. The degree to which independent variables were not similar was determined following Fornel and Larcker. This involved comparing the square root of the average variance extracted of each variable and the inter-construct correlation. We found that the inter-construct correlation coefficients among study variables were below the square root of each average variance explained meaning that the variables we used were suitable for the study of talent management. We further used the heterotrait-monotrait ratio (HTMT) as recommended by Henseler et al. (2015) to confirm discriminant validity.

We used Smart-PLS 4.0.9.0 to test for structural model fit and Q-squared to establish the predictive relevance of endogenous constructs. Q2 values above zero indicated that the model had predictive relevance (Hair et al., 2017). Predictions from partial least squares path model were generated and assessed following Hair et al. (2017). Two-tailed *t*-tests were used to assess all statistical tests at the 5% level of significance. The two-tailed *t*-tests allowed us to identify relationships between aspects of organisational culture and aspects of talent management.

Results

Socioeconomic Characteristics of Study Participants

Most of the tertiary institutions included in the study were from the Central region (n = 48, 55.8%), followed by the Western region (n = 20, 23.3%) and the Eastern region (n = 10, 11.6%), while a smaller number were from the Northern region (n = 8, 9.3%). Of the 573 participants, 63% were male, 45.5% were aged 35–44 years old, 60.1% had a master's degree as the highest level of education while 46.1% had served their respective institutions for 5–8 years (Table 20.1). Overall, majority (85%) of the respondents were married, followed by those who were single (Table 20.1). This means that the biggest number of respondents were dual career couples. They participate in the activities of the institution and deal with family affairs that need balancing.

Table 20.1 Socioeconomic characteristics of respondents

Variable	Category	n (%)
Gender	Male	361 (63.0%)
	Female	212 (37.0%)
Age	25–34 years	87 (15.2%)
	35–44 years	261 (45.5%)
	45–54 years	191 (33.3%)
	55–64 years	32 (5.6%)
	65 years and above	2 (0.3%)
Level of education	Diploma	9 (1.6%)
	Bachelors	94 (16.4%)
	Masters	344 (60.0%)
	PhD	126 (22.0%)
Marital status	Married	487 (85.0%)
	Single	45 (7.9%)
	In partnership	27 (4.7%)
	Separation	14 (2.4%)
Position of responsibility	Human Resource Manager	157 (27.5%)
	Head of Academic Unit	202 (35.2%)
	Dean of Academic Unit	214 (37.3%)
Years of service	1–4 years	134 (23.4%)
	5–8 years	264 (46.1%)
	9–12 years	113 (19.7%)
	13–16 years	35 (6.1%)
	17–20 years	12 (2.1%)
	21 and above	15 (2.6%)

Table 20.2 Two-tailed *t*-tests used to identify relationships between variables of interest

	B	μ	δ	*t*-statistic	*P*-value
Direct effects					
Organisational culture → talent management	−0.10	−0.11	0.12	0.86	0.39
Total effects					
Organisational culture → talent management	0.08	0.07	0.12	0.68	0.50

Perception of Organisational Culture and Talent Management Practices

Results of the two-tailed *t*-tests used to identify relationships between variables of interest show that the relationship between organisational culture and talent management was not significant (Table 20.2).

Discussion

Although studies elsewhere have found that organisational culture influences talent management, our findings if validated would imply that talent recognition, open communication, and teamwork as attributes of organisational culture are not directly related to talent management in Uganda's tertiary institutions. These findings, however, remain inconclusive and efforts are underway to conduct more in-depth analyses. The researchers who noted that organisational culture affects talent management include Kontoghiorghes (2016), Paais and Pattiruhu (2020) and Saleh and Atan (2021). If validated, the findings would suggest that culture alone might not influence the way top talent is identified, maintained and retained, implying that other factors are more important like the way institutional systems work. Managers may be having good intentions, and encourage sharing of information but fail to provide what professors need like international visibility or research opportunities, that in the end they leave.

The findings would also be surprising because these cultural traits are usually thought to create a positive work environment. However, in complex human systems like universities, things do not always work in a straight line. Open communication or teamwork may exist, but if the systems and structures for mentoring and rewarding talent are weak or missing, culture alone cannot produce the desired results. This supports the complex adaptive systems (CAS) theory which suggests that organisations are not predictable and that they are made up of many interacting parts in changing ways. Thus, the CAS theory assists us understand that what works in one place may not work in another, and that change often happens from the bottom-up, not top-down, plans. So even when staff recognize talent and communicate openly, talent management may not improve unless there are strong feedback loops, adaptive leadership, and a system that learns and adjusts over time (Uhl-Bien et al., 2007). This shows that managing talent in complex environments requires not only culture but a whole system that can adapt and support people in meaningful ways.

References

Anlesinya, A., Amponsah-Tawiah, K., & Dartey-Baah, K. (2019). Talent management research in Africa: Towards multilevel model and research agenda. *African Journal of Economic and Management Studies, 10*(4), 440–457.

Collings, D. G., Vaiman, V., & Scullion, H. (2022). *Talent management: A decade of developments* (pp. 1–18). Emerald Publishing Limited.

Cronbach, L. J. (1951). Coefficient alpha and the internal structure of tests. *Psychometrika, 16*(3), 297–334.

Fahim, M. G. A. (2018). Strategic human resource management and public employee retention. *Review of Economics and Political Science, 3*(2), 20–39.

Gallardo-Gallardo, E., Thunnissen, M., & Scullion, H. (2020). Talent management: Context matters. *The International Journal of Human Resource Management, 31*(4), 457–473.

Hair, J. F., Hult, M., Ringle, C. M., & Sarstedt, M. (2017). *A primer on partial least squares structural equation modeling (PLS-SEM)*. Sage.

Hair, J. F., Jr., Howard, M. C., & Nitzl, C. (2020). Assessing measurement model quality in PLS-SEM using confirmatory composite analysis. *Journal of Business Research, 109*, 101–110.

Henseler, J., Ringle, C. M., & Sarstedt, M. (2015). A new criterion for assessing discriminant validity in variance-based structural equation modeling. *Journal of the Academy of Marketing Science, 43*, 115–135.

Kemethofer, D., Weber, C., Brauckmann-Sajkiewicz, S., & Pashiardis, P. (2022). Examining the trident: How data from the PISA study can be used to identify associations among context, school leadership and student outcomes. *Journal of Educational Administration, 61*(2), 162–167.

Kontoghiorghes, C. (2016). Linking high performance organisational culture and talent management: Satisfaction/motivation and organizational commitment as mediators. *The International Journal of Human Resource Management, 27*(16), 1833–1853.

Krishnakumar, S., & Kavya Vasu, P. K. (2021). The impact of organization culture on talent management. *Dr. Srinidhi K. Parthasarathi.*

Kwon, K., & Jang, S. (2022). There is no good war for talent: A critical review of the literature on talent management. *Employee Relations: The International Journal, 44*(1), 94–120.

Matongolo, A., Kasekende, F., & Mafabi, S. (2018). Employer branding and talent retention: Perceptions of employees in higher education institutions in Uganda. *Industrial and Commercial Training, 50*(5), 217–233.

McDonnell, A., Collings, D. G., Mellahi, K., & Schuler, R. (2017). Talent management: A systematic review and future prospects. *European Journal of International Management, 11*(1), 86–128.

McMillan, E. (2008). *Complexity, management and the dynamics of change: Challenges for practice.* Routledge.

Mitchell, T., Holtom, C. B., & Lee, T. (2001). How to keep your best employees: Developing an effective retention policy. *Academy of Management Executive, 15*, 96–108.

Paais, M., & Pattiruhu, J. R. (2020). Effect of motivation, leadership, and organisational culture on satisfaction and employee performance. *The Journal of Asian Finance, Economics and Business, 7*(8), 577–588.

Patel, J. (2024). The role of adaptive culture in organizational success: A comprehensive review. *Journal of Organisational Culture Communications and Conflict, 28*(S2), 1–3.

Podsakoff, P. M., MacKenzie, S. B., & Podsakoff, N. P. (2012). Sources of method bias in social science research and recommendations on how to control it. *Annual Review of Psychology, 63*, 539–569.

Putri, S. M., Rivai, H. A., & Syahrul, L. (2023). The effect of talent management and organizational culture on employee performance with job satisfaction as a mediating variable. *Enrichment: Journal of Management, 1*(1), 41–54.

Sackey, J. A., & Selassie, E. A. (2022). The role of human resource management practices and employee job satisfaction in predicting organizational commitment in tertiary institutions. *Malaysian Business Management Journal, 1*(1), 14–18.

Saleh, R., & Atan, T. (2021). The involvement of sustainable talent management practices on employee's job satisfaction: Mediating effect of organizational culture. *Sustainability, 13*(23), 13320.

Schophuizen, M., Kelly, A., Utama, C., Specht, M., & Kalz, M. (2022). Enabling educational innovation through complexity leadership? Perspectives from four Dutch universities. *Tertiary Education and Management, 29*, 471–490.

Uhl-Bien, M., Marion, R., & McKelvey, B. (2007). Complexity leadership theory: Shifting leadership from the industrial age to the knowledge era. *The Leadership Quarterly, 18*(4), 298–318.

Vaiman, V., Cascio, W. F., Collings, D. G., & Swider, B. W. (2021). The shifting boundaries of talent management. *Human Resource Management, 60*(2), 253–257.

Part VI

Education, Digital Age and Knowledge Production in Africa

Scoping Review of Knowledge on Digital Uptake in Educational Institutions in Uganda

21

Alex Alison Tongu and Modest Kayi O'dama

Abstract

Digital literacy skills and competencies are crucial in the management of information and improvement of pedagogical practices in educational institutions. In this paper, we conducted a scoping review and present a summary of the state of existing knowledge on digital uptake among institutions in Uganda, with particular focus on educational institutions. The findings show there is a diversity of digital resources in use and challenges underpinning their promotion, including among others, elderly managers having challenges with knowledge acquisition and application of ICT in operations, resource constraints limiting educational institutions from transitioning to digital management systems. An in-depth study is required to provide information to enable design and implementation of strategies that would intensify digital uptake in the management of educational institutions in Uganda.

Keywords

Learning management systems · COVID-19 pandemic · Enabel Uganda · E-library · Secondary schools

A. A. Tongu (✉) · M. K. O'dama
Uganda Martyrs University, Kampala, Uganda
e-mail: tongu.alison@stud.umu.ac.ug

Introduction

In line with the global Sustainable Development Goals (SDGs), the government of Uganda established a Digital Vision 2040. This vision was meant to enable universal integration, sustainable development, economic progress and eradicating poverty by becoming digitally empowered (Gillwald et al., 2019). This was followed by the formulation of the ICT policy in 2014, which designated ICT as an industry that enhances economic growth (Bowman, 2019). But according to Andema et al. (2013), a big gap exists between the digital policy requirements and the reality on the ground in Uganda. In addition, although the 2014 Digital Policy proposed establishing computer studies, e-learning and e-administration, only the former has been fully implemented. Generally, head teachers in public and private schools still manage information and pedagogy conventionally. In this paper we provide an overview of the state of digital uptake in Uganda, with particular emphasis on educational institutions. We provide sectional overview of what exists in the literature generally.

Digital uptake in schools refers to the increasing adoption and integration of digital technologies into the teaching and learning process within educational institutions (Moore & Pratt, 2023). This includes the use of digital tools, online plat-

P. Nyadoi et al. (eds.), *Sustainable Development Goals in Africa*, Sustainable Development Goals Series, https://doi.org/10.1007/978-3-032-20742-5_21

forms and interactive content to enhance the learning experience and to prepare students for a digitally driven world. As reported by Bowman (2019), digital uptake requires significant investment in hardware, software and internet infrastructure. In educational institutions globally, it entails establishing digital instructional materials, e-libraries, online learning and internet-based laboratories. Proponents of digital uptake in educational institutions recognize the positive impact on the management this makes despite the costs involved (Nahamya & Tumwebaze, 2014; Snoussi, 2019). For example, digital uptake could enable managers to share, keep and archive information and make it accessible to every stakeholder for effective management (Adekunle et al., 2024). In many countries, Uganda inclusive, use of technology became a necessity during the outbreak of COVID-19 pandemic. Many schools, both public and private, were forced to rely on digital technology as a strategy to support containment of the pandemic.

History and Progress of Digital Uptake in Uganda

In Uganda the use of modern technologies including telephones, telegrams and printing presses dates back to the late colonial era (Summers, 2015). The country has experienced a significant increase in the adoption of digital technologies in recent years, particularly in areas like mobile money and e-commerce (Bowman, 2019). This growth has been driven by government initiatives, private sector innovation and increased internet penetration. The government has been allocating significant investments and securing funding from public and private sources to support the implementation of digital initiatives in urban and rural areas (Bowman, 2019).

Uganda is one of the countries in sub-Saharan Africa with a policy that supports digital uptake (Asongu & Odhiambo, 2019). This policy has been embedded in Uganda's Vision 2040, which aims to transform the country into a competitive, upper middle-income nation by 2040. In addition to the digital policy, an Education Digital Agenda (EDA) was formulated in 2020 and was aimed at leveraging and scaling up the impact of digitalization in improving the activities in the education sector (Ssenyonga et al., 2022). However, school administrators and teachers have not been given sufficient support to acquire the necessary digital competences. When Enabel Uganda (a Belgian agency for international cooperation) organised a training for teacher trainees, it was noted that a big gap exists between the digital policy requirements and the knowledge, skills and competences of secondary schools' teachers (Mwogeza, 2018). On the other hand, Lutalo and Bisaso (2020) reported inability to learn and acquire ICT knowledge and skills among some school managers despite Continuous Professional Development (CPD) trainings.

Diversity and State of Digital Uptake in Uganda

Digital facilities including computer labs, interactive whiteboards, e-books and educational software are in short supply in many Ugandan schools. The major digital components that enable information processing, storage, transfer and retrieval are the internet-based and web-based tools (Udochukwu Ukanwa & Chimdiya Chiemeka, 2021). In most Ugandan secondary schools, basic files including student, staff and inventory records keep hard copy manual records which would easily be managed electronically (Mwadulo & Odoyo, 2020). Schools would do well to invest more in digital systems to organise, store and manage student and administrative records (Munemo & Sibanda, 2021).

Despite the fact that digital libraries offer several advantages over physical libraries, Juma et al. (2016) found that lack of investment in digital facilities has made most schools to rely heavily on physical libraries. There is thus need for schools to invest more in hardware and software to support modern teaching and learning (Asuman et al., 2018; Kyomuhendo et al., 2024). Available literature indicates that schools that invest in digital facilities have higher enrolments (Syahnita, 2021; Schellnack-Kelly et al., 2021). As suggested by Kahiigi et al. (2008), schools and parents can take advantage of the affordable and accessible internet offered by telecommunication companies.

Conclusion

Findings show there is a diversity of digital resources in Uganda's educational institutions. The main challenges limiting the potential benefits of digital learning include inadequate training of teachers and lack of investment in the necessary computers, reliable internet access and updated software. An in-depth study is required to inform the design and implementation of strategies for increasing adoption and integration of digital technologies in educational institutions.

References

Adekunle, P., Aigbavboa, C., Akinradewo, O., Ikuabe, M., & Otasowie, K. (2024). Towards the uptake of digital technologies for construction information management: A partial least squares structural equation modelling approach. *Buildings, 14*(3). https://doi.org/10.3390/buildings14030827

Andema, S., Kendrick, M., & Norton, B. (2013). Digital literacy in Ugandan teacher education: Insights from a case study. *Reading & Writing, 4*(1), 1–8. https://doi.org/10.4102/rw.v4i1.27

Asongu, A. S., & Odhiambo, M. N. (2019). *Enhancing ICT for quality education in Sub-Saharan Africa*. AGDI Working Paper No WP/19/007, Africa, Government and Development Institute (AGDI).

Asuman, B., Khan, M. S. H., & Clement, C. K. (2018). Integration of web-based learning into higher education institutions in Uganda: Teachers' perspectives. *International Journal of Web-Based Learning and Teaching Technologies, 13*(3), 33–50. https://doi.org/10.4018/IJWLTT.2018070103

Bowman, W. M. (2019). Technological distribution in Uganda: Information and communications technology and the state in an eastern African nation. *Review of Policy Research, 36*(6), 835–865. https://doi.org/10.1111/ropr.12358

Gillwald, A., Mothobi, O., Tusubira, F., & Ndiwalana, A. (2019). *The state of ICT in Uganda* (Policy Paper No. 8; Series 5: After access—Assessing digital inequality in Africa). Research ICT Africa. https://researchictafrica.net/wpcontent/uploads/2019/05/2019_After-Access-The-State-of-ICT-in-Uganda.pdf

Juma, I. W., Wamukoya, J., & Wekullo, C. (2016). The role of digital libraries in bridging the knowledge gap in Africa. *Mousaion: South African Journal of Information Studies, 32*(1), 64–83. https://doi.org/10.25159/0027-2639/1700

Kahiigi, K. E., Ekenberg, L., Hanson, H., Danielson, M., & Tusubira, F. F. (2008). Explorative study of e-learning in developing countries: A case of the Uganda education system. In *MCCSIS'08—IADIS Multi Conference on Computer Science and Information Systems; Proceedings of e-Learning 2008*, (Vol. 2(June 2014), pp. 195–199).

Kyomuhendo, T. K., Mugizi, W., Rwothumio, J., & Waiswa, M. M. (2024). School resources and effective implementation of e-learning in secondary schools in Kasese Municipality, Uganda. *Interdisciplinary Journal of Education Research, 6*, 1–16. https://doi.org/10.38140/ijer-2024.vol6.01

Lutalo, F., & Bisaso, S. M. (2020). Information communication technology integration in school systems and management of secondary schools in Nakawa Division, Uganda. *Islamic University Multidisciplinary Journal, 7*(4), 136–149.

Moore, J., & Pratt, M. K. (2023). *What is digital transformation? Everything you need to know*. www.techtarget.com/searchcio/definition/digital-transformation

Munemo, P., & Sibanda, A. (2021). Users' perceptions and use of electronic resources in academic libraries: A case of ARRUPE College library, Zimbabwe. *GSJ, 9*(8), 1856–1878.

Mwadulo, M. W., & Odoyo, C. O. (2020). ICT adoption in the educational management of primary schools in Kenya. *Universal Journal of Communications and Network, 8*(1), 1–5. https://doi.org/10.13189/ujcn.2020.080101

Mwogeza, R. (2018). *Report on the baseline study of technology-enabled learning at Uganda Management Institute*. https://oasis.col.org/server/api/core/bitstreams/62490d74-134a-4793-912e-907db478f4da/content

Nahamya, W., & Tumwebaze, H. (2014). The role of digital literacy in continuous based assessment in technical institutions in Uganda. *International Journal of Vocational Education and Training*.

Schellnack-Kelly, I., Nwagwu, W. E., & Dubale, F. F. (2021). Use of electronic information resources in a special library in Ethiopia. *Library Philosophy and Practice, 2021*, 1–24.

Snoussi, T. (2019). Learning management system in education: Opportunities and challenges. *International Journal of Innovative Technology and Exploring Engineering, 8*(12S), 664–667. https://doi.org/10.35940/ijitee.l1161.10812s19

Ssenyonga, R., Sewankambo, N. K., Mugagga, S. K., Nakyejwe, E., Chesire, F., Mugisha, M., et al. (2022). Learning to think critically about health using digital technology in Ugandan lower secondary schools:

A contextual analysis. *PLoS One, 17*(2), e0260367. https://doi.org/10.1371/journal.pone.0260367

Summers, C. (2015). Slander, buzz and spin: Telegrams, politics and global communications in the Uganda Protectorate, 1945-55. *Journal of Colonialism and Colonial History, 16*(3), 1–13.

Syahnita, R. (2021). *Information communication technology integration and performance of library in public universities in Nairobi City County*. Kenyatta University.

Udochukwu Ukanwa, G., & Chimdiya Chiemeka, E. (2021). Utilization of ICT for management and administration in education sector. *International Journal of Advances in Engineering and Management (IJAEM), 3*(3), 987. https://doi.org/10.35629/5252-0303915920

Teaching and Learning Practices at Four East African Universities: Lessons and Experiences from the TESCEA Project

22

Jacinta Bwegyeme and Josephine Namuli

Abstract

Over time, teaching and learning practices in higher education have been characterized by didactic, teacher-centered approaches. Subsequently, the need to use active learning methods that meet the twenty-first-century skills expectations led to the birth of the *Transforming Employability for Social Change in East Africa* (TESCEA) Project. The aim of the project was to promote transformative learning that, among others, enhances teachers' knowledge about critical thinking concepts, teaching for critical thinking, and gender responsiveness. Experiences and lessons that emerged from the project show that active learning strategies create a reflective environment that challenges previously held assumptions. We recommend a long-term study to assess the impact of applying active learning strategies, such as those that were promoted by TESCEA, on teachers' ability to promote transformative learning.

Keywords

Beliefs · Pedagogy · Lecturers · Teaching methods · Uganda · Higher institutions of learning

J. Bwegyeme (✉) · J. Namuli
Uganda Martyrs University, Kampala, Uganda
e-mail: jbwegyeme@umu.ac.ug

Introduction

The need to use teaching and learning approaches that engage students and address twenty-first century skills such as collaboration, critical thinking, and problem-solving has been documented by various scholars (Abdella and Fataar, 2021; Morrison et al., 2015). Several studies indicate that teachers' style of teaching in many institutions of higher learning especially in Africa has remained didactic and teacher centered (Abdella and Fataar, 2021; Snyder, 2017). Embracing change from these didactic approaches is often hindered by teachers' attitude and resistance to innovations (Snyder, 2017). The realization of the need to address the existing teaching/learning challenges led to the partnership of several institutions leading to the emergence of Transforming Employability for Social Change in East Africa (TESCEA) project. The TESCEA project has been involved in the training of academic staff selected from various faculties in four universities in East Africa. The aim is to enhance skills that promote transformative learning, critical thinking, and problem-based learning. Face-to-face workshops were the mode of instruction before the COVID-19 lockdowns and then later, utilization of the online mode was adopted. The following East African universities were involved in the project: Uganda Martyrs University, Gulu University, University of Dodoma, and Mzumbe University. Of significance, at TESCEA's incep-

P. Nyadoi et al. (eds.), *Sustainable Development Goals in Africa*, Sustainable Development Goals Series, https://doi.org/10.1007/978-3-032-20742-5_22

tion, the partners to the project acknowledged that although several institutions had established trainings for a few lecturers, the existing initiatives had not adopted a sufficiently systemic approach to transform pedagogy. Hence the partnership proposed a need to equip academic staff with new pedagogical approaches in a bid to transform teaching and learning.

Transformative learning theory, proposed by Jack in 1990, and experiential learning, proposed by Kolb in 1984, were utilized to elucidate the modern pedagogies that involve transformative learning, critical thinking, and problem-based learning. In the TESCEA project, transformative learning was operationalized in the context of modern teaching and learning practices as a process through which individuals critically examine and, ultimately, shift their deeply held beliefs, values, and assumptions (Sahin et al., 2018). This change does not simply involve the acquisition of new information but constitutes a fundamental reconstruction of one's ingrained habits, thought patterns, and ways of understanding the world. Problem-based learning is an instructional method that centers on learners working in groups to address complex, real-world problems (Yu and Mohamed Zin, 2023), while critical thinking is the disciplined process of actively and skillfully conceptualizing, analyzing, synthesizing, and evaluating information gathered from observation, experience, or communication.

Transformative learning theory posits that adults rely on previously established combinations of habits of mind, beliefs, and values that shape how they interpret experiences (Mezirow, 1990; Fleming, 2018). These frames are built up over years of social, cultural, and personal experiences and tend to remain stable until significantly challenged. The lived experience forces the learner to question previously unquestioned assumptions. The theory describes the process by which individuals critically examine and potentially revise their existing beliefs, assumptions, and perspectives through reflection and discourse. This theory has significantly influenced adult education by emphasizing the role of critical reflection in facilitating profound personal and societal change (Fleming, 2018).

Cranton (2023) opines that fostering group ownership, providing experiential activities, and encouraging critical reflection can facilitate transformative learning in adult education settings to create a transformation in employees. Cranton (2023), however, highlights the significance of emotions and relationships in transformative learning. Transformation is not solely a cognitive process but also involves affective and interpersonal elements. However, Mezirow's theory prioritizes cognitive processes while neglecting emotional and spiritual dimensions of learning (Carter & Nicolaides, 2023). *The transformative theory is also criticized for overemphasis on rationality*, prioritizing cognitive processes while neglecting emotional and spiritual dimensions of learning. Furthermore, the transformative theory is criticized for its Eurocentric perspective, potentially limiting its applicability across diverse cultural contexts and ignoring the role of context in shaping transformative learning experiences (Fleming, 2018). Rodríguez Aboytes and Barth (2020) note that although transformative learning theory is superficially used in many studies, it can contribute to the design and implementation of educational interventions and assessments of learning toward sustainability if carefully studied. As such, integrating the Kolb experiential learning theory becomes pivotal in elucidating the gaps of affective and interpersonal elements that are neglected under transformative learning.

Kolb's experiential learning theory offers a cyclical model that encompasses four stages: concrete experience, reflective observation, abstract conceptualization, and active experimentation. This model complements transformative learning theory in several ways. While transformative learning emphasizes the critical reappraisal of assumptions through cognitive reflection, Kolb's model grounds learning in actual experiences by engaging learners in concrete activities and observations. Kolb's model ensures that learning is not only reflective but directly tied to lived experiences. Kolb's approach accommodates the non-linear nature of learning. The continuous cycle of acting, reflecting, conceptualizing, and then experimenting provides an iterative structure that contrasts with the more staged model of transformation. Further, by incorporating reflective observation as a distinct phase, Kolb's theory allows for both the analysis of experience and the integration of personal insights. When combined with transformative learning's call for critical reflection on one's core assumptions, this approach helps balance the rational and emotional aspects of personal change. Hitchcock (2017) and Murawski (2014) elucidate that the process of thinking critically involves problem identification and analysis, clarification of meaning, gathering and assessing of evidence, inferring conclusions, considering other relevant information, and making an overall judgment. Halpern's (2019) situates critical thinking skills within real-world contexts, advocating for instruction that integrates cognitive strategies with metacognitive monitoring.

Cultivating independent thinking involves transformative learning that facilitates critical thinking (Murawski, 2014). Thus, critical thinking is the ability not just to acquire knowledge as advocated in the transmission of knowledge approach but also to make sense of new information, other than accumulation of knowledge. Further, Hitchcock (2017) differentiates critical thinking from the logical appraisal of arguments by proving that the former extends beyond a single argument, has a creative component, and involves critical assessment of evidence. Whereas critical thinking concept refers to a purposeful, self-regulated reflective process, teaching for critical thinking refers to application of structured approaches that encourage students to engage in reasoning, problem-solving, and reflective thinking (Srivastava, 2020). It is the responsibility of the educator/teacher to create and sustain the type of environments that encourage critical reflection rather than the didactic transmission of knowledge (DiPasquale, 2017). Hitchcock (2017) proposes the following learning environment in a stand-alone course: adaptation to the students' situation, communication of the course goals, motivation of students, use of a checklist as a course framework, fostering a critical spirit, preference of depth to breadth, use of bridging, taking advantage of key issues, use of realistic examples, giving students lots of guided practice with feedback, checking for understanding, encouraging metacognition, thinking about context, avoiding empty use of technical terms, and designing multiple-choice items carefully if they should be used.

DiPasquale (2017) and Berg et al. (2019) proposed the strategies that should be integrated into a course to promote critical thinking and these include: elaboration, organizing, metacognitive self-regulation and peer learning, students' encouragement, students' participation and interaction during the course, class presentation, and providing detailed critiques on written assignments, group projects, and fictitious problem scenarios. A study by Changwong et al. (2018) in Thailand established that students who master the ability to think critically and insightfully perform better academically and become creative thinkers and problem solvers. Higher institutions of learning are vessels for building skills and competencies for economic and social development in the volatile environment of the twenty-first century (Zeleza, 2018). If these demands are to be addressed, the education system must go beyond validation of vocational and technical skills to embracing critical thinking (Zeleza, 2018). Abrami (2015) in 117 studies established that teaching critical thinking explicitly by providing strategies, modeling, and guided practice improved students' experiences.

Gender-Responsive Concept and Teaching for Critical Thinking

Gender responsiveness refers to "taking action in order to correct gender bias and discrimination in order to ensure gender equality and equity" (FAWE, 2020). Aligning people's gender identity and expression with their biological sex may mislead organizations to assume that they are gender responsive (Tarrayo et al., 2021). Gender-responsive pedagogy goes beyond simply acknowledging gender differences. It involves designing and facilitating learning experiences that actively dismantle gender stereotypes and create an inclusive classroom environment

(Ananga, 2021). Examples of gender-responsive practices implemented in teacher education include: using diverse learning materials, facilitating reflective discussions, and tailoring teaching methods to address diverse learner needs. Integrating gender responsiveness and teaching for critical thinking fosters an inclusive environment and encourages students to critically question existing gender norms, thus strengthening critical thinking skills (Ananga, 2021). When educators marry gender-responsive pedagogy with teaching for critical thinking, they create a dynamic learning environment that addresses both the cognitive and the social dimensions of learning (Ananga, 2021). For example, by exposing students to texts, case studies, and images that challenge stereotypical representations, teachers encourage learners to question biases and to develop alternative interpretations. Such learning environments prompt students to engage in reflective dialogue, a practice that further nurtures critical thinking skills.

Several studies have documented that critical thinking skills do not emerge in a vacuum; they are influenced by learners' socialization, cognitive style, and prior experiences which, in turn, are often shaped by gender. For instance, Sukarma et al. (2019) suggest that traditional gender roles and societal norms may influence how students process and analyze information. Similarly, research by Albarracín-Vivo et al. (2024) examined the written compositions of primary education students in contexts where controversial gender issues were highlighted. Their qualitative results revealed that female students often exhibited higher levels of critical engagement when reflecting on gender-related content. The results suggest that early educational experiences framed around gender sensitivity may foster more robust critical thinking by challenging conventional norms and encouraging distinctive evaluation of social contexts (Albarracín-Vivo et al., 2024). The integrated approach is particularly effective as it recognizes that the ability to think critically is influenced by the cultural and gendered context of learning. By fostering an environment where gender differences are acknowledged and addressed, teachers promote a more equitable distribution of participatory roles in classroom discussions and collaborative problem-solving sessions (Shubina & Kulakli, 2019). In doing so, students learn not only to evaluate evidence rigorously but also to appreciate the role of diverse perspectives in enriching the learning process.

Lecturers' Beliefs Toward Facilitation Techniques, Teaching for Critical Thinking and Gender Responsiveness

Attitudes and beliefs drive classroom actions and influence teachers' processes of change (Murawski, 2014). Teachers' beliefs fall under two types: (1) the direct transmission beliefs and (2) the constructivist beliefs about learning and instruction. The direct transmission view supports teacher-directed methods of learning (Murawski, 2014). In contrast, a constructivist view emphasizes development of thinking and reasoning processes and supports use of active facilitation methods. Abrami et al. (2015), through meta-analysis, established that instructors who believed critical thinking was a transferable skill were more likely to explicitly teach it across subjects. In contrast, those who saw it as discipline-specific tended to neglect teaching it directly.

Semerci & Elaldı (2021) who conducted a qualitative study in Turkey established that while most lecturers supported critical thinking in theory, only a minority used strategies like problem-based learning or Socratic questioning in practice. Their hesitation stemmed from beliefs that students lacked readiness or that institutional exams did not reward critical reasoning. Yuan (2017) suggests that teachers' beliefs are shaped by professional development training and new pedagogical orientations. Also, the different personal experiences contribute to the formation of strong enduring beliefs about teaching and learning. Despite all the literature on the shortfalls of the direct transmission models, the dominant method used in universities has remained lecturing (Snyder, 2017). Katende (2023) reported that most faculty in Uganda's higher institutions of learning believe critical thinking is important but often default to lecture-heavy methods which do not strongly support the development of critical thinking skills in learners. A large-scale study by Ertmer et al. (2015) found that many higher education instructors held on to traditional, teacher-centered views despite advocating student-centered facilitation. Their beliefs, particularly about control, often limited their use of inquiry-based or dialogue teaching strategies. It is explicit that under no circumstance can such teachers be able to churn out students who are critical thinkers.

Namatende-Sakwa et al. (2022) found that pre-service teachers who have experienced schools with strong gender roles, combined with poor training, are more likely to continue old gender stereotypes in their future classrooms. The authors suggest completely revising the curriculum to include more in-depth, gender-sensitive topics so that future teachers can effectively challenge inequality from the start. Ismail et al. (2021) established that early childhood teachers in South Africa held stereotypical beliefs about gender roles until exposed to structured gender-responsive training. According to UNESCO (2023), empirical studies point to the persistence of gender biases among teachers, especially in *Science, Technology, Engineering, and Mathematics* (*STEM*) where low female participation is notable. INASP (2021) also found that lecturers' resistance to gender-responsive practices was less about ideology and more about lack of exposure and training. The studies suggest that lecturers' beliefs act as filters for both pedagogical decisions and responsiveness to learners' needs. In addition, studies indicate that the teachers can be exposed and trained in gender responsiveness.

The TESCEA intervention sought to introduce transformational approaches that would change the beliefs of teachers, hence the need to review literature on formation of beliefs. Whereas research reveals that cognitive beliefs develop gradually through education and experience (Ferguson 2020), there are significant research findings that indicate that both long- and short-term interventions can lead to changes in an individual's views of knowledge and knowing (Kienhues et al., 2016; Lunn Brownlee et al. 2017). For example, targeted interventions have been used to introduce cognitive conflict in pre-service teachers (Gill et al., 2017), by inducing text-based materials that introduce and then demerit commonly held myths about teaching knowledge (Gill et al., 2017). Furthermore, Parkinson and Maggion (2017) highlight the role of reflection in belief change where reflection can be introduced in a guided approach. Interventional research by Braasch et al. (2013) explores students' engagement in group discussions based on instructional materials and demonstrates how different ways of thinking enhance reflection and change in belief.

Teachers' Attitudes Toward Teaching for Critical Thinking and Gender Responsiveness

According to Le and Nguyen (2024), teachers' attitudes toward teaching for critical thinking have been affected by two models, namely the constructivist view and the direct transmission view. The literature reviewed indicates that although most teachers seem to understand the importance of teaching critical thinking skills, they still do not practice teaching for critical thinking. The survey by Vincent-Lancrin et al. (2019), which was conducted across 11 countries, found that 90% of teachers regarded critical thinking as a key goal but were uncertain about how to embed it meaningfully into their everyday classroom practices. Another study reported that some teachers neglect critical thinking skills and claim that they do not have time (Asgharheidari & Tahriri, 2015). Such teachers still believe that their responsibility is to cover all subject matter through lectures and the students' responsibility is to memorize the knowledge transmitted to them. In Moroccan secondary schools, a survey of English teachers revealed generally positive attitudes toward critical thinking where 82% believed critical thinking skills were essential for language learning but only 30% reported routinely using critical thinking. Therefore, encouraging a positive attitude toward critical thinking and active facilitation techniques is essential.

Research indicates that teachers who hold traditional gender stereotypes may inadvertently reinforce inequities. Carlana (2019) demonstrates that teachers' implicit gender stereotypes lead to differential expectations and behaviors in the classroom, thereby reinforcing inequities such as attributing disruptive behavior more readily to boys or setting lower expectations for girls. Conversely, teachers with gender-sensitive attitudes are more likely to foster inclusive classroom environments that challenge stereotypical roles, encouraging both girls and boys to explore a broader range of academic and career-related possibilities (Niones, 2022). A few studies were reviewed to establish how teachers' attitudes can be influenced. A study by Bicchieri et al. (2019) investigated how social norms influence attitude change. The research suggested that attitudes shift when individuals perceive a change in collective behavior or expectations within their social group. Another study by Entman (2017) highlights how media framing affects attitudes by shaping how information is presented. This applies to political persuasion, advertising, and public health messaging. Findings show that the use of narratives and storytelling can enhance attitude change more effectively than purely factual reporting. Furthermore, Nabi and Myrick (2019) show how emotional appeals can facilitate attitude change. Their study suggests that positive emotions like inspiration and hope lead to more sustainable behavioral changes compared to fear-based messaging.

The TESCEA Project

The TESCEA interventions were designed according to the transformative learning theory of Mezirow (1990), whereby critical thinking and problem-solving were emphasized in teaching and learning. Before commencement, a baseline survey was conducted to establish the knowledge and capacity of project beneficiaries with respect to aspects that the project aimed to address. The beneficiaries (university lecturers) were given several trainings from AFELT. The course content included formative and summative assessments, the theory of transformative learning, the Kolb cycle of experiential learning, and application of technology in learning. We used active learning strategies like role-plays, group discussions, debates, simulations, puzzles, case studies, and peer teaching and reviews. Besides, individual readings before the class were encouraged to create an environment for thinking and reflection as individuals.

The trainings and readings helped lecturers to reflect on the flaws of traditional teaching and exposed them to transformative learning and active learning strategies. In collaboration with JAG, universities redesigned selected courses, and in doing so, the lecturers underwent a process of rethinking their courses and reorganized the content and methodology of delivery. The lecturers were also introduced to active learning methods of teaching and learning that they applied in their classes. Peer teaching and evaluation were encouraged to allow them to learn from each other. To establish lessons learned, the perceptions of project participants were assessed before and after the interventions to determine trends and effects. The report was written using qualitative data obtained from the project beneficiaries (90 lecturers who were trained by the project in transformative learning approaches). The data was obtained about the project beneficiaries' knowledge (understanding critical thinking, teaching for critical thinking, and gender responsiveness), their beliefs (views regarding teaching techniques, critical thinking, and gender-responsive pedagogy implementation), and their attitudes toward teaching and learning (perspectives on teaching for critical thinking and gender responsiveness).

The Effect of TESCEA Interventions

Analysis of quantitative and qualitative data obtained through online questionnaires to 86 project beneficiaries using the Survey Monkey tool revealed that most beneficiaries acknowledged positive outcomes in critical thinking, the level and understanding of teaching for critical thinking, and gender-responsive pedagogy. Similarly, most beneficiaries by the end of the project believed that teaching for critical thinking was important. It was, however, not possible for the beneficiaries to apply critical thinking in the hard sciences. They did not have to answer all questions from their students; the mastery of concepts was more important than students covering every small fact in the curriculum.

We found that most of the project beneficiaries by the end of the intervention believed that students working in groups and using case study analysis enhanced learning outcomes and that just listening and taking notes did not provide the required skills. The beneficiaries also believed that role-plays, debates, team-based learning, project-based learning, and peer teaching techniques were appropriate methods for teaching for critical thinking. Similarly, the project beneficiaries' beliefs on gender-responsive pedagogy changed and by the end of the project interventions, most of them believed that adopting gender-responsive pedagogy was important for all classes/subjects. They mentioned that it is the role of a teacher to equally engage male and female students. Similarly, by the end of the project, most of the beneficiaries reported that their attitude had changed about teaching and learning and that they were changing from the traditional approach to teaching for critical thinking. They believed that engaging students in small groups to discuss issues and make presentations, doing group project work over time, teachers not being obsessed with the idea of syllabus completion but concept clarity, teachers using active learning techniques such as role-play, fish bowl, debate, and peer teaching, using more real-world scenarios to engage learners in problem-solving, using complex scenarios to stimulate debates in class, and using blended learning were important.

Discussion

The findings revealed that teachers' knowledge of the concept of critical thinking was minimal before the intervention but improved markedly after the intervention. The beneficiaries mentioned that they had mastered what critical thinking is, its benefits, and how to apply it. As observed from the data, the trainees reported that they could develop and integrate critical thinking skills in their teaching. Although different initiatives had been implemented to improve teaching in the universities involved, the baseline survey indicated that most universities were still using traditional approaches to teaching and learning. According to Abrami et al. (2015), developing students' critical thinking skills requires first ensuring they understand what critical thinking entails, along with providing them with clear, practical strategies for engaging in it. In alignment with this view, TESCEA trainers endeavored to emphasize these strategies to the participants. Similarly, Paul and Elder (2014) emphasize that critical thinkers engage in practices such as questioning, clearly defining problems, scrutinizing evidence, recognizing assumptions and biases, steering clear of emotional reasoning, avoiding oversimplification, exploring alternative viewpoints, and embracing ambiguity. Most of these key principles were introduced early on by AFELT trainers to the lecturers. As they explored the literature on critical thinking, the lecturers came to value its significance and became more aware of how to integrate it effectively into their teaching.

Vansteenkiste et al. (2018) observed that enhancing the personal relevance and perceived significance of tasks boosts individuals' propensity to engage in critical thinking. Grounded in self-determination theory (SDT), their study altered instructional approaches to underscore the meaningfulness and importance of tasks. The results revealed that learners who viewed the tasks as aligning with their personal interests or aspirations showed greater engagement and invested more cognitive effort in analyzing the material critically. In a similar vein, AFELT facilitators encouraged active participation among lecturers by guiding them through processes, such as developing learning outcomes aligned with their university's mission, envisioning the desired graduate profile, consulting industry stakeholders, redesigning curriculum content, and completing real-world tasks. This hands-on involvement fostered a deeper level of engagement, enhancing their capacity for critical thinking.

Aston (2023) emphasizes that moments of surprise often serve as catalysts for the critical thinking process by creating a sense of cognitive disequilibrium. The study explores how unexpected events can either disrupt or deepen critical reflection, noting that when learners encounter conflicting ideas, they are more likely to question assumptions and engage thoughtfully with new perspectives. This dynamic was evident during the presentation of the "banking concept" a framework describing the traditional, transmission-based model of teaching to university lecturers. For many, this moment prompted a profound realization: they became aware of the unintended harm perpetuated through conventional pedagogical approaches, recognizing that they themselves had been shaped by the same educational model. This critical awakening spurred deeper reflection and a renewed commitment to transformative teaching practices. Dellantonio and Pastore (2020) contend that exposing learners to contradictions between established scientific understanding and their own intuitive or misguided beliefs triggers cognitive conflict. This tension not only encourages critical thinking but also supports the revision of ingrained yet inaccurate assumptions. During the intervention, instructors highlighted the contrast between the traditional "banking" model of education and more transformative teaching approaches. This comparison prompted lecturers to critically examine long-held beliefs about teaching and learning, leading to a powerful realization: in the conventional model, educators—rather than empowering students—can unintentionally adopt the role of oppressors, thereby undermining the very purpose of education as a liberating force. Clearly, the teaching and learning environment that was created by the intervention through methodology of delivery and exposure to transformative learning motivated the teachers to start teaching for critical thinking.

The TESCEA project interventions enabled university lecturers to believe in transformative learning that emphasizes critical thinking. The findings are in agreement with Mezirow's (1990) theory of transformative learning and Festinger's (1957) theory of cognitive dissonance, which stress that the "existence of the misfit between the new experience and previously held assumptions produces a disequilibrium in the learner that triggers the dilemma to be solved." Clearly, the conflicting beliefs between new experiences on critical thinking and previously held assumptions of traditional methods produced a disorientation to solve the problem; hence, the teachers' beliefs had to change. Gill et al. (2017) and Pinillos (2023) explain a change in beliefs in terms of the introduction of cognitive doubt surrounding prior beliefs, thus invoking the experience of dissatisfaction with existing beliefs. The study materials that were availed to the lecturers and different presentations made by the facilitators introduced new and preferred beliefs in teaching and learning and highlighted the drawbacks of the traditional direct transmission approaches, hence the introduction of cognitive conflict. Further, the creation of a reflective environment through collaborative work, Socratic questioning, and guided group discussions supported change in beliefs as similarly supported by previous studies that show that reflection when introduced using a guided approach can change teacher beliefs (Valanides and Angeli, 2005; Parkinson and Maggion, 2017).

The TESCEA project findings show the potential for teachers to shift from traditional ways of teaching to the critical thinking approach. This finding is in agreement with the literature on attitude change which shows that attitudes are partly memory based and partly constructed on the spot as a hybrid model that allows for attitude stability and change as observed by Albarracin et al (2018). The situated activities that were carried out during the intervention motivated on-the-spot attitude change. For instance, the course redesign activities gave the lecturers an opportunity to think through their courses and formulate fresh learning outcomes. The TESCEA project exposed the beneficiaries to critical thinking content through individual reading, group reflection, discussions, and appropriate feedback from industry experts about university courses; hence, attitude change was registered. The active learning methods that were used to deliver the content created impact as the results reflect, in line with the cognitive response view of Carfora et al. (2018), who note that as people gain information about the attitude object, new or altered beliefs may emerge as a result of that exposure. Interaction between groups or people has an impact on attitude change as seen in Lemmer and Wagner (2015). In the TESCEA intervention, the lecturers had contact with peers from other faculties, universities, and countries who underwent similar experiences, and this may explain the observed change. Markedly, the attitude change of lecturers can also be explained by cognitive dissonance whereby a large body of research using different experimental paradigms has supported the role of dissonance theory in attitude change (Festinger, 1957; Kaaronen, 2018). Regarding Festinger's theory, it proposes that inconsistency among beliefs or behaviors causes an uncomfortable psychological tension that stimulates change to reduce the dissonance. Nevertheless, the lecturers were aware and believed that it was their responsibility to propel learning, but through the TESCEA trainings it was clear that they were the ones hindering learning through the use of the traditional approach. Hence, this cognitive dissonance prompted the lecturers to change their current practice of facilitation so that their practice/attitude could align with what they believed.

Conclusions

Overall, although exploratory in nature, effects of the TESCEA project interventions pinpoint the pathway for promoting transformative learning approaches. Application of these approaches in pedagogy could enhance learners' experiences and the learning outcomes. Our findings are therefore valuable as potential pointers to what can be done to empower teachers with modern pedagogical approaches, although not sufficient to allow definitive conclusions. A comprehensive longitudinal study to establish the impact of application of active learning strategies on teachers' ability to promote transformative learning is thus worthwhile.

References

Abdella, A. S., & Fataar, A. (2021). Teaching styles of educators in higher education in Eritrea. *Journal of Higher Education in Africa, 19*(1), 45–62. https://www.jstor.org/stable/48645902

Abrami, P. C., Bernard, R. M., Borokhovski, E., Waddington, D. I., Wade, C. A., & Persson, T. (2015). Strategies for teaching students to think critically: A meta-analysis. *Review of Educational Research, 85*(2), 275–314.

Albarracin, D., & Shavitt, S. (2018). Attitudes and attitude change. *Annual Review of Psychology, 69*, 299–327. https://www.annualreviews.org/doi/full/10.1146/annurev-psych-122216-011911

Ananga, E. D. (2021). Gender Responsive Pedagogy for Teaching and Learning: The Practice in Ghana's Initial Teacher Education Programme. Creative Education, 12(4).

Albarracin D, Shavitt S. Attitudes and Attitude Change. Annu Rev Psychol. (2018) 4;69:299-327. https://doi.org/10.1146/annurev-psych-122216-011911. Epub 2017 Aug 25. PMID: 28841390; PMCID: PMC12479097.

Asgharheidari, F., & Tahriri, A. (2015). A survey of EFL teachers' attitudes towards critical thinking instruction. *Journal of Language Teaching and Research, 6*(2), 388. http://academypublication.com/issues2/jltr/vol06/02/20.pdf

Aston, K. J. (2023). 'Why is this hard, to have critical thinking?' Exploring the factors affecting critical thinking with international higher education students. *Active Learning in Higher Education, 25*(3), 537–550. https://doi.org/10.1177/14697874231168341

Berg, C., Philipp, R., & Taff, S. D. (2019). Critical thinking and transformational learning: using case studies as narrative frameworks for threshold concepts. *Journal of Occupational Therapy education., 3*(3) https://encompass.eku.edu/cgi/viewcontent.cgi?Article=1162&context=jote

Bicchieri, C., Dimant, E., Gächter, S., & Nosenzo, D. (2019). *Observability, social proximity, and the erosion of norm compliance* (CeDEx Discussion Paper Series, No. 2019-08). The University of Nottingham, Centre for Decision Research and Experimental Economics (CeDEx).

Braasch, J. L., Braten, I., Stromso, H. I., Anmarkrud, O., & Ferguson, L. E. (2013). Promoting Secondary School students' evaluation of source features of multiple documents. *Contemporary Educational Psychology, 38*(3), 180–195. https://www.researchgate.net/publication/236631399_

Carfora, V., Catellani, P., & Conner, M. (2018). Narrative persuasion and belief change: The role of information elaboration. *Frontiers in Psychology, 9*, 1398. https://doi.org/10.3389/fpsyg.2018.01398

Carlana, M. (2019). Implicit stereotypes: Evidence from teachers' gender bias. *The Quarterly Journal of Economics, 134*(3), 1163–1224. https://doi.org/10.1093/qje/qjz008

Carter, P. L., & Nicolaides, A. (2023). Transformative learning: An emotional (r)evolution. *New Directions for Adult and Continuing Education, 177*, 25–36. https://doi.org/10.1002/ace.20476

Changwong, K., Sukkamart, A., & Sisan, B. (2018). Critical thinking skill development: Analysis of a new learning management model for Thai high schools. *Journal of International studies, 11*(2), 37–48.

Cranton, P. (2023). *Understanding and promoting transformative learning: A guide to theory and practice* (3rd ed.). Routledge Publishers. https://www.routledge.com/Understanding-and-Promoting-Transformative-Learning-A-Guide-to-Theory-and Practice/Cranton/p/book/9781620364123

Dellantonio, S., & Pastore, L. (2020). Ignorance, misconceptions and critical thinking. *Synthese, 198*(8), 7473–7501. https://doi.org/10.1007/s11229-019-02529-7

DiPasquale, J. P. (2017). *Transformative learning and critical thinking in asynchronous online discussions*. Research Thesis. University of Ontario Institute of Technology.

Dimant E. Bicchieri, Dimant, Gächter & Nosenzo (2020). Observability, Social Proximity, and the Erosion of Norm Compliance. 2020.

Domingo Albarracín-Vivo & Eduardo Encabo-Fernández & Isabel Jerez-Martínez & Lourdes Hernández-Delgado. (2024). Gender Differences and Critical Thinking: A Study on the Written Compositions of Primary Education Students, Societies, MDPI, vol. 14(7), 1–12.

Entman, R. M. (2017). Framing bias: Media in the distribution of power. In K. Kenski & K. H. Jamieson (Eds.), *The Oxford handbook of political communication* (pp. 295–310). Oxford University Press.

Ertmer, P., Anne, O.-L., & Tondeur, J. (2015). Teacher beliefs and uses of technology to support 21st century teaching and learning. In International handbook of research on teachers' beliefs (pp. 403–419). New York, NY, USA: Routledge.

Ferguson, L. E. (2020). Teacher beliefs in higher education. *Laering Om Laering, 4*(1).

Festinger, L. (1957). *A theory of cognitive dissonance (2)*. Stanford University Press.

Fleming, T. (2018). Attachment theory and transformative learning: rethinking the origins of meaning making. In M. Welch, V. Marsick, & D. Hope (Eds.), *Building transformative community: Proceedings of the XII Transformative Learning Conference* (pp. 288–294). Teachers College, Columbia University.

Forum for African Women Educationalists (FAWE). (2020). *Gender responsive pedagogy: A toolkit for teachers and schools* (2nd edition, updated) https://www.unicef.org/esa/media/6726/file/GRP

Gill, D. L., Williams, L., & Reifsteck, E. J. (2017). *Psychological dynamics of sport and exercise*. Human Kinetics.

Halpern, D. F. (2019). *Critical thinking across the curriculum: A brief edition of thought and knowledge* (4th ed.). Routledge.

Hitchcock, D. (2017). Critical thinking as an educational ideal. In *On reasoning and argument* (pp. 477–497). Springer. https://doi.org/10.1007/978-3-319-53562-3_30

INASP. (2021). https://www.inasp.info/publications/gender-responsive-pedagogy-higher-education-framework

Ismail, M., Molefi, L., & Dhlamini, N. (2021). *Gender responsive pedagogy for early childhood education*. VVOB, FAWE & Dept. Basic Education South.

Kaaronen, R. O. (2018). A theory of predictive dissonance: Predictive processing presents a new take on cognitive dissonance. *Frontiers in Psychology, 9*, 2218.

Katende, E. (2023). Critical thinking in curriculum content: the case of Uganda. *Journal of Education and Practice, 7*(8), 1–18. Available at: https://www.carijournals.org/journals/index.php/JEP/article/view/1564/1931. Accessed: 25 June 2025

Kienhues, D., Ferguson, L. E. and Stahl, E. (2016). Diverging information and epistemic change. In J.A. Greene, W.A. Sandoval and I. Braten (Eds). *Handbook of Epistemic Cognition* pp 318–330. New York Routledge Https://www.taylorfrancis.com/chapters/edit/10.4324/9781315795225-24/diverging-information-epistemic-change-dorothe-kien

Le, H. V., & Nguyen, L. Q. (**2024**). Promoting L2 learners' critical thinking skills: The role of social constructivism in reading class. *Frontiers in Education, 9*, 1241973. https://doi.org/10.3389/feduc.2024.1241973

Lemmer, G., & Wagner, U. (2015). Can we really reduce ethnic prejudice outside the lab? A meta-analysis of direct and indirect contact interventions. *European Journal of Social Psychology, 45*(2), 152–168. https://doi.org/10.1002/ejsp.2079

Lunn Brownlee, J., Ferguson, L. E., & Ryan, M. (2017). Changing teachers' epistemic cognition: Anew conceptual framework for epistemic reflexivity. *Educational Psychologist, 52*(4), 242–252.

Mezirow, J. (1990). Transformative learning: theory to practice. *New directions for adult and continuing education.*

Morrison, J., Roth McDuffie, A., & French, B. (2015). Identifying key components of teaching and learning in a STEM School. *School Science and Mathematics, 115*(5), 244–255. https://doi.org/10.1111/ssm.12126

Murawski, L. M. (2014). Critical thinking in the classroom and beyond. *Journal of Learning in Higher Education, 10*(1), 25–30. https://files.eric.ed.gov/fulltext/EJ1143316.pdf

Nabi, R. L., & Myrick, J. G. (2019). Uplifting fear appeals: Considering the role of hope in fear-based persuasive messages. *Health Communication, 34*(4), 463–474. https://doi.org/10.1080/10410236.2017.142284

Namatende-Sakwa, L., Ezati, B., Kariisa, H. and Ndawula, S. (2022). *Equipping pre-service teachers to champion gender equality in Uganda: Context matters research report*. Retrieved from https://code.ngo/wp-content/uploads/2022/09/Context-Matters-Research-Report-Final-Sakwa-August-2022.pdf

Niones, S. M. (2022). Teachers' gender sensitive attitudes towards their students in relation to the parameters of gender sensitivity. *International Journal of English Literature and Social Sciences, 7*(2). https://doi.org/10.22161/ijels.72.17

Parkinson, M., & Maggion, L. (2017). The potential of course interventions to change pre-service teachers' epistemological beliefs. In G. Schraw, J. Brownlee, L. Olafson, & M. Vander Veldt (Eds.), *Teachers' personal epistemologies: Evolving models for transforming practice*. Information Age Press.

Paul, R., & Elder, L. (2014). Critical thinking: intellectual standards essential to reasoning well within every domain of human thought, Part 4. *Journal of Developmental Education, 37*(3), 34.

Pinillos, N. Á. (2023). *Why we doubt: A cognitive account of our skeptical inclinations*. Oxford University Press.

Rodríguez Aboytes, J. G., & Barth, M. (2020). Transformative learning in the field of sustainability: A systematic literature review (1999–2019). *International Journal of Sustainability in Higher Education, 21*(5), 993–1013.

Sahin, N. T., Keshav, N. U., Salisbury, J. P., Vahabzadeh, A. (2018). Safety and Lack of negative effects of wearable augmented-reality social communication aid for children and adults with autism. J. Clin. Med.7:e188. https://doi.org/10.3390/jcm708018

Semerci, Ç., & Elaldı, Ş. (2021). A qualitative study on lecturers' attitudes and practices toward critical thinking: the case of Turkey. *International Journal of Educational Research Open, 2*(1), 100–112.

Shubina, I., & Kulakli, A. (2019). Critical thinking, creativity and gender differences for knowledge generation in education. *Literacy Information and Computer Education Journal, 10*(1), 3086–3093. https://doi.org/10.20533/licej.2040.2589.2019.0405

Snyder, R. R. (2017). Resistance to change among veteran teachers: providing voice for more effective engagement. *International Journal of Educational Leadership Preparation, 12*(1) https://files.eric.ed.gov/fulltext/EJ1145464.pdf

Srivastava, R. P. (2020). Critical thinking in higher education: Understanding its perspectives. *Journal of Education and learning, 9*(3), 93–104.

Sukarma, I. K., Budayasa, I. K., & Suwarsono, S. (2019). Gender, cognitive style and its relationship with critical thinking. *IOSR Journal of Research & Method in Education (IOSR-JRME), 9*(5), 62–67. Available at: https://www.iosrjournals.org/iosr-jrme/papers/Vol-9%20Issue-5/Series-4/E0905046267.pdf

Tarrayo, V. N., Potestades, R. R., & Ulla, M. B. (2021). Exploring the gender perspective in English Language Teaching (ELT): Voices from ELT practitioners in Philippine higher education institutions. *Sexuality and Culture, 25*(5), 1652. https://doi.org/10.1007/s12119-021-09840-x

UNESCO. (2023). *Strengthening gender responsive pedagogy for STEM in Uganda*. UNESCO. Retrieved from https://www.unesco.org/en/articles/strengthening-gender-responsive-pedagogy-stem-uganda

Valanides, N., & Angeli, C. (2005). Effects of instruction on changes in epistemological beliefs. *Contemporary Educational Psychology, 30*(3), 314–330. https://www.academia.edu/download/32336071/CEP.pdf

Vansteenkiste, M., Aelterman, N., De Muynck, G. J., Haerens, L., Patall, E., & Reeve, J. (2018). Fostering personal meaning and self-relevance: A self-determination theory perspective on internalization. *The Journal of Experimental Education, 86*(1), 30–49.

Vincent-Lancrin, S., Sainsaulieu, G., & Sjöholm, E. (2019). *Fostering students' creativity and critical thinking: what it means in school. Educational Research and Innovation.*, OECD Publishing. https://doi.org/10.1787/62212c37-en

Yu, L., & Mohamed Zin, Z. M. (2023). The critical thinking-oriented adaptations of problem-based learning models: a systematic review. *Frontiers in Education, 8*, 139987. https://doi.org/10.3389/feduc.2023.1139987

Yuan, E. R. (2017). Exploring University–Based teacher educators' teaching beliefs and practice: A Hong Kong study. *Teaching in Higher Education, 22*(3), 259–273. https://doi.org/10.1080/13562517.2016.1248393

Zeleza, P. T. (2018). Positioning universities as engines of innovation for sustainable development and transformation. *Journal of Higher Education in Africa, 15*(1), 45–62.

23 Perspectives on Uganda's Education Sector Services and Education Quality

Charles Patrick Namisi, Aloysius Byaruhanga, Veronica Munuve, Priscilla Nyadoi, and Patrick Kyamanywa

Abstract

Quality education has ripple effects on many aspects of life, including success or not in employment. The government of Uganda has had interventions such as universal free primary and secondary education and the sector liberalization, allowing private institutions to offer educational services. Despite these interventions, the quality of education remains a challenge, as findings from secondary data review and analysis reveal, among others, limitations related to teachers' competencies, instruction and curriculum design, cost of education, and school completion rates. Our review finding also pinpoints the need for teachers' capacity building on transformative learning, curriculum and policy review, and research to identify innovative strategies for education financing and servicing to deliver desired standards and quality.

Keywords

Educational cost burden · Educational Review Commission · Educational Psychology and Philosophy · National Teachers Policy · Modules

C. P. Namisi (✉) · A. Byaruhanga · V. Munuve · P. Nyadoi · P. Kyamanywa
Uganda Martyrs University, Kampala, Uganda
e-mail: cnamisi@umu.ac.ug

Introduction

Education has a ripple effect on all aspects of life and is the critical factor for socioeconomic development. This perhaps is the reason why sub-Saharan Africa, with the lowest youth literacy rate globally (72% compared to 91%), is also the world's poorest region (about seven times higher than the global average extreme poverty rate) (UNESCO World Higher Education Conference, 2022; World Bank, 2024). For Uganda, the literacy rate is additionally affected by high dropout rates of 34%, 16%, 05%, and 03% by Primary Leaving Education (PLE), Ordinary Certificate of Education (UCE), Advanced Level Education Certificate (UACE), and Higher Education Certificates (HE), respectively (Network for Education and Multidisciplinary Research Africa (MoES, 2024).

Elsewhere, underfunded education systems have been documented to cause poor education services that also lead to poor-quality graduates who lack employability skills required in the job markets (UBoS, 2021). Improving the quality of education therefore requires changing the learning content and experiences for graduates to suit the job market demands. This necessitates government and private sector investments in policies and strategies that support curriculum designs and instructional practices. It necessitates offering personal learning, domain-specific knowledge, and transferable skills to learners

P. Nyadoi et al. (eds.), *Sustainable Development Goals in Africa*, Sustainable Development Goals Series, https://doi.org/10.1007/978-3-032-20742-5_23

(Dewey, 1896; World Economic Forum, 2020). This necessitates governments objectively identifying, developing, and scaling up transformative educational approaches that ensure that quality educational services are provided and learners graduate as fit for the job market from all the levels of the education system.

In this paper, we reviewed literature and made summaries on the state of affairs regarding educational quality and underlying factors. To complement this, we talked to and documented perceptions of key stakeholders in the education sector. In particular, 5 groups of teachers and their learners were involved in a focus group discussion setting during a 5-day mental health training. Sources of literature reviewed included the Constitution of the Republic of Uganda, the Government White Paper on the Education (The Policy Review Commission, 1992), the National Teachers Policy (2019), the National Curriculum Development Act (2000), The National Council for Higher Education Framework (2014), Uganda's Vision (2040), the Uganda Bureau of Statistics Report (UBOS, 2021), and the Education Policy Report of the Finnish Government (2020), among others. The emerging summary from these is as follows:

Access to Education in Uganda

Although the Constitution of the Republic of Uganda provides that the *state shall promote free and compulsory basic education, the* situation currently is that early childhood education is 100% offered by private institutions in Uganda and is not free of charge. The International Standard Classification of Education (ISCE) defines basic education as corresponding to the first nine years of formal schooling (Statistics, 1976).

Although the free universal primary and secondary education intervention by the government of Uganda led to an increase in school enrolment, the school dropout rate remains high (UBOS, 2024). To further improve access and quality of education to learners, the Ministry of Education and Sports (MoES) began a competency-based curriculum for secondary schools through the National Curriculum Development Centre (NCDC) in 2020 and teachers were trained to implement it (National Curriculum Development Centre, 2021; MoES, 2024). Additionally, the National Teachers' Policy that among others required that teachers have a minimum of *a Bachelor's Degree in Education* was enacted (MoES, 2019).

The implementation of free primary and secondary school education, the upgrading of teachers to degree holders and other reforms the government instituted to increase access and ensure quality of education in Uganda, however, remain problematic. For example, there are limitations due to insufficient financing of the sector by government (MoES, 2024). Conversely, a majority (73–93%) of Ugandan families are not able to afford pre-primary (early childhood education) that is not free and offered largely by private educational institutions. For the children of such families, access to early childhood education is not possible.

Quality of Education in Uganda

Early childhood education is known to improve learning abilities for the learners (Brunette et al., 2017; UBoS, 2024) and therefore learners that miss out are challenged. The teachers too are reportedly part of the largely demoralized workforce in the country. They receive low pay for their services and, coupled with limitations on pedagogy approaches (as we explain a little more below), deliver poor-quality services to the learners. The latter in turn graduate and lose out in the job market because of insufficient job-market-required skills (MoES, 2024).

Although, the Uganda National Teacher Policy 2019 provides that "Teachers must embrace new pedagogies and transform pedagogical practices" as recommended in the Sustainable Development Goal number 4 (SDG4), not much strides have been realized on this front, for similar implementation challenges as denoted for interventions to do with access to education above. Indeed, knowledge and skills are constantly emerging and changing, and hence the need for the sector interventions to be ahead and timely. The country's education system currently largely ignores measures of technical performance, such as practical skills, ethics, and moral achievement, and focus is put more on grades, with teachers forced to rush through to complete the syllabi to enable their learners achieve desired grades (performance). For instance, one of the key informants we engaged in this study had this to say: "We hire the experts who mark national examinations to coach our students." Evidently, the passing of examinations is decoupled from mastery of subjects and acquisition of practical skills, moral values, social skills, and innovation ability on the part of the learners. While this is the state of affairs with regard to pedagogical approaches, the need for reform in Uganda's education sector was recognized decades ago, even as stipulated then in the Government White Paper on Education (1992).

The Government White Paper on the Education (Policy Review Commission, 1992) recommended that primary education be progressively practical, beginning from P.1. This would promote transformative learning and experiences to the learners, enabling them to achieve quality education. However, it has been 33 years since the Policy Review Commission's Report and this recommendation is yet to be implemented. The quality of education remains low, characterized by low literacy and numeracy levels, and this is in addition to the high school dropout rate we discussed above under access to education in Uganda. Reportedly, 43% of the employed Ugandans are undereducated in skills required for their jobs (UBoS, 2021). To improve the situation, the country needs to move fast with its competency-based curriculum implementation and, in particular, make curriculum and pedagogical reforms that would, among others, encourage originality and interdisciplinary teaching approaches that allow learners to see, learn, and make connections across subjects as espoused elsewhere (Bereiter et al., 2023). There is need for curriculum review to integrate crosscutting interdisciplinary core learning concepts in foundational modules at all levels of education and be aligned with the learner-centered philosophy.

From the voices still, it is established that the teacher–learner interaction time in Uganda currently is 8:00 a.m.–5:00 p.m., meaning learners spend most of their time at school and have very little time left for play, domestic work engagements, and experiential learning. The teacher–learner interaction time in school needs to be reduced and homeschooling and experiential learning should be encouraged more and integrated into policy. This will allow time for the learners to play so as to have good mental health, gain experiential learning, and engage in income-generating activities for their households. This kind of intervention would potentially enable an increment in school enrolment and reduce the school dropout rate, as the learners would have time to engage in livelihood projects. Free time boosts the morale, social, and entrepreneurial skills of learners, families, and the teachers.

Conclusion

The UN Sustainable Development Goal 4 calls for accessible and inclusive learning that is open to all (SDG4, 2030). Every detail at each school and informal learning must be planned through the lens of inclusion, right from the curriculum, learning materials, and teachers, among others. Our findings show that for Uganda, the education sector is still falling short of the recommendation of SDG4. The country needs to invest in training its teachers to offer transformative education, review its educational policy and curriculum, invest funds in and commit to standards in order to deliver quality service to enable achievement of quality graduates. Comprehensive research in the educational sector is needed, and the results will inform revision of the sector policy and strategies for service delivery.

References

Bereiter, C., Bogouslavsky, M., Tsuji, W., & Scardamalia, M. (2023). Knowledge building: Improving ideas, improving writing. In *The Routledge international handbook of research on writing* (pp. 225–235). Routledge.

Brunette, T., Crouch, L., Cummiskey, C., Dick, A., Henny, C., Jordan, R., et al. (2017). Repetition of primary 1 and pre-primary education in Uganda. International Development Working Paper, no. 2017-02. *International Journal for Research Trends and Innovation*.

Dewey, J. (1896). Psychology of number. *Journal of Science, 3*(60), 286–289.

Education Policy Review Commission. (1992). *Government white paper on implementation of the recommendations of the report of the Education Policy Review Commission entitled "Education for national integration and development"*. Uganda Ministry of Education and Sports. Retrieved from https://unesdoc.unesco.org/ark:/48223/pf0000172298

MoES. (2019). *The National Teacher Policy*. Uganda Ministry of Education and Sports.

MoES. (2024). Planning, budgeting and implementation guidelines for local governments for the education, sports and skills development sub sector (FY 2024–2025).

National Curriculum Development Centre. (2021). *National Curriculum Development Centre (Amendment) Act 2021 Uganda*. Uganda Ministry of Education and Sports.

Statistics, U. O. O. (1976). *International Standard Classification of Education (ISCED)* (Vol. 1). UNESCO.

UBoS. (2021). *The national labour force survey main report*. Uganda Bureau of Statistics (UBOS).

UBoS. (2024). *National population and housing census 2024 final report*. Uganda Bureau of Statistics (UBoS).

World Economic Forum. (2020). Platform for shaping the future of the new economy and society. The promise of platform work: Understanding the ecosystem. In World Economic Forum White Papers.

24 Visualization Strategies and Teaching Geography to Learners with Hearing Impairment in Secondary Schools: Lessons Learnt from Mbale and Lira Districts in Eastern and Northern Uganda

Peter Mwesigwa and Mark Kiiza

Abstract

This study investigated the use of visualization strategies in teaching geography to students with hearing impairment in secondary schools in Mbale and Lira Districts in Uganda. Results of the data analyzed for themes and descriptive statistics in Microsoft Excel revealed that 82% of secondary school students with hearing impairments in these two districts struggled to understand geographical concepts. Application of role-play and visualization in teaching methods enhanced their comprehension, with a 57% improvement in understanding. These results demonstrate the effectiveness of role-play in teaching strategies in enhancing learning among learners with hearing impairments and the need for policy underpinnings to this effect.

Keywords

Hearing impairments · Visualization · Strategies · Inclusive education · Policies

Introduction

Inclusive education is essential for creating equitable learning environments, yet learners with hearing impairments in Uganda's Eastern Region face significant barriers. They experience isolation and reduced educational opportunities due to resource limitations, inadequate teacher training, and societal stigma (Agrawal & Books, 2024;Chandra, 2024). Visualization strategies, including the use of flashcards, illustrated charts, and digital media, have shown potential in bridging this gap, offering an accessible means of understanding concepts through visual cues (Phillips, 2022). The current study investigated the effectiveness of visualization strategies in teaching geography to the deaf and hard-of-hearing (DHH) learners in Mbale and Lira Districts in Uganda. We pursued the following objectives.

Specific Objectives

(i) Examined the influence of visualization strategies in teaching geography concepts to deaf and hard-of-hearing students in Lira and Mbale Districts in Uganda.

P. Mwesigwa · M. Kiiza (✉)
Uganda Martyrs University, Kampala, Uganda
e-mail: pmwesigwa@umu.ac.ug; mkiiza@umu.ac.ug

P. Nyadoi et al. (eds.), *Sustainable Development Goals in Africa*, Sustainable Development Goals Series, https://doi.org/10.1007/978-3-032-20742-5_24

(ii) Assessed the effect of role-playing and visualization/demonstration in teaching strategies on the deaf and hard-of-hearing students' learning of geography concepts.

Methodology

This study was anchored on dual-coding theory (Kanellopoulou et al., 2019), which recognizes that images and words have different cognitive representations and that the brain uses separate memory systems for different types of information. They state that verbal memory is related to the language system, while image memory includes graphics and tastes. Verbal information is acquired from sensory memory to visual processors (Agwu & Ogochi, 2019). We adopted the mixed-method approach that enabled the collection of quantitative and qualitative data cross-sectionally (Netexplo & France, 2019; Engebretsen & Kennedy, 2020) and to allow for generalizations (Taylor, 2023). We collected both primary and secondary data on the aspects of our investigation. A total of 55 research participants were engaged in this study, including 8 teachers, 45 students with hearing impairments, and 2 administrators, from 2 secondary schools. We used interviews and observation for data collection, approaches that elsewhere have been reported to suit studies on visualization strategies for DHH learners (Chandra, 2024). The data obtained from the respondents were sorted, coded, and entered into a Microsoft Excel spreadsheet and into SPSS. Descriptive statistics, including frequency and percentages, were interpreted. Content analysis was conducted for the qualitative data, and emerging themes were identified.

Results

The results indicate that visualization aids are highly effective in supporting the comprehension and engagement of hearing-impaired learners. Among the teachers surveyed, 78% reported that visual aids such as flashcards, illustrated charts, and digital projectors significantly improved students' comprehension of concepts. The findings further indicate that there is a great impact of role-play/demonstration methods of teaching DHH learners. According to test 2, 82% of the learners were below the average of understanding the geographical concept after they were taught theoretically; however, in test 2 after the use of role-play/demonstration, only 9% of the students were below the average level. This implied that 63% of the learners were able to understand the concept to the average level of understanding, thus reducing the number of students below the average to an average and above the average.

Secondly, in test 1, there were over 7% of DHH students at the average level of understanding geographical concepts, and in test 2, the number increased to 64%. This implies that there was a growth in the students' understanding of the concept by 57% as a result of using role-play and visualization. The table also shows that during test 1, only 11% of the DHH students were above the average level of understanding geography concepts. That is to say, they could apply, match, identify, and explain. While in the post-test, the DHH learners increased to 27% above the average level of understanding geographical concepts. Thus, the results also show that there was an increase of 16% of the DHH learners in understanding the concepts above an average level.

Effect of Using both Graphic Design Visualizations and Role-Play/Demonstration in Teaching DHH Students in Learning Geography Concepts in North Eastern Uganda

The findings indicate that there is a great impact of combined use of different visualization methods in teaching DHH learners (both graphic design visualizations and role-play/demonstration). According to test 2 (affiliated to role-play/demonstration), only 9% of the learners were below the average of understanding the geographical concept; however, after the test 3 (affiliated to both graphic design and role-play/demonstration), none (0%) of the students was below the average level. This implies that the 9% of learners who were below average during test 2 managed to understand the concept to the next level of "average" after the researcher used combined strategies to teach the DHH learners. In the second phase of average level, in test 2, there were over 64% of DHH students at the average level of understanding the geographical concepts, and in test 3, the number reduced to 4% (Fig. 24.1). Since there was no student below the average, this proves that the difference in the percentage moved to the next level. This implies that there was a growth in the students' understanding of the concept by 60% as a result of using both graphic designs and role-play to teach DHH learners. The third phase, tagged "above average" in the table, also shows that during test 2, only 27% of the DHH students were above the average level of understanding geography concepts. That is to say, they could apply, match, identify, and explain. While in test 3, the percentage of DHH learners increased overwhelmingly to 96% above the average level of understanding geographical concepts. Thus, the results also show that there was an increase of 69% of the DHH learners in understanding the concepts up to an average level.

The study reveals that the visualization strategies like role-play and demonstration have a great positive effect on students' understanding of geography concepts among the DHH learners in north-eastern Uganda. In addition to this, these results prove that visual materials together with sign language are more effective in teaching geographical concepts that are both abstract and concrete to learners with hearing impairment than using solely sign language and lecture methods of teaching. This is because the statistical findings indicated that using visual materials together with the combined use of graphic designs and role-play is more effective than using only the lecture methods of teaching DHH. This is because results collected from test 2 and 3 after using visual materials outcompeted the result one collected after teaching learners theoretically/using lecture methods. Therefore, visualization technique provides better and more permanent learning than using only verbal/sign language instruction for DHH learners.

DHH learners learn and remember/understand better when they are introduced to new vocabulary items by using visual aids like graphics, role–play, and real objects, which can be demonstrated with the test results of the study. Similarly, the research has shown that teachers of DHH learners have inadequately used visualization in teaching DHH, which resulted in the low level of knowledge on geographical concepts due to limited skills to use them and lack of knowledge about deafness. Further, the research has shown that secondary schools of north-eastern Uganda lacked adequate funding to purchase teaching/learning resources such as maps, globes, and textbooks for use by teachers during lessons to be used. This research has also shown that visual-aid teaching/learning resources such as TV, CDs,

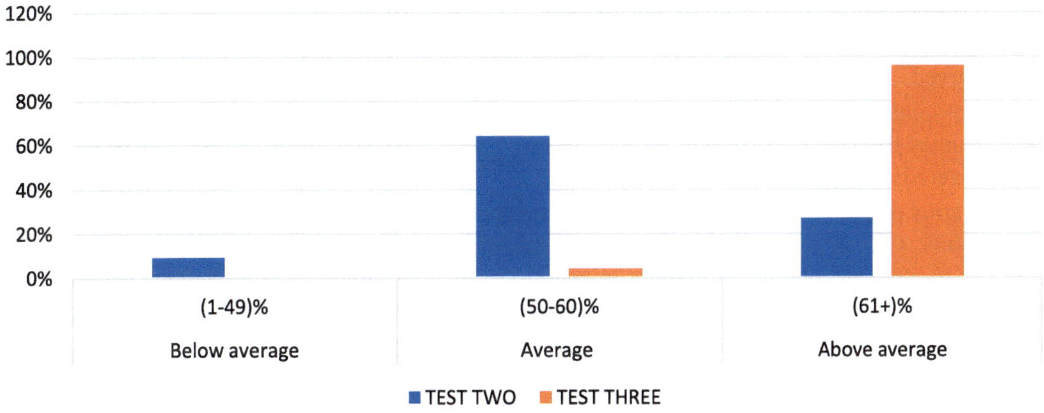

Fig. 24.1 Effect of the use of graphic designs and role-play in teaching students with hearing impairment

computers, recording tapes, and radios were not in schools and therefore could not be used by both teachers and students for teaching and learning, respectively.

Discussion

The findings align with recent literature on the potential of visualization strategies to enhance educational services delivery and learning outcomes for hearing-impaired learners (Lane et al., 2023). The effectiveness of visualization in promoting comprehension and engagement, as observed by 78% of teachers in our study, is in agreement with findings from elsewhere on the same issues, for example, reported by Patay-Horváth (2023) and Simunich et al. (2024).

Our findings also show that the use of visualization strategies in the teaching of the DHH learners enhanced their understanding of geography concepts and this needs to be integrated in teaching methods of this group of learners. However, effective use of visualization learning aids in teaching requires specific skills and familiarity with different strategies, underscoring the importance of professional development programs. Therefore, we recommend training in special education techniques for teachers engaged in teaching of the DHH. The teachers, as documented elsewhere, need to be equipped with knowledge and skills for diverse learning needs (Olusanya et al., 2023).

In our observations, we noted that more than half of the hard-of-hearing learners felt isolated socially due to negative societal attitudes. This, we observed, also affected their classroom participation and overall confidence. Elsewhere, other studies have also shown that promoting a supportive educational environment for learners with hearing impairments is necessary (Kaluri et al., 2024; Chiedu & M, 2024). Our findings thus underscore the need for community engagement to foster a more inclusive atmosphere, aiding the social integration of hearing-impaired students. The community could be engaged in resource mobilization to provide visualization aids for the students. These aids, according to our results and in the findings of other scholars—for example, Edyburn (2020), Sanders and White (2023), and Sengupta (2023)—have been found to enhance their comprehension and learning.

Conclusion

This study concludes that visualization could enhance teaching and learning for the DHH learners and hence its use should be integrated in the teaching approaches for this group of students. We highlight the transformative potential of visualization strategies in improving educational outcomes for hearing-impaired learners. By enhancing comprehension and classroom engagement, these aids may enable hearing-impaired students to participate more fully in the learning process.

We recommend review and revision of existing education policies to ensure they enforce use of visualization strategies in the teaching strategies for DHH learners and that teachers are supported to acquire the requisite knowledge and skills in the application of these tools as learning aids. Additionally, communities need to be sensitized and mobilized to support the teaching and learning of DHH through provisioning of a supportive learning environment and supply of visualization-strategy-integrated learning aids.

Additional Recommendations

- The discussion and conclusions of the study findings deduce significant recommendations based on theoretical and practical implications that contribute to a deeper understanding of the school experiences of learners with hearing impairments. These implications can inform future educational practices and interventions aimed at improving the teaching and learning process for learners with hearing impairments in the region.
- Teachers teaching DHH geography concepts and other related subjects like agriculture, biology, and chemistry, which are scientific

and abstract in nature should endeavor to use a visualization strategy to suit the learning ability of DHH learners, since their way of processing information to the brain for meaningful ideas is through the use of their sight to see and interpret, since they are unable to process audio information.

- Teachers teaching DHH geography concepts and other related subjects like agriculture, biology, and chemistry, which are scientific and abstract in nature, should use graphic designs to break down the concepts to enable the DHH learners to easily accommodate, process, and connect the information given.
- The study recommends ongoing teacher training and provision. Professional development programs should focus on inclusive deaf reaching strategic techniques, equipping teachers with the skills necessary to apply visualization strategies effectively.
- To ensure there is increased funding for provision of visual aids, the government should prioritize funding for special needs education, particularly in rural and underserved regions of North-eastern Uganda.
- Educational authorities should embrace proactive policies, and the government should invest in the integration of digital resources, such as tablets, laptops, and interactive boards, to provide hearing-impaired learners with more personalized learning experiences.

References

Agrawal, C., & Books, N. (2024). *TNPSC-Tamilnadu Combined Technical Services exam-research assistant exam-research aptitude subject practice sets*. Chandresh Agrawal.

Agwu, S. N., & Ogochi, M. A. (2019). Assessing the effect of visual aids on secondary school students' achievements in learning English language in Agbani education zone of Enugu state, Nigeria. Advance Journal of Education and Social Sciences, 4(10),1–2.

Chandra, C., & Harish. (2024). *Social inclusion tactics for people with intellectual and developmental disabilities*. IGI Global.

Chiedu, E., M, D., & Boitumelo. (2024). *Perspectives on career transitioning of students with hearing impairments*. IGI Global.

Edyburn, D. (2020). *Rapid literature review on assistive technology in education: Research report*. Knowledge by Design, Incorporated.

Engebretsen, M., & Kennedy, H. (2020). *Data visualization in society*. Amsterdam University Press.

Kaluri, R., Mahmud, M., Gadekallu, T. R., Rajput, D. S., & Lakshmanna, K. (2024). *Applied assistive technologies and informatics for students with disabilities*. Springer.

Kanellopoulou, C., Kermanidis, K. L., & Giannakoulopoulos, A. (2019). The Dual-Coding and Multimedia Learning Theories: Film Subtitles as a Vocabulary Teaching Tool. Education Sciences, 9(3), 210. https://doi.org/10.3390/educsci9030210.

Lane, D., Catania, N., & Semon, S. (2023). *Progress toward agenda 2030: A mid term review of the status of inclusive education in global contexts*. Emerald Group Publishing.

Netexplo (France), & Information, U. A. D.-G. for C. (2019). *Human learning in the digital era*. UNESCO Publishing.

Olusanya, B. O., Hadders-Algra, M., Nair, M. K. C., & Lynch, P. (2023). *Optimizing school readiness for children with developmental disabilities*. Frontiers Media SA.

Patay-Horváth, A. (2023). *Transformations of Pelops: Myths, monuments, and cult reconsidered*. Taylor & Francis.

Phillips, F., & Tamara. (2022). *Exploring ethical problems in today's technological world*. IGI Global.

Simunich, B., Gregg, A., & Ralston-Berg, P. (2024). *High-impact design for online courses: Blueprinting quality digital learning in eight practical steps*. Taylor & Francis.

Taylor, R. (2023). *Kielhofner's research in occupational therapy: Methods of inquiry for enhancing practice*. F.A. Davis.

25 Perspectives on the Challenges to Timely Progress and Completion of Study Programmes Among Doctoral Students

Priscilla Nyadoi, Joseph Ssekandi, Jimmy Spire Ssentongo, Simon Omaada Esibo, Aloysius Byaruhanga, John Bosco Lokolimoi, Edward K. Ssegawa, Laura Ariko Otaala, and Patrick Kyamanywa

Abstract

This chapter presents perceptions of students and staff on factors underpinning timely progress and completion of doctoral studies. Overall, doctoral students perceived that supportive activities such as conferences, publications in university outlets, timely feedback, and frequent interaction with supervisors/managers help them progress. Staff, however, perceived that late or non-payment of supervisors and examiners, poor quality students, and burdensome workload are problems that impede the timely completion of doctoral studies. Together, doctoral students and staff perceived that unharmonized timetabling and teaching in graduate and undergraduate programmes, insufficient supervisor–supervisee interactions, and limited ICT application in pedagogy lead to delayed progress and completion. While an in-depth investigation is still needed, these perspectives provide useful pointers to strategies that may be needed to improve progress and completion rates of doctoral candidates.

Keywords

Mentorship · University press · Disappearing students · Finance · Scholarships · Professors

Introduction

The challenges facing educational institutions world over are reported to be multi-dimensional, multilevel, dynamic, and often dictated by the institution's educational model, mission, objectives, standards, programmes, and interests of the stakeholders, including students, labour market, society, and government (Choudaha and De Wit, 2014; Mayanja, 2020). For Uganda, although the country has made efforts to improve access through universal primary and secondary education and allowed private institutions to offer education services, the quality of education has remained a matter of concern. Taken from key policy documents, recent scholarly debates and literature, persisting challenges to quality education in the country include teachers' competencies, instruction and curriculum designs, cost of

P. Nyadoi (✉) · J. Ssekandi · J. S. Ssentongo · S. O. Esibo · A. Byaruhanga · J. B. Lokolimoi · E. K. Ssegawa · L. A. Otaala · P. Kyamanywa
Uganda Martyrs University, Kampala, Uganda
e-mail: pnyadoi@umu.ac.ug

P. Nyadoi et al. (eds.), *Sustainable Development Goals in Africa*, Sustainable Development Goals Series, https://doi.org/10.1007/978-3-032-20742-5_25

education, and school dropout, among others (National Teachers Policy, 2019; Brunette et al., 2017; UBoS, 2024). For a public university in Uganda, Banoti (2011) categorizes the challenges as being of economic, political, quality (of students and faculty members) in nature. Related, Rwakenda (2016) identified them to include pedagogy and curriculum development, administration and management, educational equipment, leadership, budget, technology, quality, ethics, and human resources management issues.

Both private and public educational institutions face financial constraints that lead to poor remuneration (for staff making their retention hard) and high tuition cost burden for students, which leads, among others, to non-enrolment and low completion rate or dropout of programmes (Bunoti 2011; Mulhern et al., 2015; Halabieh et al., 2025). There are often fewer qualified staff (professors and doctoral-level instructors) for doctoral programmes (Wapman et al., 2022). Institutions often face conflicting values and lack of commitment among leaders (Asiimwe & Steyn, 2013). There are pedagogical and academic challenges that include outdated teaching methods that emphasize memorization rather than critical thinking and problem-solving, limited academia and industry interaction, leading to limited opportunities for building skills and competences (Almulla, 2023; Eslit 2023). There is a disconnect with communities characterized by the failure of educational institutions to connect with community needs and contribute meaningfully to solving problems at the household level. There are also quality challenges, whereby students graduate with skills gaps and hence face unemployment (El Baradei & Kadry, 2022). The other challenges include increasing competition for students and funding among institutions of higher learning, shifting students' learning style expectations, rapid technological advancements that require universities to constantly adapt their teaching and learning methods, and political pressures threatening their autonomy and academic freedom (Wapman et al., 2022).

Mushemeza (2016) and Malunda et al. (2024) give detailed accounts of challenges higher education institutions in Uganda face. These include inadequate training of the administrators, planning based on government ideology and policies, complex and unpredictable human behavior, high financial dependence and inadequate releases of resources to drive institutional mandates that lead to restriction on staff recruitment, research and dissemination use, quality assurance and compliance, staff development, and response to uncertainties that are aimed at enhancing efficiency and effectiveness. The consequences of these restrictions included high workload stress for the academic staff, low-quality research output, exposure and aggravation of the social-economic inequalities, and declining morale of staff. These in turn determine the success or failure of initiatives, policies, or programmes of the higher education institution (Malunda et al., 2024).

In his study of private universities in Uganda, Echel-Ochwa (2016) reports funding, lack of permanent academic staff, a curriculum that does not meet economy or market needs, and poor infrastructure as the outstanding challenges to the institutions. Conversely, Bunoti (2011), in a case study of one public university, reported that academic staff were poorly remunerated, overloaded with work, and stressed. Consequently, they offer fewer services than desired to their students because of low motivation. This has far-reaching impacts on the students; they receive poor-quality education and consequently become poor-quality graduates. The students, however, were burdened by the cost of university education. Bunoti found that the students, due to inability to afford certain courses, enrol for what they can afford and not necessarily what they desired.

Bunoti (2011) mainly attributed the challenges facing higher education managers and learners in Uganda and overall Africa to education commercialization, an increasing student population in need of higher education, limited funding to educational institutions, and policy. Conversely, Echel-Ochwa (2016) and Malunda et al. (2024) posit that private universities align their teaching, learning approaches and strategies to their visions, missions, objectives, programmes, and values. Meaning that strategies for addressing management and academic challenges differ among private universites. Therefore, each institution may need to investigate its ecosystem to identify solutions for its challenges while in general, adding on to the body of knowledge relevant to managers in higher institutions of learning. Aligned to this perpetual need, for this paper, we explored students and staff perspectives on challenges/factors they felt impede progress and therefore need to be addressed in order to improve study progression and completion among doctoral students.

Methods: Data Collection and Analysis

The staff (faculty deans, heads of departments, associate deans, and administrators) and students were from a faith-oriented private university in Uganda. Information on the subject was obtained from them following the focus group discussion approach that a similar study by Bunoti in 2011 also adopted (Bunoti, 2011). Focus group discussion is an approach that is known to be suitable for studies focused on purposively identifying peoples and communities—shared perceptions, attitudes, knowledge, and experiences on a subject matter under investigation (Creswel, 2013; Adekola & Olumati, 2023). For this study, the focus group discussions were conducted in a meeting setting with agenda outlined and purposively selected participants involved. The participants were selected due to their roles in the doctoral programme training, which placed them in a position to have the knowledge, experiences, and therefore hold perspectives on the subject that was being investigated. The focus group discussions were held at faculty level, with at least 3–5 members who included the deans, associate deans, heads of departments, and administrators of the individual faculties, a total of six meetings with the six faculties and overall, about 30 participants were involved. The focus group discussion with academic staff involved in doctoral student supervision was held once in a one whole day physical meeting with about 30 participants (doctoral student supervisors).

Separate from the staff, a focus group discussion that was held with the doctoral students was done once and online for one day. This involved about 30 students, covering the ten cohorts of PhD students, from the 2016 academic year to 2024. In the online meeting with the students, they were asked to submit both their group and individual perceptions on the subject matter. For ethical consideration, participation in the study was based on roles played in the institution. The study was also commissioned by management. Participation in the study was by invitation, and providing information in response to questions advanced to respondents was voluntary. The timing and mode for the meetings were agreed on by the participants, and the processes involved ensured no harm was posed and/or caused to the respondents.

The information we obtained from respondents as described above was summarized and descriptively analyzed to identify students and staff perspectives on factors that cause progress and completion delay in doctoral programmes. The entire exercise was conducted between January and March 2025, within the normal procedures of the case study university and was in part commissioned to help the institution elucidate and find solutions to the challenges constraining better performance in its doctoral programmes. This was after the institution's tracer study revealed that the completion rate was just about 40%.

Results

Factors Staff Perceived to Delay Doctoral Students' Study Progress and Completion

The staff (deans, associate deans, heads of departments, administrators, and doctoral student supervisors) from six faculties offering graduate programmes in the case study university mentioned 18 factors they perceived to cause delay of progress and completion of studies among doctoral students (Fig. 25.1).

To address the identified challenges, the staff gave about 22 recommendations they felt would help improve students' progress and timely completion of doctoral study programmes and from these themes emerged as follows:

Financing
Finance department to prioritize and make timely payment to supervisors and examiners.

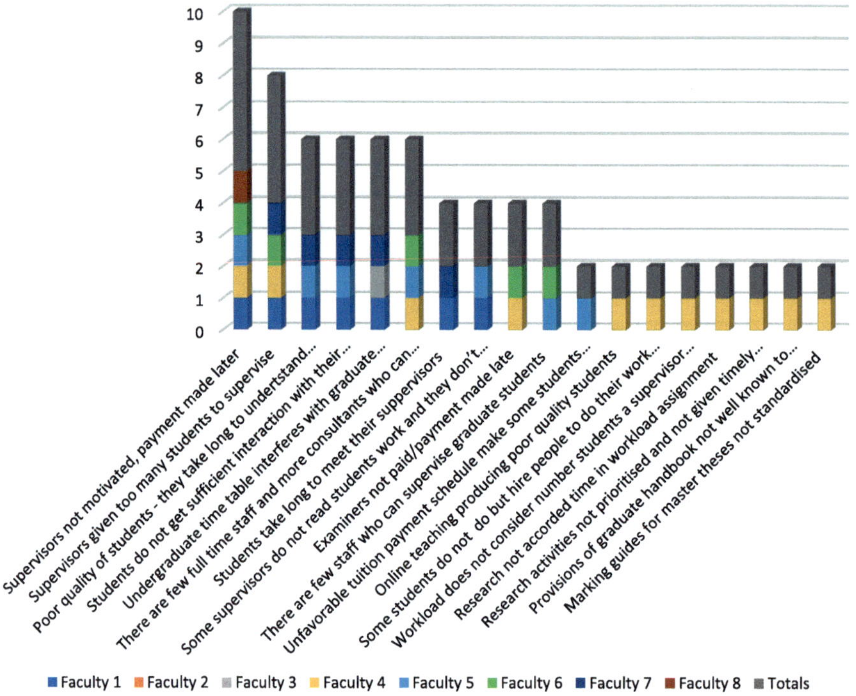

Fig. 25.1 Factors university staff perceived to delay progress and completion of studies among doctoral programme students

Human Resource

- The human resource department plans for and recruits qualified staff for graduate programme teaching and supervision.
- Assign an appropriate workload to staff.
- Give contracts to the consultants before they commence duty.

Research

- Faculty members to develop research ideas as a group.
- Provide access criteria and award staff research fund.

Pedagogy (Teaching and Learning Approaches)

- Deans, heads of departments, and research-methods lecturers are to take students through the graduate handbook during orientation.
- Faculties requisition their ICT Facilities needs to the ICT Directorate.
- Convene graduate seminars for doctoral students.
- Limit the number of students that can be assigned for supervision by one supervisor.
- Harmonize graduate and undergraduate programme lectures timetabling.

Policy, Regulations, and Standards

- Revise the graduate handbook to have sections for faculty-specific requirements (for admission, research, and examination processes and procedures).
- Convene awareness-raising meetings for deans, heads of departments, coordinators of graduate programmes, lecturers of research methods, supervisors of graduate students and the students, on the provisions and requirements of the graduate handbook.
- Print and distribute copies of the graduate handbooks to faculty deans, heads of depart-

ments, supervisors, coordinators, examiners, and lecturers of research methods in faculties offering graduate programmes and to the students.

Doctoral students mentioned the factors they perceived to delay progress and made suggestions for improvement as presented in Table 25.1.

Table 25.1 The factors students perceived to delay doctoral study progress and completion

Factors that delay study progress and completion	Strategies for improvement of study progress and completion
Supervisors who never respond to students' emails, messages, and phone calls and take very long to provide feedback on thesis drafts submitted for their guidance	The graduate school should follow up on the commitment of the supervisors
Late issuance of teaching timetables and lack of adherence to the time table schedule on the part of lecturers, leading to delay in programme completion	Timely preparation, issuance, and adherence to schedule in the timetables by the graduate programme academic staff
Lack of practical skills for data collection, analysis, journal article preparation and publication on the part of students	The graduate school conducts periodic training on data collection, analysis, and scholarly writing and supports students in publishing their work
Slow response to students' concerns by different units of the university	Timely response to students' concerns by the different units of university
High tuition	Need for the university to offer scholarships to reduce the tuition charges
Some faculties of the university are struggling to produce PhD graduates, while others are generally doing well	Need for the university to audit and build the capacity of all faculties offering doctoral Programmes, to function well
Transport and related costs students meet, so as to reach their supervisors for supervision meetings and guidance	Need for relevant ICT deployment in doctoral programme teaching, learning, and supervision

Discussion

Evidently, the factors perceived to impede progress and timely completion of doctoral studies are double-edged, recognized by both the students and staff involved. Some of the factors we found to be impeding progress of students in their academic programmes had been reported in earlier studies by a number of researchers such as Bunoti (2011), Mayanja (2020) and Malunda et al. (2024). These factors our analysis reveal continue to include, financial constraint in form of high tuition (for the students) and low remuneration (for staff); pedagogical approaches (in our case, conflicting timetables, lack of ICT facilities that would easy teaching and learning, insufficient teacher–learner interactions, which in our case is supervisor–supervisee interaction) and human resource constraints (this in our case being few qualified staff, complaining of work overload in addition to low pay).

Doctoral students complained of supervisors' delay in giving them feedback for improvement of their thesis, while the supervisors complained of poor-quality students who take long to grasp research concepts and meet their supervisors. The students ask for capacity-building support in scholarly writing, publications, data collection, and analysis skills, while the staff ask for scholarships for colleagues to do PhD so that human resource capacity improves. Also emerging from our exploratory study is the need for staff and students to be educated on the provisions of the graduate handbook, as unawareness of this is also perceived to affect the progress and completion of doctoral programmes. Because of the exploratory nature of our study, it is not possible to ascertain to what extent each of the factors the staff and students viewed as challenges may be, if at all, contributing to delay in progress and completion of doctoral study programmes. The perspectives, however, aligned with findings of past studies conducted in private and public universities in Uganda. For example, low-pay/underpaid, work-overloaded staff deploying rudimentary tools and delivering low-quality education to their students were reported both in Uganda and elsewhere (Choudaha & De Wit, 2014; Bunoti, 2011;

Rwakenda, 2016; Malunda et al., 2024). In Uganda, the commercialization of higher education is reported to have translated into an education cost burden for students to the extent that they opt to do courses they can afford rather than what they originally wanted to do at the university (Bunoti, 2011; Mayanja, 2020; Malunda et al., 2024).

Financial burdens, low motivation, work overload, and supervision challenges in doctoral study programmes, emerging as themes in this paper, are not problems unique to Uganda. These challenges have been reported elsewhere in other continents outside Africa (Mulhern et al., 2015; Acharya, 2024; Haya et al., 2025). Acharya and his colleagues talk about institutional, supervisory, and personal demands (ambiguity in doctoral research roles, inadequate resources for doctoral programmes, interpersonal conflict that threatens the well-being of doctoral students), workload, role responsibility, and role complexity as some of the challenges for doctoral programme students and staff. They recommended an approachable institutional ecosystem, the vital role of supervisors, and work–life balance as necessary solutions for addressing the challenges. These are in agreement with the themes emergent from our study from both the staff and the students, for example, the needs for timely supervision, feedback, and interaction between the supervisors and their supervisees (the doctoral student), ICT use in pedagogy affirming the need for universities to adaptively deploy technology in pedagogy earlier reported (Rwakenda, 2016; Almulla, 2023; Eslit, 2023; Malunda et al., 2024). Others are personal demands for capacity building support (to publish and for tuition reduction on the part of the students), and scholarships for PhD studies and higher remuneration for staff.

Conclusions

Perspectives from our exploratory study align with most documented challenges for staff and students in doctoral programmes globally. Nevertheless, they are pointers to the need for in-depth individual institutional studies, in order to identify relevant endemic solutions. Such solutions when deployed would lead to lessons and best practices scalable for the improvement of strategies for delivering doctoral training across institutions.

Acknowledgment The authors thank the management of the university for allowing this case study to be done, and the graduate programme students and staff who provided information and participated in planning and implementing activities that contributed to realization of this manuscript.

References

Acharya, M. (2024). Exploring learning difficulties among the students in public schools in Kirtipur municipality. *Orchid Academia Siraha, 3*(1), 95–104.

Adekola, G., & Olumati, E. S. (2023). Focus group discussion: A research method in community development. *International Journal of Research and Innovation in Social Science (IJRISS), 7*(05), 392–399. https://doi.org/10.47772/IJRISS.2023.70533

Almulla, M. A. (2023). Constructivism learning theory: A paradigm for students' critical thinking, creativity, and problem solving to affect academic performance in higher education. *Cogent Education, 10*(1). https://doi.org/10.1080/2331186X.2023.2172929

Asiimwe, S., & Steyn, M. G. (2013). Obstacles hindering the effective governance of universities in Uganda. *Journal of Social Sciences, 34*(1), 17–37.

Brunette, T., Crouch, L., Cummiskey, C., Dick, A., Henny, C., Jordan, R., Weatherholt, T. (2017). Repetition of primary 1 and pre-primary education in Uganda. International Development Working Paper, no. 2017-02. *J RTI International.*

Bunoti, S. (2011). *The quality of higher education in developing countries needs professional support.* Retrieved on July 19, 2025 from http://www.intconfhighered.org/FINAL%20Sarah%20Bunoti.pdf.

Choudaha, R., & De Wit, H. (2014). Challenges and opportunities for global student mobility in the future: A comparative and critical analysis. In S. Bernahard (Ed.), *Internationalisation of higher education & global mobility* (pp. 19–34). Symposium Books.

Creswel, W. J. (2013). *Qualitative enquiry and research design. Choosing among five approaches* (3rd ed.). SAGE Publications.

Echel-Ochwa, J. (2016). Private universities in Uganda: Issues and challenges. *International Journal of Education and Social Science.* www.ijessnet.com, *3*(3).

El Baradei, L., & Kadry, M. (2022). *Bridging the employability skills gap in higher education. Policy paper. The public policy HUB.* The School of Global Affairs and Public Policy (GAPP). The American University in Calro (AUC).

Eslit, R. E. (2023). 21st century teaching: Updates. *Challenges, and Best Practices.* https://doi.org/10.13140/RG.2.2.21864.65284

Halabieh, H., Hawkins, S., Bernstein, A. E., Lewkowict, S., Unaldi Kamel, B., Fleming, L., & Levitin, D. (2025). The future of higher education: Identifying current educational problems and proposed solutions. *Education in Science, 2022*(12), 888. https://www.researchgate.net/deref/https%3A%2F%2Fdoi.org%2F10.3390%2Feducsci12120888?_tp=eyJjb250ZXh0Ijp7ImZpcnN0UGFnZSI6InB1YmxpY2F0aW9uIiwicGFnZSI6InB1YmxpY2F0aW9uIiwicG9zaXRpb24iOiJwYWdlQ29udGVudCJ9fQ

Malunda, N. P., Atwebembeire, J., & Bazanye, K. N. (2024). Institutional planning and unintended consequences in public higher education institutions in Uganda. *International Journal of Management (IJM), 15*(4), 102–112. Article ID: IJM_15_04_008.

Mayanja, S. C. (2020). Participatory monitoring and evaluation for quality programs in higher education: What is the way for Uganda? *International Journal of Educational Administration and Policy Studies, 12*(1), 52–59.

Mulhern, C., Spies, R. R., & Wu, D. (2015). The effects of rising student costs in higher education: Evidence from public institutions in Virginia. *Research Report.* https://doi.org/10.18665/sr.221021

Mushemeza, D. E. (2016). Opportunities and challenges of academic staff in higher education in Africa. *International Journal of Higher Education, 5*(3).

Rwakenda, B. B. H. (2016). Challenges faced by leaders of institutions of higher learning in Uganda: Case study of Mbale public and private universities. *South American journal of academic Research Special Edition May 2016, 1.*

UBoS. (2024). *National Population and housing census 2024 final report*. Uganda Bureau of Statistics (UBoS).

Wapman, K. H., Zhang, S., Clauset, a., & Larremore, B. D. (2022). Quantifying hierarchy and dynamics in US faculty hiring and retention. *Nature, 610*, 120–127. https://doi.org/10.1038/s41586-022-05222-x

Index

P. Nyadoi et al. (eds.), *Sustainable Development Goals in Africa*, Sustainable Development Goals Series, https://doi.org/10.1007/978-3-032-20742-5